Advances in Sustainability Science and Technology

Series Editors

Robert J. Howlett, Bournemouth University & KES International, Shoreham-by-sea, UK

John Littlewood, School of Art & Design, Cardiff Metropolitan University, Cardiff, UK

Lakhmi C. Jain, University of Technology Sydney, Broadway, NSW, Australia

The book series aims at bringing together valuable and novel scientific contributions that address the critical issues of renewable energy, sustainable building, sustainable manufacturing, and other sustainability science and technology topics that have an impact in this diverse and fast-changing research community in academia and industry.

The areas to be covered are

- Climate change and mitigation, atmospheric carbon reduction, global warming
- Sustainability science, sustainability technologies
- Sustainable building technologies
- Intelligent buildings
- Sustainable energy generation
- Combined heat and power and district heating systems
- Control and optimization of renewable energy systems
- Smart grids and micro grids, local energy markets
- Smart cities, smart buildings, smart districts, smart countryside
- Energy and environmental assessment in buildings and cities
- Sustainable design, innovation and services
- Sustainable manufacturing processes and technology
- Sustainable manufacturing systems and enterprises
- Decision support for sustainability
- Micro/nanomachining, microelectromechanical machines (MEMS)
- Sustainable transport, smart vehicles and smart roads
- Information technology and artificial intelligence applied to sustainability
- Big data and data analytics applied to sustainability
- Sustainable food production, sustainable horticulture and agriculture
- Sustainability of air, water and other natural resources
- Sustainability policy, shaping the future, the triple bottom line, the circular economy

High quality content is an essential feature for all book proposals accepted for the series. It is expected that editors of all accepted volumes will ensure that contributions are subjected to an appropriate level of reviewing process and adhere to KES quality principles.

The series will include monographs, edited volumes, and selected proceedings.

More information about this series at http://www.springer.com/series/16477

Neha Sharma · Santanu Ghosh ·
Monodeep Saha

Open Data for Sustainable Community

Glocalized Sustainable Development Goals

Springer

Neha Sharma
Analytics and Insights
Tata Consultancy Services Ltd.
Pune, Maharashtra, India

Santanu Ghosh
Analytics and Insights
Tata Consultancy Services Ltd.
Kolkata, West Bengal, India

Monodeep Saha
Analytics and Insights
Tata Consultancy Services Ltd.
Bengaluru, Karnataka, India

ISSN 2662-6829 ISSN 2662-6837 (electronic)
Advances in Sustainability Science and Technology
ISBN 978-981-33-4311-5 ISBN 978-981-33-4312-2 (eBook)
https://doi.org/10.1007/978-981-33-4312-2

This Springer imprint is published by the registered company Springer Nature Singapore Pte Ltd.
The registered company address is: 152 Beach Road, #21-01/04 Gateway East, Singapore 189721, Singapore

Foreword

I am pleased to write this foreword, because of my close association with the themes of 'Data Centricity' and 'Sustainability Ecosystems' that is relevant for businesses, communities and individuals. In my current role as the Global Head of Analytics and Insights, Tata Consultancy Services, I come across various customer execs, policy makers, government officials and industry leaders who are aspiring to make their organizations more purpose driven and adaptive, attain business growth but at the same time make a strong positive contribution towards environment and communities. My association with industry bodies, academia and strategic think tanks has led me to a view that growth, development and transformations are only meaningful if they are sustainable. The power of data, analytics and AI and the related digital technologies play a major role in this journey which is also vetted by the guidance provided by the United Nations Sustainable Development Goals. The world is seeing democratization of technologies and data resulting in newer possibilities. Open data is one of the levers for achieving those.

In order to further the sustainability agenda, there needs to be a strong collaboration among governments, corporates, academia and citizens with resulting ecosystem capabilities. This publication stands at the intersection of such an ecosystem and an excellent compilation of research projects based on open data with outcomes having a bearing on the community. The next generation is a very important stakeholder of this ecosystem as they would foster innovation to shape up a better tomorrow. Such a publication will encourage the students and academic institutions to focus on sustainability as a theme and contribute to the ecosystem. Each of the work outlines interesting possibilities which will be of interest to business enterprises.

Open Data for Sustainable Community is a result of joint interests between the students, data science communities and the TCS Analytics and Insights team. They have been instrumental in terms of topic selection, open data resources and researcher orientation. It explores and presents a part of the open datasets from government institutions to achieve the sustainable objectives at local level, in turn contributing towards global mission. Reading through the book, you will find some of the specific issues in the areas of environment, Indian agriculture and health care seen through the lens of data science, which has deep relevance in today's world.

I would encourage students, researchers and practitioners to contribute to the sustainability ecosystem and build further on the good work that has been done.

Dinanath Kholkar
Vice President and Global Head
Analytics and Insights
Tata Consultancy Services Ltd.
Pune, India

Preface

By and large, we are at the crossroad of the 4th Industrial Revolution, where phy-gital systems are going to play a massive role. This transformation is cutting across every known sphere to mankind. The world will become globally localized marketplace. COVID-19 has convoluted the entire space-time fabric, and there is a massive paradigm shift. We are looking the world through the new lenses where technological transformation via machine learning and artificial intelligence is the new norm. We are at the cusp of the future where AI/ML will be imbibed in day-to-day activities via cloud platforms. Doing business in a greener way is going to be norm for us if we intend to sustain life force on this planet. Fighting against natural calamities like drought, pandemics and pollution needs proactive intervention, clear vision, ground-level implementable and scalable technology. The transformation of thought process at an individual level will help to achieve the same. Even the policies and strategies have to be top down and the implementation needs to start bottom up and most importantly at grass-root level.

Since the last 300 years, industrial revolutions have game changing impact on societies. As our topic suggests, we are looking at some of these like health care, agriculture and environment through the lens of AI. Demographic-level analysis and GIS analysis are novel methods that are used in this field. The intent is to explore into these areas and identify cracks through which deeper in roads can be made. With the above background given, we will go in the deeper waters and explore the content of this book.

Health care—This chapter intends to dig on the Sustainable Development Goal 3. With that bigger picture as a vision, a study is conducted on COVID-19. The kind of impact coronavirus has on our society is at a massive scale and there is no geography which is untouched. There is no geography which is untouched. A cohort study on the indeterminate contagion pattern of COVID-19 is done, and effort is made to map the same with the potential features. The demand and supply sides both are mapped and identified with some of the state-of-the-art ML techniques which are used on the data to create analysis to derive insights from them. Though these are early days, these are initial steps in that direction.

Medical fraternity along with researchers are working to fast track a vaccine for the COVID-19. Nations and pharmaceutical organizations are joining forces and trying to gain a handle on situation carrying joint clinical trials. This book is a first-hand attempt to provide a consultative approach of looking at the demand and supply sides using federated data using open-source technology adoption. The data-driven thought leadership shown in the book ensures a detailed outlook. While pharmaceutical and healthcare organizations across the globe are in the process of doing deep research around application of machine learning on federated data, the opportunities are limitless. This approach of looking at catastrophic events, which has the ability to shock both demand and supply metrics, is a perfect experimental set-up and would be leveraged in future as framework for analysis. This study can further be utilized in the financial services industry to analyse the perturbation effect and how individual demographics are impacted. Utilities industries also can benefit by bringing in another angle of pricing. This base framework in the chapters provides the detail set-up of this experiment, data staging, analysing the data, steps involved, peeling various layers of data, using algorithms to derive insights. Apart from the results, the setting of the process will result in value.

Agriculture—The study is being done on farmer contact centre by Government of India queries. The intent of the study is to reduce the false positives and identify, define and create recommendations to remove these process deficiencies which can reduce time debt on the government and in turn resolve queries for the farmers faster. This solution provides and opens up the doors for usage of natural language process and use of automating. The idea is novel and can be used in setting up of a command centre which is in line with the revolution of Industry 4.0. Smart farming alarm system can be used to build early warning systems, which can map the grievances with solutions. One potential use case which can be tapped in is creating a database for common grievances and mapping them demography-wise, which when mapped with the loan data provided to farmers can become strong indicators. Those who have operated with micro-finance institutions will find this information extremely valuable. What this book aims is to create a methodology in terms of how to stage the data and convert it into a goldmine.

Environment—This section highlights the use of data democratization to identify correlations and patterns between air pollution and green cover for Pune City. Different techniques are fused together to create analysis which are technically intensive and extremely data driven to derive key insights from the data. They not only establish the age-old fact that the trees are key to the societal development but also help in forming strategy regarding where and how these plantations need to be done for it to have maximum impact. Reduction in carbon footprint is one major goal of all the corporate houses. Automobile sector and industries using fossil fuel for their energy requirements have to relook from a different lens. The Pune City is taken as a use case to highlight the fact that in case we want to develop technology solutions to take us on the road of creating smart cities, then it can not happen without a green cover. The book attempts not only using the available data but also looking at environment barometer by using open-source technology.

The three domains mentioned above are the intended social sample that weaves the social fabric by enabling interaction between individuals, technology and governance. Our study of sustainable development by means of artificial intelligence will help us understand these dynamics. The intent with which this book is written is very close to our hearts and is to provoke the thought process which draws out various possibilities in which our society becomes a better place for living. The caveat to be drawn here is that all the three frameworks in finer way are monetizable. The three domains and pillars of this book call out the essence of **process excellence**, use of **federated data** and **open-source technology**.

In the new normal, these three trends are not going to go anywhere soon.

The book will cater to the wide range of readers including professionals from sustainable development goals, social scientists, data scientists and machine learning experts.

This book is an early attempt towards that process. The book presents a prototype of thought process about taking the initial steps of harnessing data openly available and how to craft a solution. The authors of this book have come together from different walks of life with one common goal—a strong sense of commitment and a burning desire for betterment of society utilizing their technical skills. A sincere effort from the entire team which includes the authors, publishing house and the students. A special mention to our family members, friends and colleagues who kept us sane and focused during the entire journey.

Pune, India — Neha Sharma
Kolkata, India — Santanu Ghosh
Bengaluru, India — Monodeep Saha

Contents

Authors and Contributors

About the Authors

Neha Sharma is working with Tata Consultancy Services and is a Founder Secretary, Society for Data Science. Prior to this she has worked as Director of premier Institute of Pune, that run post-graduation courses like MCA and MBA. She is an alumnus of a premier College of Engineering and Technology, Bhubaneshwar and completed her PhD from prestigious Indian Institute of Technology, Dhanbad. She is an ACM Distinguished Speaker, a Senior IEEE member and Executive Body member of IEEE Pune Section. She is an astute academician and has organized several national and international conferences and published several research papers. She is the recipient of "Best PhD Thesis Award" and "Best Paper Presenter at International Conference Award" at National Level. She is a well-known figure among the IT circle, and well sought over for her sound knowledge and professional skills. Neha Sharma has been instrumental in integrating teaching with the current needs of the Industry and steering students towards their bright future.

Santanu Ghosh is a IT veteran with over 26 years of experience and a Sustainability Evangelist. He leads the Sustainability Ecosystem for TCS A&I unit. He is responsible for the exploration and launch of Data & Analytics-driven sustainability initiatives in the areas of Environment, Climate, Health and Biodiversity for both commercial and community enterprises. His endeavors align with UN's Sustainable Development Goals and its mapping to Enterprise and Community KPIs. He opines that data plays a critical role in creating data-driven environment-friendly solutions and creating a circular economy by creation of a knowledge ecosystem. He has been part of many fora promoting his views on Environment, Climate, Community Impacts and their relationships with the SDGs.

Mr. Monodeep Saha is currently working as Project Lead in TCS for BFSI clients. He has a rich and diverse experience of over a decade in Financial services Analytics industry, with consulting experience of major financial institutions/FinTech's in Asia pacific/US/EMEA region. Analytics/Machine Learning expert with industry expertise in Risk and Marketing Analytics space. Mr. Saha has master's in physics from IIT Madras with focus on advanced mathematics and computing. In his spare time, he spends time with family, loves to read and write. He authors short writeup at various forums covering fiction and non-fictional topics. This book is an attempt to converge his professional skills for greater good of humanity and society at large.

Contributors

Sudhanshu Bhoi is a computer engineering student who loves to think through problems and build creative solutions and an enthusiast in the field of AI; currently, he is studying machine learning and applying models to create a better world.

Sanket Bijawe is an undergraduate student at VIT, Pune, and an Exe-com member at Student Branch, VIT, Pune. He is interested in the field of deep learning and NLP.

Arjun Ghose is pursuing M.Sc. in applied statistics from Symbiosis Statistical Institute, Pune. He has graduated with a B.Sc. degree in statistics from St. Xavier's College, Mumbai.

Chaitanyasuma Jain is a computer engineering undergraduate student. She has previously worked on software development and data science projects. She is an active participant in women-based STEM communities and has a keen interest in machine learning. She believes in ideation and wishes to use her technical skills for meaningful social causes.

Monisha Jhamb is currently pursuing masters in applied statistics at Symbiosis Statistical Institute, Pune. She is a data science engineer with a post-graduate programme in data science and engineering from Great Lakes Institute of Management. She is also Six Sigma Green Belt certified by KPMG.

Shreyas Joshi is currently pursuing his graduation in electronics and telecommunication engineering at P. E. S. Modern College of Engineering, Pune. His area of interest includes embedded systems and Python for speech recognition.

Anirudh Kolwadkar is a dynamic, team-spirited and performance-driven computer engineering student. He is CS sophomore passionate about technology, Android development and artificial intelligence. His interests include coding and writing technical blogs.

Aboli Marathe is a third-year student studying computer engineering at Pune Institute of Computer Technology. She is a machine learning researcher and is working towards transforming the idea of artificial intelligence for social good. Her research interests include causal inference, explainable artificial intelligence and deep learning.

Vaishnavi Nair is currently pursuing her undergraduate course in electronics and telecommunication engineering from P. E. S Modern College of Engineering, Pune-05. She is currently serving as the CoSSR of the IEEE Pune Section.

Aman Pande is pursuing M.Sc. in applied statistics from Symbiosis Statistical Institute. He has graduated with a B.Sc. degree in statistics from Sir Purshurambhau College of Pune.

Manoj Patil, is currently pursuing his graduation in electronics and telecommunication engineering at P. E. S. Modern College of Engineering, Pune. His area of interest includes IoT and Python programming.

Shivraj Patil is currently pursuing bachelor's degree and has developed a keen interest in the field of machine learning, especially in computer vision. He wishes to work on more amazing projects in the future. As a beginner in reinforcement learning, his current goal is to train an agent (bot) to play tennis.

Atreyee Saha is pursuing M.Sc. in applied statistics from Symbiosis Statistical Institute. She has completed B.Sc. in statistics honours from Bethune College affiliated by Calcutta University, Kolkata.

Siddharth Saoji is pursuing CS Undergrad at VIT, Pune. He is CoSSR of IEEE Pune section and interested in finding solutions to real-world problems using data science and full stack development.

Madhavi Shamkuwar is currently working as an assistant professor in Zeal Education Society, Zeal Institute of Business Administration, Computer Application and Research **(ZIBACAR)**, Pune, Maharashtra, India. She is pursuing Ph.D. from Department of Management Sciences (PUMBA), Savitribai Phule Pune University, Pune, in wildlife management using information technology tools. She had published several research papers and participated in conferences, seminar and FDPs. She had 11 years of teaching experience, and her areas of interest are data mining, system analysis and design, algorithms and big data.

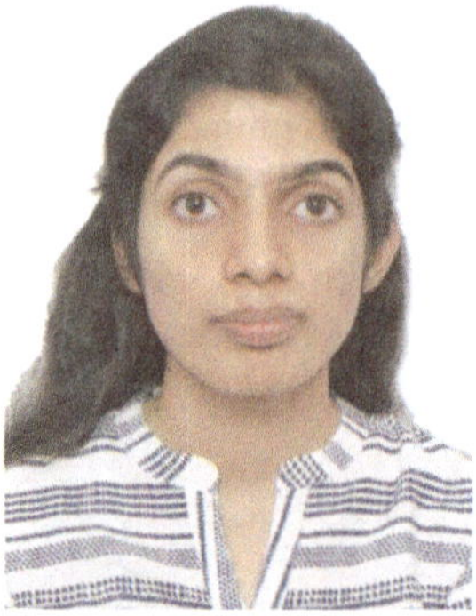

Mishita Sharma is currently pursuing master of science in applied statistics at the Symbiosis Statistical Institute, Pune. She is having a strong background in mathematics; her interests lie in data mining, statistical analysis and machine learning techniques.

Arpan Sil is currently a student of applied statistics in Symbiosis Statistical Institute. He is a bronze medallist in electrical engineering from Maulana Abul Kalam Azad University of Technology for his outstanding academic achievements at undergraduate level. Besides taking an avid interest in data sciences and allied fields, he is passionate about football and writing and has published several articles in reputed sports websites. A through and through optimist, he envisages a better future for humankind and is dedicated to make it happen.

Vineet Tambe is a budding E&TC engineer with a keen interest in robotics and AI. He works with embedded systems and robotics and aims to build advanced secure robotic technology and hardware systems for the future using various artificial intelligence techniques.

Part I
Environment—A Fact-Based Study using Tree Census and Air Pollution Data

(SDG13—Climate Action and SDG15—Life on Land)

Chapter 1
Inching Towards Sustainable Smart Cities—Literature Review and Data Preparation

1.1 Introduction

Conservation of environment is critical for sustainable survival of diverse species on this planet [1–3]. Hence to create the awareness among the global ecosystem, and the term 'biodiversity' was coined by Conservation International (CI) in the year 1998 [4]. Biodiversity implies coexistence of variety of living organism (terrestrial and marine) and plants in a particular region to create a favourable environment and to lift the productivity of the ecosystem [4]. Earth is in natural possession of biodiverse regions, mainly in tropical areas which covers 10% of Earth's surface but is a habitat of 90% of world's species [5–7]. The endeavour should be to preserve species in biodiverse hotspots as well as conserve environment to ensure natural sustainability for all life forms [8–10]. However, in reality, of more than 300,000 known species of plants, the International Union for Conservation of Nature (IUCN) has evaluated only 12,914 species, finding that about 68% of evaluated plant species are threatened with extinction [11, 12]. Many such precarious situations forced United Nations to present a blue print of 17 global goals (also known as Sustainable Development Goals—SDGs) in the year 2015 to ensure social development, environmental protection as well as economic growth and to be achieved by 2030 [13–15].

Conservation International listed 17 countries that are mega-biodiverse and are mostly located in the tropical or subtropical region. India is at 16th position in the world with 21% of forest and 5% protected area covered around the total land area of the country [16]. But deforestation and other harmful human activities pose a great threat to these existing biodiverse forests [17]. Therefore, in keeping with the sustainable development goals related to the environment (SDG13-Climate Action), the Government of India has adopted a National Action Plan on Climate Change as well as a National Mission for Green India, to address this issue directly. These national schemes are complemented by a host of specific programmes on solar energy,

N. Sharma et al., *Open Data for Sustainable Community*,
Advances in Sustainability Science and Technology,
https://doi.org/10.1007/978-981-33-4312-2_1

enhanced energy efficiency, sustainable habitats, water, sustaining the Himalayan ecosystem, and to encourage strategic knowledge exchange towards climate change. Another important Sustainable Development Goal towards environment is SDG15 (Life on Land), which calls for urgent action to halt the degradation of natural habitats, to end poaching and trafficking of animals, as well as to integrate ecosystem and biodiversity values into local planning and development processes. Two important targets for SDG15 are to conserve, restore and sustain land and water ecosystems and its resources like forests, wetlands, mountains and drylands and to promote sustainable management of forests, halt deforestation and increase afforestation as well as reforestation globally [18].

Trees constitute a major part of the ecosystem and play a key role in sustainable development. Through photosynthesis, they provide the oxygen we breathe as well as the food we eat and are thus the foundation of most life on Earth. Trees absorb carbon dioxide (CO_2) we release in the atmosphere and store it in the form of carbon in its branches, leaves, trunks, roots and in the soil. Carbon is sequestered in soil by plants through photosynthesis and can be stored as soil organic carbon [19–22]. However, due to pollution and deforestation, stored carbon is released into the atmosphere, mainly as CO_2, resulting in global warming and many health hazards. Smart cities thus must provide carbon-neutral atmosphere by intelligently using data and technology towards implementing green infrastructure, smart grids, carbon-free public transport, energy-efficient homes, etc. [23, 24]. The key objectives of the Smart City mission in India revolve around making cities more sustainable and citizen friendly, and Pune Smart City mission is fully aligned with that. One of the major steps in this process of transforming cities is using the available open data and analyzing them appropriately for determining the best way to plan the modifications to be carried out [25–27].

In this unit, we present a case study using democratization of green cover data and air pollution data for Pune City in Maharashtra State to find important correlations as well as insightful patterns using data science techniques as shown in Fig. 1.1. The data used in the study is the open data related to tree census and air pollution created by Pune Municipal Corporation [28]. Along with NGOs and social organizations, the civic body has been involved in conducted various tree plantation drives in the city on a large scale. Due to this, they have helped increase the green cover by 20 lakh trees and documented their progress in tree census surveys [29–32]. Also, the air pollution dataset gave us the air pollution characteristics (AQI, concentration of SO_2, NO_x, PM_{10}) in five areas of Pune city (Swargate, Pimpri, Nalstop, Bhosari and Karve) from January 2018 to March 2019. We can observe the environment, measure the biodiversity, find the correlation with major pollutant levels, integrate data analytics and suggest effective sustainable practices to build the smart cities.

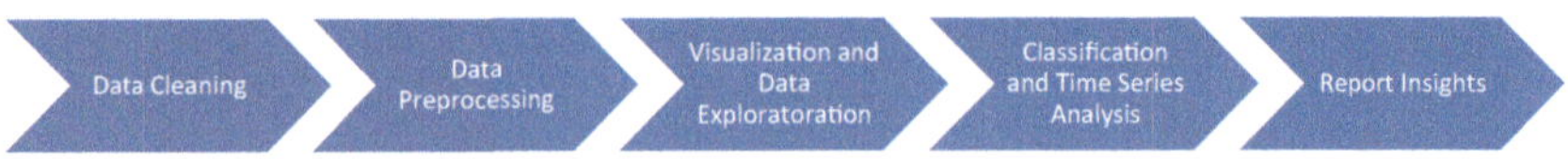

Fig. 1.1 Data science pipeline for handling open data and generating insights

In this study, we are confirming the positive changes in the environment, correlated with afforestation, for the city of Pune.

The main objectives of the Unit I are as follows:

1. To study the features, distribution and contents of the green cover (tree census) and air pollution dataset.
2. To find correlations between the above mentioned datasets, so as to confirm the positive environmental effect of tree density on a region.
3. To suggest improvements in existing city flora, so as to curtail the alarming air pollution levels.

In this chapter, an attempt has been made to understand the green cover (tree census) and air pollution dataset in detail and present the detailed process of data preparation. The rest of the chapter is organized as follows: Section 1.2 discusses various literatures in the similar domain, Sect. 1.3 deliberated on green cover (tree census) and air pollution dataset, respectively, Section 1.4 presents the data cleaning and preparation process, and Sect. 1.5 concludes the chapter.

1.2 Literature Review

According to the National Geographic, around 70% of Earth's land animals and plants live in forests and many cannot survive the deforestation which destroys their home [33]. Cutting trees results in the habitat loss for animal species, which in turn damages ecosystems. Likewise, the Amazon rainforest, a tropical rainforest, contributes a cycle of evaporation and rainfall. Loss of such huge ecosystems could result in warmer and drier climates near the tropics, eventually resulting in habitat destruction for the species of plants and animals living there. Interestingly, trees provide many more services and benefits, [34] which may or may not be visible to the common man. They also have a lot of economic value and a potential to conserve, clean and maintain other natural resources. They even have psychological benefits like reduction in air temperature, pollution removal, building energy use, reduction in ultraviolet radiation, conserving wildlife populations and so on.

In 2003, the Parks and Recreation Department of Little Rock, Arkansas, took an initiative that encouraged people to adopt a natural setting and promote healthy lifestyle choices. The Medical Mail Trail, a part of the Arkansas River Trail system which runs for eighty miles, is health inspired and supported by the medical community. It includes picnicking areas, playgrounds, sculptures, spray grounds and a mural wall with the underlying message to inspire fitness and nutrition among the people.

Also, urban forests mitigate a variety of health-related issues, ranging from respiratory diseases to melanoma, promoting an active lifestyle, which can help in reducing obesity. Many studies done by the health professionals and government agencies have found that being beset by natural settings (specifically urban forests) have a tremendous impact on mental health. A 20-min walk or sitting in a forested park in a busy metropolis can help in reducing mental stress for people with attention-deficit/hyperactivity disorder (ADHD), thus improving concentration and promoting creativity.

Talking about canopy cover and its importance, Scott Trimble in his blog posted on 2 April 2019 [35] explained that the canopy cover was the layer formed by the branches and crowns of plants or trees. This cover can be either continuous (for primary forests) or discontinuous—with gaps (for orchards). In tropical and temperate forests, canopies serve as very important habitats for many animals and plants species. A dense canopy cover generally lets in little light which can reach the ground and subsequently work in lowering the temperatures. The canopy also protects the ground from the force of rainfall and somewhat moderates the wind force. Thus, the ground habitat conditions are affected by a considerable degree by the canopy cover. There are differences in forest canopies and their effects on the immediate ecology.

David J. Nowak, USDA Forest Service, Syracuse, NY 2002 [36], in his research paper, had talked about the effects the urban trees on air quality. Tree transpiration and tree canopies hugely affect air temperature [37], radiation absorption, heat storage, wind speed, relative humidity, turbulence, surface albedo and surface roughness. These changes in local meteorology can go a long way in altering pollution levels in the urban areas.

Trees generally remove gaseous air pollutants primarily via leaf stomata. Inside the leaf, gases get diffused into the intercellular spaces and may end up getting absorbed by the water films to form acids or react with the surfaces of the inner leaf. Trees also reduce pollution levels by intercepting the airborne particles. Some particles might be absorbed into the tree, though most of the particles which get intercepted are retained on the surface of the plant. The intercepted particles often get resuspended back into atmosphere, get washed off by rain or get dropped to the ground with leaf falls. As a result, vegetation is only some sort of transient retention site for most of the atmospheric particles. In 1994 [38], a study found that trees in the New York City removed an estimated 1821 metric tons of air pollutants at an estimated value of $9.5 million to the society. Air pollutants removal by the urban forests in the city of New York was far greater than in Atlanta (1196 t; $6.5 million) and Baltimore (499 t; $2.7 million), but pollution removal per m^2 of canopy cover was fairly along the same lines for these cities (New York: 13.7 g/m^2/year; Baltimore: 12.2 g/m^2/year; Atlanta: 10.6 g/m^2/year). These standardized pollution removal rates differ among cities in accordance to the exact amount of air pollution, the length of leaf in season, the precipitation and some other important meteorological variables. Larger and healthier trees which are greater than 77 cm in diameter remove more or less around 70 times more air pollution (1.4 kg/year) annually than the smaller healthy trees which are less than 8 cm in diameter (0.02 kg/year). The total air quality improvement in the city of New York due to the pollution removal by the trees during daytime of the in-leaf season averaged around 0.30% for nitrogen dioxide, 0.47% for the particulate matter (PM10 and PM2.5), 0.43% for sulphur dioxide, 0.45% for ozone, and 0.002% for carbon monoxide. Air quality genuinely improves with the increase in percentage of tree cover and decreased mixing-layer heights. In urban areas with around 100% tree cover (i.e. contiguous forest stands), short-term improvements in air quality (one hour) from pollution removal by trees were near around 15% for ozone, 14% for SO_2, 13% for particulate matter, 8% for NO_x and 0.05% for CO.

David, in his another paper, 'Assessing the Benefits and Economic Values of Trees', [37] discussed 'i-Tree' software. i-Tree is a state-of-the-art software suite developed and used by the USDA Forest Service which provides analysis of urban and rural forestry along with various assessment tools. The i-Tree tools can help in strengthening the forest management and related advocacy efforts by quantifying the forest structure and the associated environmental benefits that trees can provide [38].

An article published by The Hindu Newspaper on 22 June 2019 and updated on 27 June 2019, [39] described about the tree census that was conducted in Bangalore. Some of the key points discussed were regarding the increment of green cover in a locality, the necessary things to do and the important points to be kept in mind. Taking a good stock of the trees that share space in the selected area and preparing an inventory should be the first priority. In December 2016, the last census was conducted by the students in the city of Bangalore in order to take stock of the losses due to Cyclone Vardah.

Pauline Deborah R., an associate professor of the Department of Plant Biology and Plant Biotechnology of WCC, who had led these tree censuses, explained the significance and benefits of conducting a tree census. She also highlighted that the key to any census is planning the time and date of the census, selection of volunteers and requisite permission from concerned authorities. Training of the volunteers and a tree identification guide are essential to derive the maximum output from a tree census. Height, girth, canopy diameter, stress, flowering, fruiting season, name and frequency of distribution and the health of the tree are to be collected. Suitable measures should be taken if a rare tree is identified. The removal of precariously standing trees and appropriate pest treatment is also essential for the unhealthy and the infected ones.

The paper 'The urban forest in Beijing and its role in air pollution reduction' written by Yang et al. [40] focused on the study of a proposal given by the municipal government as a measure to alleviate the air pollution levels in the Chinese capital, Beijing. It was based on the analyses done with the help of the satellite images and requisite field surveys to study the defining characteristics of the urban forest in the central part of Beijing. The Urban Forest Effects Model studied the influence of urban forests on the air pollution level of the designated area. The results showed that there are around 2.4 million trees in the central part of Beijing. Small diameter distribution trees were more prominent, and the urban forest was mostly dominated by a few species. About 29% of trees in the central part of Beijing were classified as being in poor condition. With respect to air pollution removal levels, the trees around 1261.4 tons of pollutants from the air in 2002. PM10 was the pollutant most reduced by the tree cover; the reduction amounted to around 772 tons. Also, the carbon dioxide (CO_2) stored in biomass formed by urban forests amounted to around 0.2 million tons.

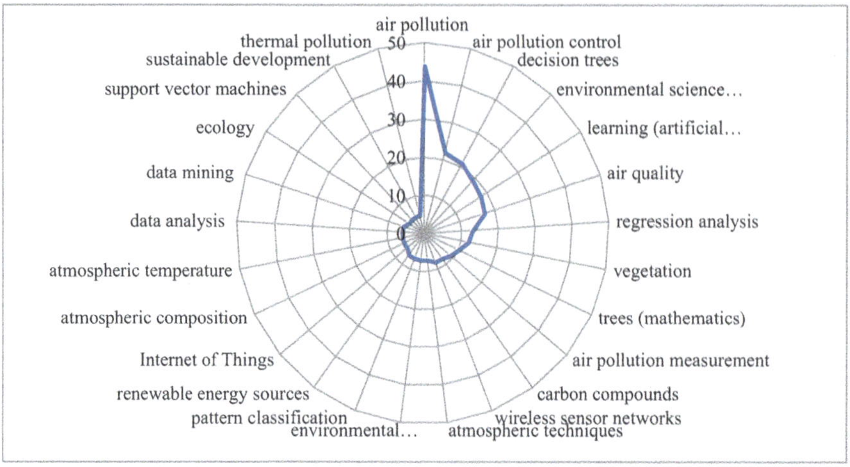

Fig. 1.2 Top 10 publication terms related to air pollution and green cover

In this section, our endeavour was to review the literature related to tree census and air pollution. Finally, a systematic literature review is carried out using IEEE Explore database as shown in Fig. 1.2.

IEEE Explore database is a standard research database, which was studied to know more about digital dimensions of the aforesaid topics. The 'advanced search' was made using the terms 'tree' and 'air pollution', for a span of last ten years, i.e. from 2009 till date. The top ten 'publication terms' related to 'tree' and 'air pollution' are given in Fig. 1.2, which indicates that the most popular terms used in research are decision trees, environmental science computing, learning (artificial intelligence), regression analysis, trees (mathematics), etc.

1.3 Data Preparation: Tree Census Data

The process of preparing data involves many sub-processes like identification of data source(s), data acquisition, data cleaning and data wrangling, so that dataset is ready to be explored, analysed and provide amazing insights.

1.3.1 Understanding the Importance of Tree Census

"Plant more Trees!" Time and again we have heard this phrase. What is the value of these trees? Is all the effort and money that goes into planting and maintaining them worth it? The benefits of the green cover of a city are not limited to their contribution to the environment and aesthetic value alone. Jack Payne, Senior Vice

President for Agriculture and Natural Resources and Professor of Wildlife Ecology and Conservation, University of Florida, in his blog titled 'What is the economic value of your city's trees?' stated that cities routinely rake up tens of millions of dollars from their urban forests annually in ways that are not always obvious [41]. The urban green cover has a great potential to conserve, clean and maintain other natural resources. Besides, it provides a large number of environmental and economic benefits, and some of them are discussed below:

1. *Trees conserve energy*: If the trees are strategically placed around a house they can significantly reduce demand for energy used in cooling during summer by reducing the need for air conditioning by about 50%. This when done at a larger level can reduce the emission of carbon dioxide (CO_2) and other pollutants from power plants [34].
2. *Trees combat climate change*: Excess carbon dioxide in the atmosphere is a major reason for climate change. Trees absorb CO_2, remove and store the carbon (C) and release oxygen (O_2) back into the atmosphere. In one year, an acre of mature trees have the potential to absorb the same amount of CO_2 produced by a car when driven for 26,000 miles [34].
3. *Trees save water and prevent water pollution*: The shade from a tree canopy slows the evaporation of water from lawns. Most of the trees which are newly planted need only 15 gallons of water a week. Also, as trees transpire, i.e. the exhalation of water vapour through the stomata, they add to the atmospheric moisture [34].
4. *Trees help to prevent the erosion of soil*: The rain water which falls with a great speed and force on the ground has the potential to cause runoff of the upper layer of the soil carrying away all the essential minerals that maintain the fertility of the soil. Rahman and Ennos, in their paper titled 'What we know and don't know about the surface runoff reduction potential of urban trees' [42], describe that some of the rain is intercepted by vegetation and eventually evaporates from the canopy back into the atmosphere. This significantly reduces the water that hits the soil. Thus, only some of the water reaches the soil by either falling through the canopy or by running down the trunks, leaving only a small portion of water that usually gets stored in puddles or infiltrates into the soil. The compaction of soil and it being covered by impermeable buildings and roads in the urban areas greatly reduces the amounts of rainfall that gets intercepted, evaporated or stored. The majority of tree covers and vegetation in urban areas grows in patches of soil, such as parks, gardens, and small areas of urban woodland, that are disconnected from the drainage network of the cities and act hydrologically in parallel with it [42]. Figure 1.3 shows the difference between the percentage of evapotranspiration and infiltration of water from natural ground cover, i.e. soil, and the man-made surfaces like the roads and impermeable surfaces of the roofs of buildings. For the natural surfaces, the runoff of the excess water due to watering of plants or heavy rain leads to only 10% of the storm water to runoff, whereas a large 55% of the runoff takes place from the roads and impermeable surfaces which also have various pollutants on them that carried away in the

Fig. 1.3 Natural and impervious cover diagrams

runoff and enter the drains, eventually reaching and polluting the water bodies like lakes, rivers, sea, etc.

5. *Trees heal*: Studies have shown that patients that have a view of trees out of their windows in hospitals/homes show a faster rate of recovery and with less complications. Also the children with ADHD, when given access to nature, have shown fewer symptoms. Taking a 20-min walk or sitting in a forested park within a bustling metropolis can reduce mental stress and fatigue for people with ADHD, improve concentration and promote creativity. Exposure to trees and nature has the potential to improve an individual's concentration level by reducing mental fatigue [34].
In addition to healing, urban forests reduce the risk of many other health issues, such as respiratory diseases and skin cancer, and promote an active lifestyle. They also improve the mental health of the residents. A report for the European Commission [43] discusses that the increasing stress in working life and living in urban environments exposes citizens to noise and pollution along with visual disturbances. As per Grahn et al. and Korpela et al., there are sufficient evidence to show that the more people get exposed to green areas, the better their mood and mental—wellbeing is [43].
6. *Trees create economic opportunities*: Trees provide a large number of economic and ecosystem services that produce benefits to a community [37]. The yield from trees, for example, wood, fruits, flowers, medicinal products, essential oils, etc. from community orchards can be sold, thus providing income. There are several business opportunities created in green waste management and landscaping, when cities value mulching, i.e. the process of applying a layer of material over a soil surface in order to conserve soil moisture, improve fertility and health of the soil, and its water-saving qualities. There is a name for jobs in this sector, the 'green jobs'. Vocational training is provided to the people who are interested in jobs in this sector [34].

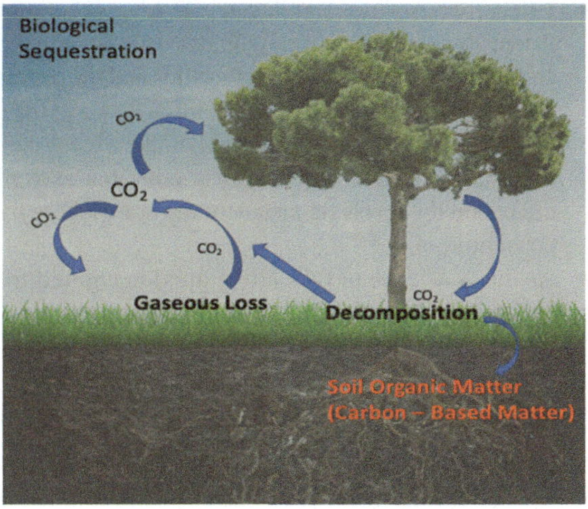

Fig. 1.4 Process of carbon sequestration by a tree

7. *Trees increase the value of a property*: The beauty of a well-planted property, its surroundings and neighbourhood can raise its value by as much as 15% [34].
8. *Trees increase business traffic*: Several studies have shown that the more trees and landscaping a business district has, the more business flows in [34]. People actually have shown a tendency to spend more time, in a shopping street or avenue if it has well maintained trees planted properly than the ones which don't. Several cities around the world have accepted that there doesn't have to be a constant battle between infrastructure development and increasing green cover. Both the goals can be achieved together with effective urban planning.
9. *Carbon sequestration*: It is a process of long-term removal, capture or sequestration of CO_2 from the atmosphere as shown in Fig. 1.4.
   Biological sequestration helps to slow down and reverse the atmospheric pollution due excessive contents of CO_2 in the atmosphere. Thus, in the long run, it aids the mitigation of global warming. In order to examine and estimate the carbon sequestration of an urban tree, factors like carbon storage in biomass, the amount of avoided carbon emission by energy conservation, and carbon emissions that are associated with tree maintenance and decomposition must be considered [44]. CO_2 in the atmosphere gets absorbed by the tree trunks, branches, foliage and roots, which then decompose carbon dioxide into carbon and oxygen. The carbon is stored in the soil and the trunk and branches of the tree, and the oxygen gets released back into the atmosphere. The figure also shows gaseous loss from patches of land without a significant green cover which ultimately causes global warming.

Important facts about carbon sequestration:

- About 25% of carbon emissions have historically been captured by Earth's forests, farms and grasslands. Scientists and land managers are working to keep landscapes vegetated and soil hydrated for plants to grow and sequester carbon.

- Urban forests play a significant role when it comes to the sequestration of carbon. A tree absorbs carbon throughout its life. But the type of vegetation in the city also calls for attention from the authorities and the citizens. The results of urban forests can have varying effects on the carbon levels in the city's air, as the vegetation that is being used to function as a sink to the excess carbon dioxide in the atmosphere may start acting as a source of it emission as well. The vegetation has indirect effects on the levels of carbon in the atmosphere by reducing the need for energy consumption.
- As much as 30% of the carbon dioxide emitted from the burning fossils fuels is absorbed by the upper layer of the ocean, which raises the water's acidity. Ocean acidification makes it harder for marine animals to build their shells. Scientists and the fishing industry are taking proactive steps to monitor the changes from carbon sequestration and adapt efficient fishing practices [45].

One of the ways in which a city can keep track of its green cover, and know where it is lacking and what areas need more attention, is by conducting a census of the trees in the city. The Pune Municipal Corporation completed its last tree in August 2019.

1.3.2 Introduction to Tree Census Open Data

The tree data has been taken from the Open Data Portal of the Pune Municipal Corporation (PMC) [46]. This data gives the records of the 4,009,623 trees of Pune that were counted in 2019 tree census project of PMC with the help of the geo-tagging technology. According to the Maharashtra Protection and Prevention of Trees Act 1975, it has been made mandatory to carry out tree census for every civic body once in every five years. The original data contains information on 28 attributes corresponding to every tree record.

The review of original tree census open data is carried out using Python programming language, and it started with the first step of importing the required Python libraries.

```
[]
Import pandas as pd
Import numpy as np
Import matplotlib.pyplot as plt
Import seaborn as sns
Import matplotlib as mpl
%matplotlib inline

mpl.rc('axes', labelsize = 15)
mpl.rc('xtick', labelsize = 12)
mpl.rc('ytick', labelsize = 12)
```

Followed by importing and reading the five .csv (Comma Separated Files) files that together contain the 4,009,623 tree records followed by converting each .csv file into a separate pandas data frame named df1, df2, df3, df4 and df5.

```
[]
# Import and convert the five files into pandas data
frames
df1 = pd.read_csv('/content/drive/My Drive/Tree
Census Data/p1.csv', dtype = types)
df2 = pd.read_csv('/content/drive/My Drive/Tree
Census Data/p2.csv', dtype = types)
df3 = pd.read_csv('/content/drive/My Drive/Tree
Census Data/p3.csv', dtype = types)
df4 = pd.read_csv('/content/drive/My Drive/Tree
Census Data/p4.csv', dtype = types)
df5 = pd.read_csv('/content/drive/My Drive/Tree
Census Data/p5.csv', dtype = types)
```

The following Python code 'concatenates' or joins the five data frames into a single large data frame. The joining takes place over the last row of each of the first four data frames provided that the positions and number of columns match.

```
[]
# Concatenate the five data frames into one
list = [df1, df2, df3, df4, df5]
tree = pd.concat(list, axis = 0, ignore_index =
True, sort = False)
```

The following code provides the structure of the data frame, i.e. the number of rows (records) and columns (fields) in the data frame.

```
[In] tree.shape
[Out] (4009623, 28)
```

Thus, the final data frame or dataset has a total of 4,009,623 rows and 28 columns.

1.3.3 Data Profiling of Tree Census Open Data

Data profiling is a step that necessitates knowing the data before cleaning and wrangling it. It describes the dataset from various perspectives like its data structure, number of null records present, outliers in the dataset, junk data (if any), and possible issues regarding data quality, etc. A comprehensive examination of each attribute of dataset can help determine if an open data source is worthy of inclusion in the data transformation effort, is there any potential data quality issues, and the amount of wrangling required to transform the data for analytics to be performed.

The tree census dataset has 28 attributes (also known as feature or column), and each attribute is described in detail in this sub-section.

- **FID**: Tree geo-tag Id of the general form trees_display.fid--1672c0a_16c67XXXXXX_XXXX (all unique values). No missing values were found for this attribute.
- **id**: Tree Id, unique 8 digit numerical ID. No missing values were found for this attribute.
- **oid**: This attribute was completely empty
- **sr_no**: This attribute was completely empty
- **girth_cm**: This attribute keeps a record of girth of a particular tree. The blog posted by S. Agarwal titled 'How to measure the diameter, height and volume of a tree' states that the girth of the trees (in centimetres), in many countries around the world especially India, is by norm measured at 4.5 feet (1.37 m) above the ground and is known as 'Diameter at Breast Height' or DBH [47]. This distance is measured along the side of the trunk from the base of the tree perpendicular to its axis.

According to the blog 'E & E: Old Trees Store More Carbon More Quickly, Than Younger Trees' posted by Tiffany Stecker on Friday, 17 January 2014, in the E & E ClimateWire, 38 researchers from 15 countries in their studies involving more than 400 tree species found that 97% of these species grew more quickly as they aged, in the process absorbing more carbon [48]. Also several studies have found that the older the tree, the greater is its potential to store carbon and thus, slow down the climate change. This is possible as the older trees have wider and thicker trunks, as well as it has denser canopy, so even though the ability of the trees to process carbon becomes less efficient with age, this effect declines their productivity only slightly as the work now is done by the leaves. The net effect is that the older trees fix more carbon in total than a newer and smaller tree and preventing it from escaping into the atmosphere which is known as carbon sequestration [48].

Diameter at Breast Height, $\text{DBH} = 2r$
Where r = radius of the trunk at breast height
Girth at Breast Height, GBH = Perimeter of Stem at Breast Height = $2\pi r$
Thus, the relationship between DBH and girth is: $\text{GBH} = \pi \times \text{DBH}$
'girth_cm' column has no null values.

- **height_m**: Height of the tree (in meters) is the vertical distance between the base of the tree and the tip of the highest branch on the tree. It cannot be accurately determined always. Tree height is not the same as the length of its trunk. The length of the trunk of a leaning tree might be greater than its height. Some advantages of taller trees are better chance for survival in the competition for sunlight and better dispersal of their pollen [47].
  This attribute has no missing values.

- **canopy_dia_m**: This attribute records the diameter of the canopy of a tree. Section 12 of 3.0 of the Phase 3 Guide—Crowns: Measurements and Sampling of the Forest Inventory and Analysis (FIA) Program of the U.S. Forest Service states that the shape of the tree canopy/crown which is measured contains all of the foliage of a tree as it grows in a stand. Abnormally long branches which extend beyond the edge of the silhouette/shape of the canopy, which is measured from branch tip to branch tip, are excluded from this measurement. The size and shape of a tree's crown vary with age and spacing and tend to be species-specific. Tree crowns tend to flatten out as they grow old and are more slender if the tree is growing in very crowded conditions [49]. Several studies have shown the crown of a tree has a marked effect on and is also strongly correlated with the growth of a tree and of its parts. Tree canopy cover is calculated by multiplying the width of the crown in the north–south direction by the width of the crown in the east–west direction.
  'canopy_dia_m' attribute contains no null values.
- **condition**: This attribute records health condition of the tree. This is a categorical variable with four unique types: Healthy, Average, Poor and Dead. This column has no missing values.
  95.6% of all the trees in Pune have been reported to be Healthy, 3.1% are in an average condition, also there is a warning that 0.22% of all the trees are in poor quality and 1.01% of the 4,009,623 trees counted in this census have been reported Dead. Tree's health is examined and compared with other tree species of the same kind with the help of signs like change in the foliage colour which is atypical, foliage density, i.e. check if there are any bare spots, sign of disease or infection and the vigour of the tree.
- **remarks**: This attribute records the visible signs of damage on the tree and the stress it is under. Because trees can only "seal" their wounds and cannot "heal" their wounds. Any physical damage done to a tree's roots, trunk, or crown affects it for the rest of its life. This column gives the information about a few other observable defects and signs of stress on a tree. The categories of this feature are: 'Mechanically cut', 'Uprooted', 'Diseased', 'On The Wall' and 'Dangerous'. On analysing further,

```
[] tree[tree['remarks'] == 'Dangerous']
```

It is found that there is only one tree corresponding to the category 'Dangerous', and this tree is a Rain Tree which is located on the Shivaji Road at 18.52617 degrees north and 73.855325 degrees east. Its society name is not available in the dataset. Although it has been reported to be balanced, however, this tree needs to be revisited and inspected at regular intervals as the risk is higher because of it being located on a road.
This attribute has approximately 44.94% missing values. These values cannot be substituted with any value. Since a null value is very ambiguous in this column. There could be trees that actually have no visible signs of damage/environmental

stress or the person who took the record might have missed it. So they are left as it is to be dealt with as an when required.

- **balanced**: This attribute records whether a tree is balanced on all sides or not. Inspecting the balancing of a tree from time to time is important. Many trees do not grow perfectly straight but if a tree starts to lean suddenly then this indicates a problem. It could be due to several factors that are linked directly to its health and strength to stand.
  This feature is a Boolean variable with the two categories 'Yes' and 'No'. It has no null values. Approximately, 99.36% of trees are balanced, i.e. True, while 0.63% of trees are not balanced, i.e. False. There are 25,571 unbalanced trees out of which only 3272 are in a poor condition or are dead, so these trees need a careful inspection which includes locating these inside the city and identifying other associated factors that might give the municipality a warning much before an accident, which may cause a loss of life or property, so that they can take an action on time.
- **special_collar**: This attribute records whether the tree possess a branch collar or not. A branch collar is the raised area that surrounds the base of every branch. There may be a wrinkled area where a branch meets the trunk. This is the branch bark ridge, and this is where the trees produce the protective callus. If this area gets damaged, then the tree is likely to get infection. Therefore, tree pruning should be done very cautiously [50]. Figure 1.5 shows the branch collar of a tree.

A tree can become hazardous if it is not taken care of properly. A tree is of great value, but it may become a liability and cause damage of large sums of money if it is not monitored properly and at regular intervals of time, especially the ones that are very large and located at places of high risk, such as beside a road or a highway, inside a parking lot, inside a children's play area, near or beside electric poles and wires, etc. A hazardous tree is the one that has a significant structural defect that may cause the tree or some portion of it to fall on someone or something of value. If a tree causes any damage, then its owner is responsible for it [51]. To prevent an accident or loss because of a hazardous tree, it is advised to pay attention to the following points during the inspection of the tree:

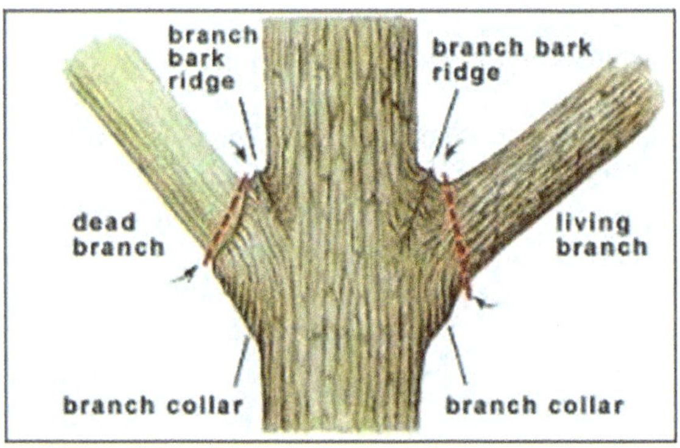

Fig. 1.5 Branch collar

- Health of the tree
- Hazardous defects in the tree
- Improper pruning
- Site conditions and targets, i.e. whether the tree can cause any damage to life or property if it falls unfortunately.

This attribute has 97.95% missing values, and therefore, this attribute is not fit to be used in the analysis.

- **other_remarks**: This is completely an empty column
- **ownership**: This attribute informs where the tree is growing or which entity owns it. This does not contain any missing values.
- **society_name**: The name of the society or the area where the tree is growing. 8.37% of its values are missing, but there are a few values that were found to have '\n' (the newline character following them). This character was stripped off from these values. Other values like 'N/A', 'Na' and 'NaN' were all converted to 'Unknown' by running the following lines of code:

```
tree [‘society_name’].replace (to_replace = ‘Na’, value = np.nan, inplace = True)
tree [‘road_name’].replace (to_replace = ‘Na’, value = np.nan, inplace = True
tree [‘society_name’].replace (to_replace = np.nan, value = ‘Unknown’, inplace = True)
tree [‘road_name’].replace (to_replace = np.nan, value = ‘Unknown’, inplace = True)
tree [‘road_name’].replace (to_replace = ‘Unknow’, value = ‘Unknown’, inplace = True)
tree [‘society_name’] = tree [‘society_name’].apply (lambda x: x.rstrip(‘\n’))
tree [‘road_name’] = tree [‘road_name’].apply (lambda x: x[:x.find(“\\n”)])
```

- **road_name**: This attribute informs about the name of the road where the tree is growing.
- **northing**: Technically, this is the projection of latitude on a two-dimensional plane, but here this column has the values same as the latitude. This column contains no missing values.
- **easting**: Technically, this is the projection of longitude on a two-dimensional plane, but here this column has the values same as the longitude. This column contains no missing values.
- **ward_name**: This attribute is same as that of column 'ward', and there are no missing values.
- **botanical_name**: The botanical name or the scientific name of a tree has three parts, which are:
    - Genus
    - Species
    - Discoverer's Name.

This is the name used in the scientific community [52]. Approximately 0.67% of the values of this attribute are missing. These cannot be imputed as well because the corresponding tree is either reported as Dead, Unidentified or Mixed Tree Growth Type trees. So expert advice is needed for the missing values of 'botanical_name' column. Therefore, these missing values are left untouched.

- **saar_uid**: These are the Tree Ids given by the IT company SAAR IT which was given the tender for the tree census of 2019. This attribute contains no missing values.
- **common_name**: Common names are the general names of the trees, i.e. the names which they are called by the common people who do not have scientific knowledge of plants and trees. The geographical area over which a particular common name is used varies. Some common names are only applicable locally, while others are virtually universal within a particular language. But some of the main limitations of the usage of common names are that they have a very local distribution and tend to change even within the same area and with time. A single species of trees can be called several common names. Conversely, a single common name is often used for multiple species of trees [53]. The 'common_name' column contains no missing values. But 14,709 common names have reported as 'Dead', another 12,021 values as 'Unidentified' and 174 occurrences of the value 'Mixed Tree Growth'.
- **local_name**: Local name of a tree is the name by which it is known locally. Local name might differ within the same city or between urban and rural areas, etc. Just like the 'common_name' attribute, it has no missing values but corresponding to the records with 'Dead', 'Unidentified' and 'Mixed Tree Growth' values for 'common_name', 'local_name' has identical values.
- **economic_i**: This attribute gives information on what is the economic utility of the yield from the tree, i.e. how is the yield from the tree used in making other products.
- **phenology**: Phenology is the time frame for any seasonal biological phenomenon, including the dates of last appearance, e.g. that of new leaves, in a tree species [54]. This attribute also contains same percentage of missing values as the 'botanical_name' column, i.e. 0.67%, and for the same records for which the 'botanical_name' values are missing. It is obvious that the trees whose 'botanical_name' is not known are all 'Unidentified' so all there properties are still is the process of discovery and therefore, have not been reported.
- **flowering**: This feature gives the information on the season in which flowers bloom on the tree. This column contains approximately 1.10% missing values and is discussed later in the data cleaning process.
- **is_rare**: This column gives the information about whether or not the occurrence of the tree species is rare in Pune. This attribute contains approximately 0.67% missing values too and is discussed later in the data cleaning process.
- **ward**: This is the number of the ward in which the tree is planted. The 'ward' attribute contains no missing values.

1.3.4 Data Cleaning and Wrangling for Analysis

Data cleaning is the process of removing corrupt and inaccurate data from the dataset, whereas data wrangling is the process of transforming the format, structure and values of data and mapping data from one raw format to another with an intent to make it more appropriate and valuable for various task. This sub-section discusses various data cleaning and wrangling techniques adopted for preparing tree census data for analysis.

Heat Map

A heat map is profusely used data visualization technique to analyse the dataset. It uses colour to illustrate degree of a phenomenon in two dimensions. The phenomenon like clustering or spread over space is demonstrated by difference in colour with respect to shade or intensity.

The heat map of a data frame shows which of its attributes or columns have missing values and gives a visual of the locations of these missing values in the respective columns. Figure 1.6 shows the heat map of the tree data frame. The light region in the heat map corresponds to the missing values, and the dark region marks the values that are known.

Fig. 1.6 Heat map of the tree census data

The following code generates a heat map of the null values of a data frame:

```
[] plt.figure(figsize = (9,9))

   sns.heatmap(tree.isnull(), cbar = False)
   plt.yticks([]);
```

From this heat map, it is very easy to see that there are some columns which do not have even one non-null value. These columns are dropped when cleaning the data as they aren't of any use for the analysis. To be more precise about the percentage of null values for each feature, the following series as shown in Fig. 1.7 is generated.

```
[] tree.isnull().sum()/len(tree) * 100
```

Values for the columns 'oid', 'sr_no' and 'other_remarks' are unknown, i.e. all their values are missing. These features, hence, would not be of any use for the analysis. Therefore, these columns are dropped from the data frame. The 'special_collar' column also has 97.95% missing values. So this column is dropped as well.

Fig. 1.7 Percentage of missing values in the original dataset

```
FID                   0.000000
id                    0.000000
geom                  0.000000
oid                 100.000000
sr_no               100.000000
girth_cm              0.000000
height_m              0.000000
canopy_dia_m          0.000000
condition             0.000000
other_remarks       100.000000
ownership             0.000000
society_name          8.367445
road_name             1.480887
northing              0.000000
easting               0.000000
balanced              0.000000
remarks              44.942280
special_collar       97.952276
ward_name             0.000000
botanical_name        0.670986
saar_uid              0.000000
common_name           0.000000
local_name            0.000000
economic_i            0.670986
phenology             1.077059
flowering             1.102622
ward                  0.000000
is_rare               0.670986
```

'FID' and 'id' columns give the Tree Ids, and they are all unique. The FID column is dropped as one column with unique Ids is sufficient. The column 'ward_name' is dropped as well, since it was found identical to the 'ward' column.

The 'geom' column is inspected, and it is observed that the values in this column have the general form:

POINT(longitude, latitude)

The longitude and latitude features were extracted from this column which led to the formation of two new columns: latitude and longitude. 'Northing' was plotted against 'latitude', and 'easting' against 'longitude' to check for any correlation between them because their values appear the same.

Scatter Plot

A scatter plot is a statistical plot to study the relationship or association between two numeric data, usually continuous variables. In case of two related variables, it gives us an idea about how one variable changes with respect to the other. Typically, if one variable is independent and the other is dependent in that case, the independent variable is plotted on the x-axis and the dependent variable on the y-axis. When both continuous variables are independent, then either type of variable can be plotted on any axis, and a scatter plot will illustrate only the degree of correlation (not causation) between two variables. Scatter plot can also be used to spot outliers.

The following lines of code generate the scatter plot shown in Fig. 1.8:

```
[]
tree['longitude'] = tree['geom'].apply(lambda x:
x[x.find("(") + 1 : x.find(")")].split()[0])

tree['latitude'] = tree['geom'].apply(lambda x:
x[x.find("(") + 1 : x.find(")")].split() [1])
```

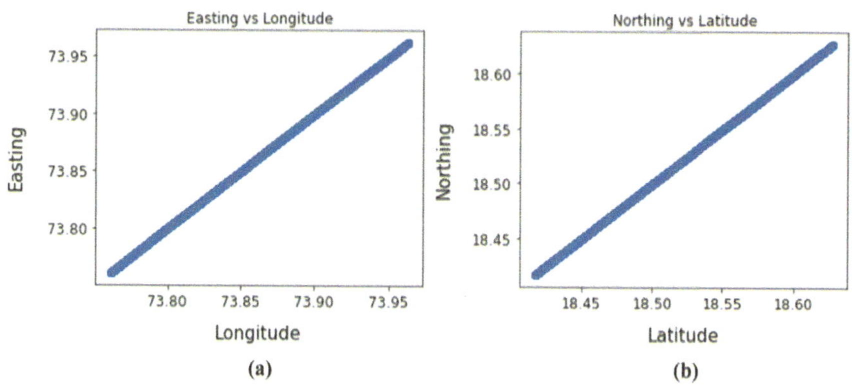

Fig. 1.8 Scatter plot of **a** easting versus longitude and **b** northing versus latitude

```
tree[['longitude', 'latitude']] = tree[['longitude',
'latitude']].astype(np.float64)

tree.drop(['ward_name', 'geom'], axis = 1, inplace =
True)

fig = plt.figure(figsize = (10, 4))

ax = fig.add_subplot(121)
plt.scatter(x = 'longitude', y = 'easting', data =
tree)
plt.xlabel('Longitude')
plt.ylabel('Easting')
plt.title('Easting vs Longitude')

ax = fig.add_subplot(122)
plt.scatter(x = 'latitude', y = 'northing', data =
tree)
plt.xlabel('Latitude')
plt.ylabel('Northing')
plt.title('Northing vs Latitude')
```

These scatter plots show that there exists a perfectly positive correlation between the pairs: easting, longitude and, northing, latitude. Although technically, northings and eastings are obtained when we project the spherical coordinates onto a two-dimensional plane, but here northings and easting hold the same values as latitudes and longitudes, respectively. So the new longitude and latitude columns are dropped as they give the values of the coordinates correct up to 14 decimal places, whereas northing and easting columns, which hold the values of latitudes and longitudes correct up to 6 decimal places, are kept. 'geom' column is dropped as well.

Removing the Inconsistencies

Few values in the dataset contradict each other and make these records a bit difficult to interpret, for example, it is found that there are some trees whose 'local_name' is 'Dead' but 'condition' has been reported as 'Healthy', 'Average' or 'Poor'. Therefore, such values were dropped by running the following lines of code:

```
[]
tree.drop(tree[(tree['local_name'] == 'Dead') &
(tree['condition'] == 'Healthy')].index, axis = 0,
inplace = True)
tree.drop(tree[(tree['local_name'] == 'Dead') &
(tree['condition'] == 'Average')].index, axis = 0,
inplace = True)
tree.drop(tree[(tree['local_name'] == 'Dead') &
(tree['condition'] == 'Poor')].index, axis = 0,
inplace = True)
```

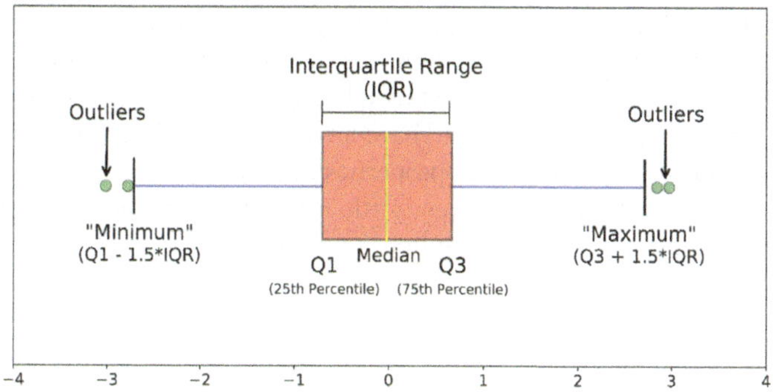

Fig. 1.9 Box-and-whisker plot

Length of the tree data frame after removing these records reduces to 40,09,504.

The 'girth_cm' feature is converted into 'girth_m' attribute so that all the features, girth, height and canopy diameter are in the same unit, i.e. meters. This is achieved by using the following code:

```
[] tree['girth_m'] = tree['girth_cm']/100
```

And then drop the 'girth_cm' column.

Box Plots

The three numerical features (girth_cm, height_m and canopy_dia_m) are found to have a large number of outliers which can be shown through box plots. Box plot is also known as a box-and-whisker plot, as shown in Fig. 1.9, is a chart which shows the distribution of numerical data with the help of a five number summary which includes minimum, 1st quartile, 2nd quartile, 3rd quartile and maximum.

The various components of a box-and-whisker plot are as follows:

- Median: The value of a numerical variable for which 50% of the observations of that variable lie below this value and the rest 50% above it is known as the median of the distribution of this feature.
- Quartiles: The distribution of every numerical variable has three quartiles which are its values that divide its entire distribution into four equal parts or Quadrants such that 25% of its observations lie below the 1st quartile (Q1), 50% below the 2nd quartile (Q2) or the median and 75% below the 3rd quartile (Q3).
- Interquartile range (I.Q.R.):

$$\text{I.Q.R.} = \text{3rd Quartile} - \text{1st Quartile}$$

All values of the variable which lie below the 'Minimum' value, Q1 – 1.5 * I.Q.R., and above the 'Maximum' value, Q3 – 1.5 * I.Q.R. are considered as outliers.

- Whiskers: The part of the box plot shown in blue.
- Outliers: The small green dots.

Here, in the following three box plots it was observed that all the three features girth_cm, height_m and canopy_dia_m of the trees have a large number of outliers. These outliers cannot even be removed or substituted as these are measurements of actual trees. Very old trees are likely to have a wider girth and canopy diameter. The height is associated with many other properties of a tree.

The box plots for the three features 'girth_m', 'height_m' and 'canopy_dia_m' are obtained by running the following lines of code:

```
[] fig = plt.figure(figsize = (10, 6))

   ax = fig.add_subplot(131)
   sns.boxplot(x = 'girth_m', data = tree, orient =
   'vertical')

   ax = fig.add_subplot(132)
   sns.boxplot(x = 'height_m', data = tree, orient =
   'vertical')

   ax = fig.add_subplot(133)
   sns.boxplot(x = 'canopy_dia_m', data = tree,
   orient = 'vertical');

   plt.tight_layout()
```

The output is obtained in the form of the three box plots shown in Fig. 1.10.

A large number of outliers exit for all the three numerical attributes, but none of these outliers can be considered as an anomaly as these are actual measurements, and that is why these outliers can't be removed cannot remove them.

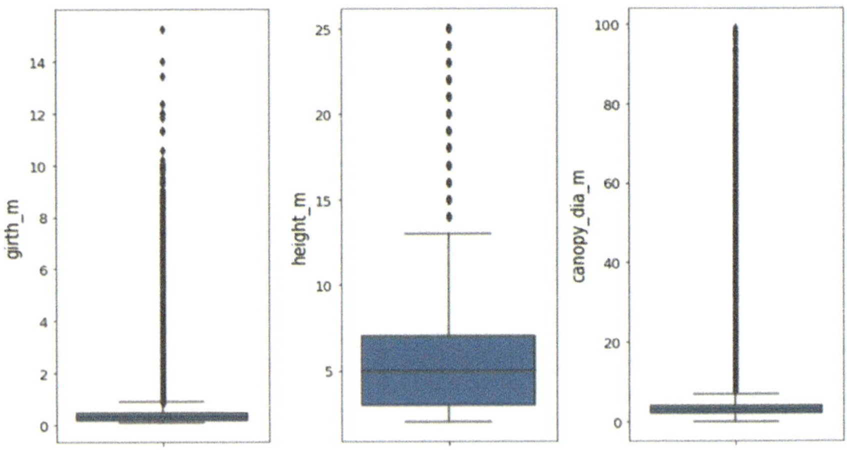

Fig. 1.10 Boxplots of girth, height and canopy diameter (in meters)

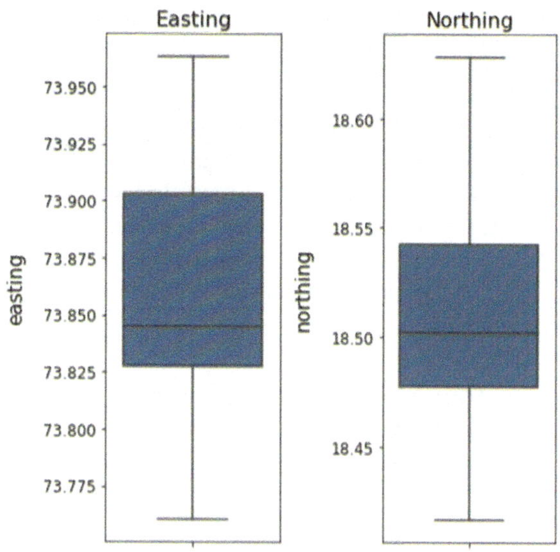

Fig. 1.11 Boxplots of easting and northing

Similarly, the box plots for the remaining two numerical attributes, 'northing' and 'easting' are also obtained by running the following lines of code. The output is in the form of the plots as shown in Fig. 1.11:

```
[]
fig = plt.figure(figsize = (6,6))

ax = fig.add_subplot(121)
sns.boxplot('easting', data = tree, orient =
'vertical')
plt.title('Easting', fontsize = 16)

ax = fig.add_subplot(122)
sns.boxplot('northing', data = tree, orient =
'vertical')
plt.title('Northing', fontsize = 16);

plt.tight_layout()
```

The above box plots of northing and easting clearly show that these features have no outliers. Hence, these are fit to be used for machine learning algorithms like clustering, which is covered in Chap. 2, Understanding Relationships between Air Pollution and Green Cover—A Quantitative Approach: Exploratory Data Analysis.

Investigating remaining attributes for cleaning and wrangling

1. ***condition***: The condition column has four unique values:

```
[In] tree['condition'].unique()
```

```
[Out] [Healthy, Poor, Dead, Average]
Categories (4, object): [Healthy, Poor, Dead,
Average]
```

2. ***ownership***: Ownership column has 15 categories which include: 'Private', 'On Road', 'Garden', 'On Foot Path', 'Govt', 'Public', 'On Wall', 'Semi Government', 'On Divider', 'Avenues', 'Industrial', 'In Well' and 'On Bridge'. A few categories were found to be similar and were given a common name. These included the pairs of categories 'Footpath', 'On Foot Path', and, 'Government', 'Govt'; which were replaced by 'On Foot Path' and 'Govt', respectively.

```
[In] tree['ownership'].unique()

[Out]array(['Private', 'On Road', 'Garden', 'On
Foot Path', 'Government', 'Public', 'On Wall',
'Semi Government', 'On Divider', 'Avenues',
'Industrial', 'In Well', 'On Bridge', 'Footpath',
'Govt'],
Dtype = object)

[In]
tree['ownership'].replace(to_replace =
['Footpath', 'Government'],
        value = ['On Foot Path', 'Govt'],inplace =
  True)

tree['ownership'].unique()

[Out] array(['Private', 'On Road', 'Garden',
'On Foot Path', 'Public', 'On Wall', 'Semi
Government', 'On Divider', 'Avenues',
'Industrial', 'In Well', 'On Bridge', 'Govt'],
dtype= object)
```

3. ***remarks***: The 'remarks' column includes 44.94% null values. For this nominal variable, two similar values were found: 'Mechanical Cut' and 'Mechanically Cut', which were replaced by the common name 'Mechanically Cut'.

```
[In]
[] tree['remarks'].replace(to_replace =
'Mechanical Cut',
Value = 'Mechanically cut',inplace = True)
```

The 44.9% missing values of the 'remarks' column cannot be imputed based on any other feature unless we go and take the recordings again. Also many of these values are missing corresponding to the Dead, Unidentified and Mixed Tree Growth type trees.

4. ***society_name, road_name***: The society and the road names 'Na' values were converted to 'NaN'. Many of these names had a trailing '\n', i.e. a new line character. These were stripped off from these string values. There are other values in this column like 'N/A', 'N/A', 'N/ A', etc. with variations which are left as it is since these can be detected and checked as when they occur.

```
[]
tree['society_name'].replace(to_replace = np.nan,
value = 'Unknown', inplace = True)
tree['road_name'].replace(to_replace = np.nan,
value = 'Unknown', inplace = True)
[]
tree['society_name'] = tree['society_name'].
astype(np.object)
tree['road_name'] = tree['road_name'].
astype(np.object)
[]
tree['society_name'] = tree['society_name'].
apply(lambda x: x.rstrip())
tree['road_name'] = tree['road_name'].apply(lambda
x: x.rstrip())
```

5. ***botanical_name***: This column also has approximately 0.67% missing values. When we tried to find out the reason for these values to not be known, by examining other features corresponding to these records, it was found that the local names and common names corresponding to all the missing values in the botanical name column were reported as either Dead, Unidentified or Mixed Tree Types.

The output of the following code gives a pandas series of the unique local names for the data points where 'botanical_name' is missing.

```
[In] tree.loc[tree['botanical_name'].isnull()]
['local_name'].unique()

[Out]

     Dead                 14590
     Unidentified         12021
     Mixed Tree Growth      174
     Name: local_name, dtype: int64
```

424 different trees in Pune were found to have unique botanical names.

Two new features, namely 'genus' and 'species', were extracted from the botanical_ name(s) using the following lines of code.

```
[In]
  bot_name = [name.split() for name in
  tree['botanical_name']]
  genus = []
  for name in bot_name:
    genus.append(name[0])
  species = []
  for name in bot_name:
    if name == ['Unknown']:
      species.append('Unknown')
      continue
  else:
    species.append(name [1])
  tree['genus'] = pd.Series(genus)
  tree['species'] = pd.Series(species)
```

It was found that Pune has trees from 260 different genera and of 352 different species.

6. ***'common_name' and 'local_name'***: There are 0.67% of missing values in this column which correspond to local_names and common_names: Dead, Unidentified and Mixed Tree Growth. These values cannot be imputed unless we recollect the data for these trees after getting trained by the experts. So they have been left as they are.
7. ***economic_i***: This feature has 12 unique categories which are obtained using the following code:

```
[In] tree['economic_i'].unique()
[Out] array(['Ornamental', 'Medicinal', 'Timber
wood', 'Fruit', 'Essential oil','Paper industry',
'Vegetable', 'Firewood', 'Spice', nan, 'Biodiesel',
'Edible oil', 'Fodder'], dtype = object)
```

This column also has approximately 0.67% missing values corresponding to the trees which have been reported as Dead, Unidentified or of Mixed Growth Type.

8. ***balanced***: This column has no null values and has two unique values.

```
[In] tree['balanced'].unique()
[Out] array([ True, False])
```

9. ***ward***: Data is available for 77 wards of Pune. When 'northing' is plotted against 'easting' in a scatter plot as presented in Fig. 1.12, a rough map of Pune is

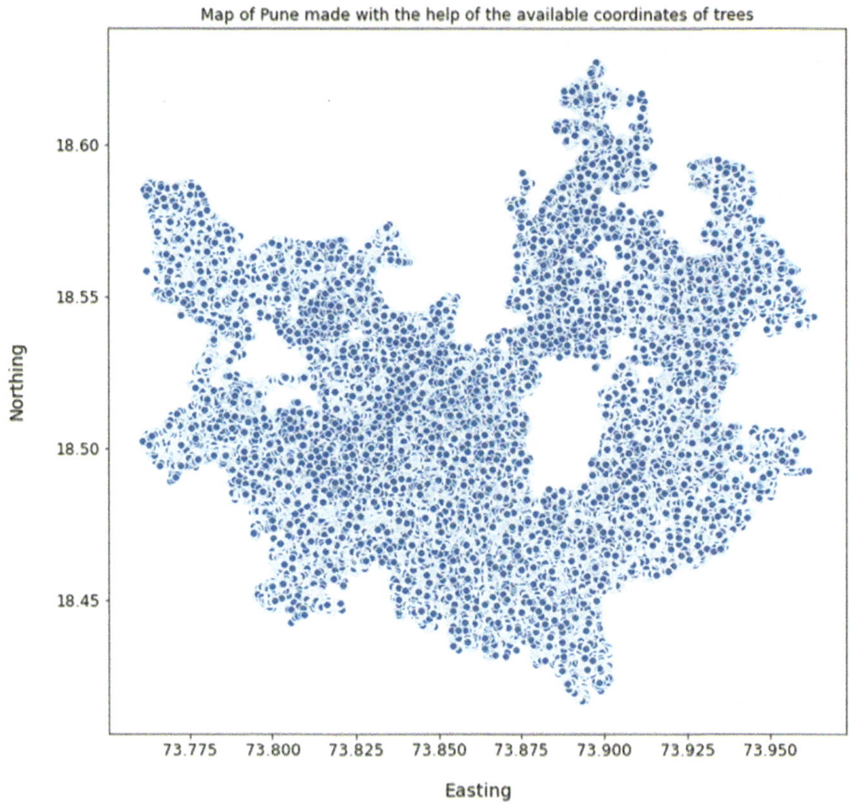

Fig. 1.12 Rough map of Pune (not to the scale)**

obtained, which clearly shows that the data is not available for the Khadki region and the Pune Cantonment area.

10. ***phenology***: This column has only two unique categories: 'Throughout Year' and 'Seasonal'.

```
[In] tree['phenology'].unique()
[Out] [Throughout year, Seasonal, NaN]
Categories (2, object): [Throughout year, Seasonal]
```

This column has 1.07% null values. The missing values corresponding to these tree records cannot be computed as major portion of the missing values were found to correspond to the three local names, 'Dead', 'Unidentified' and 'Mixed Tree Growth'. The rest of the missing values correspond to a few tree species which are listed below, for which information on neither phenology nor flowering is available.

```
[In]
tree[(tree['local_name'] ! = 'Dead') &
```

```
(tree['local_name'] ! = 'Unidentified') &
(tree['local_name'] ! = 'Mixed Tree Growth') &
(tree['phenology'].isnull())]['common_name'].unique()

[Out]
['X-Mas Tree', 'Morpankhi', 'Queen Sago', 'Agathis',
'Sago Cycad',

     'Monterey Cypress', 'Podocarpus', 'White cypress
  pine',
     'Hoop Pine', 'Pinus', 'Italian Cypress', 'Scots
  Pine',
     'Bunya Pine', 'Taxodium']
```

11. ***flowering***: This column has 98 unique values in the form of range of months, and has 1.10% of null values. These values correspond to same kind of records as mentioned for the 'phenology' column.
12. ***is_rare***: This column has two unique values, i.e. 'True' and 'False'

```
[In] tree['is_rare'].unique()
[Out] array(['false', 'true', nan], dtype = object)
```

The 0.67% missing values found in this column cannot be imputed due to same reasons as the 'phenology', 'flowering' and 'economic_i'.

Most of the missing values for botanical_name, phenology, flowering, economic_i and is_rare features are due to the fact that the local_name of the corresponding trees had been reported as Dead, Unidentified or Mixed Tree Growth or the common name of the tree belongs to the list: 'X-Mas Tree', 'Morpankhi', 'Queen Sago', 'Agathis', 'Sago Cycad', 'Monterey Cypress', 'Podocarpus', 'White cypress pine', 'Hoop Pine', 'Pinus', 'Italian Cypress', 'Scots Pine', 'Bunya Pine', 'Taxodium' for which not much information is known for these attributes.

In this step, all the 28 attributes of the dataset were critically studied, carefully cleaned and judiciously wrangled to appropriately prepare tree census data for further analysis.

1.4 Data Preparation: Air Pollution Data

Preparation of air pollution dataset involved identification of data source(s), followed by acquisition of data, and then cleaning and wrangling of data. These steps help in creating an appropriate dataset, which can further be explored, analysed and used for extracting valuable insights.

1.4.1 Understanding Air Pollution

The Earth's atmosphere is comprised of air which is a mixture of variety of gases like nitrogen (78%), oxygen (21%), and argon, carbon dioxide, neon, helium, methane, hydrogen, etc. (which together makes up for less than 1%) and some other gases in traces. Air pollution occurs due to the release of excessive quantities of harmful substances into the atmosphere. The major contributors of air pollution include but are not restricted to:

1. Ammonia
2. Carbon monoxide
3. Sulphur dioxide
4. Nitrous oxides
5. Methane
6. Chlorofluorocarbons
7. Particulates (both organic and inorganic)
8. Biological molecules.

When the concentration of the above molecules increases and exceeds a certain amount, air is said to get polluted. It can have dangerous ramifications on the entire planet and all the living beings residing in it. The major and primary sources of air pollution are the burning of the fossil fuels in power stations, aircrafts, motor vehicles, fumes from paints, aerosol sprays, waste deposition in landfills, nuclear weapons, dust, etc. The emissions from the industrial and power sector contribute immensely to the air pollution.

Two of the most dangerous consequences of air pollution are as follows:

- Health: It can cause many types of diseases including cardiovascular diseases, lung diseases, lung cancer and even deaths in humans.
- Economics: A joint study by the World Bank and the Institute for Health Metrics and Evaluation (IHME) at the University of Washington estimated that around $5 trillion economy per year is lost due to air pollution, in terms of productivity losses and overall decline in the quality of life.

Before we go on to explore a little about the key pollutants, it is important to introduce the concept of air quality index (AQI). It is a number that is used to report, study and understand the quality of air. It gives an indication of how air quality levels can affect human health. More the AQI value, the more polluted is the air. The Environmental Protection Agency (EPA) calculates the AQI for five pollutants, i.e. sulphur dioxide, nitrogen dioxide, ground level ozone, particulate matter (PM2.5 and PM10), carbon monoxide. The national standards for the above pollutants have already been established to safeguard public health [55]. At least 3 of the 5 listed pollutants should be measured, for finding the value of AQI, and it must include either PM10 or PM2.5. Figure 1.13 has been taken from airnow.gov [56] which shows the relationship between the AQI values and levels of health concern.

For a detailed explanation of how the air quality index is measured using the recorded pollutants and its reference criteria, refer to page 40 of Air Quality Report

Air Quality Index (AQI) Values	Levels of Health Concern	Colors
When the AQI is in this range:	*..air quality conditions are:*	*...as symbolized by this color:*
0-50	Good	Green
51-100	Moderate	Yellow
101-150	Unhealthy for Sensitive Groups	Orange
151 to 200	Unhealthy	Red
201 to 300	Very Unhealthy	Purple
301 to 500	Hazardous	Maroon

Fig. 1.13 AQI levels and their interpretation [56]

published by the Maharashtra Pollution Control Board (MPCB) [57] for the year 2017–18.

Now, we discuss the five major pollutants one by one.

1. **Sulphur dioxide (SO_2)**:

It is an invisible gas with a sharp and nasty smell which reacts with other substances to form acids and compounds, namely sulphuric acid, sulphurous acid and sulphate particles. Sulphur dioxide mixes with water to form sulphuric acid, which is a major constituent of acid rain. The SO_2 molecule is shown in Fig. 1.14.

Sources:

Natural: Volcanic activity, forest fires [58].

Man-made: Mostly fossil fuel combustion in industry and motor vehicles.

Effects:

Human health effects resulting from sulphur dioxide includes asthma, chronic bronchitis, affect lung function, irritation, etc.

Fig. 1.14 SO_2 molecule

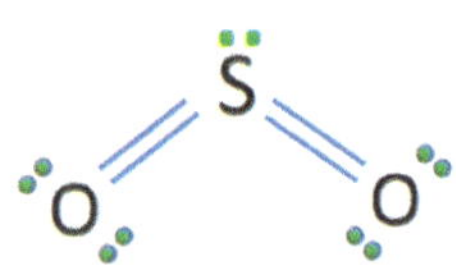

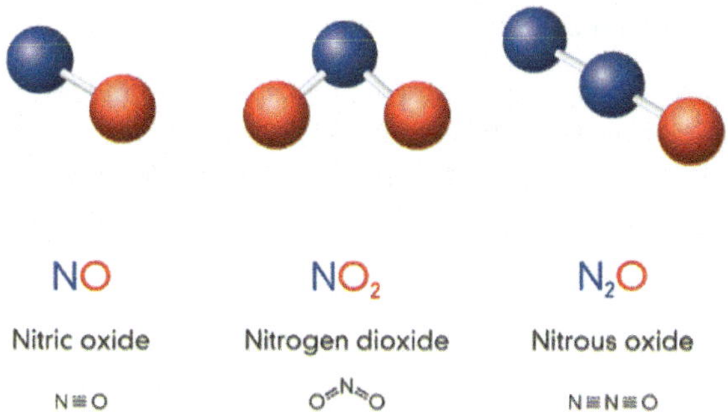

Fig. 1.15 Various nitrogen oxides in atmosphere

Recommended levels:

The recommended air quality standards (set by World Health Organization) [59] for SO_2:

- 20 $\mu g/m^3$ 24-h mean
- 500 $\mu g/m^3$ 10-min mean

According to a report released by Greenpeace on August 19 2019, India is the biggest producer of SO_2 in the world, contributing to more than 15% of global anthropogenic emissions. It also reported that 5 out of 10 global SO_2 emission hotspots from coal industry are located in India [59].

2. **Nitrogen oxides (NO_x)**:

Nitrogen reacts with oxygen to form different types of nitrogen oxides (like NO_2, N_2O, NO, etc.) shown in Fig. 1.15. Among them, nitrogen dioxide (NO_2) and nitric oxide (NO) are contributing the most towards air pollution. Nitrogen dioxide is an acidic gas, having odour and is highly corrosive. Nitric oxide is colourless and forms nitrogen dioxide upon oxidation.

Sources:

Natural sources, motor vehicles and other fuel burning sources.

Effects:

Affects the human respiratory tract and may work in increasing the severity of various respiratory infections and asthma. Chronic lung disease can be caused by long-term exposure.

Recommended levels:

The recommended air quality standards (set by World Health Organization) [60] for NO_2:

- 40 $\mu g/m^3$ annual mean
- 200 $\mu g/m^3$ 1-h mean.

3. **Ozone (O_3) at ground level**:

Ozone is colourless with a very harsh odour. Ozone at ground level is not the same as the atmospheric ozone layer. It is one of the most important constituents of photochemical smog. It is a secondary pollutant as it is formed by the reaction of 2 primary pollutants (sunlight and stagnant air). Figure 1.16 gives the molecular structure of an ozone.

Sources:

Mostly formed due to the reaction (photochemical reaction) of sunlight with nitrogen dioxide (NO_2) and volatile organic compounds (VOCs) given out by vehicles, industry, etc.

Effects:

Human health problems like asthma and various lung diseases.

Recommended levels:

The recommended air quality standards (set by World Health Organization) [59] for O_3:

- 100 $\mu g/m^3$ 8-h mean.

4. **Particulate matter (PM10 and PM2.5)**

PM10 is particulate matter having aerodynamic diameter less than 10 micrometres, whereas PM2.5 refers to particulate matter having diameter less than 2.5 μm, as shown in Fig. 1.17. These are not gaseous. They can be in suspension, liquid or solid state and having divergent composition (often called dust in local language) [61].

(Note: The aerodynamic diameter of a particle is defined as the diameter of a sphere-shaped particle that exhibits the same behaviour in the atmosphere like a dust particle which may not necessarily have a sphere shape).

Sources:

Vehicles (15–23%), biomass burning (12–22%) and dust (15–42%) are the major sources of particulate matter. Secondary particulates contribute significantly to both PM10 and PM2.5 [51].

Effects:

Health effects include heart and lung diseases, acute and chronic bronchitis, respiratory problems and may also cause premature mortality.

Recommended levels:

The recommended air quality standards (set by World Health Organization) [59] for PM10 and PM2.5 are as follows:

Fine particulate matter (PM2.5)

Fig. 1.16 Ozone molecule

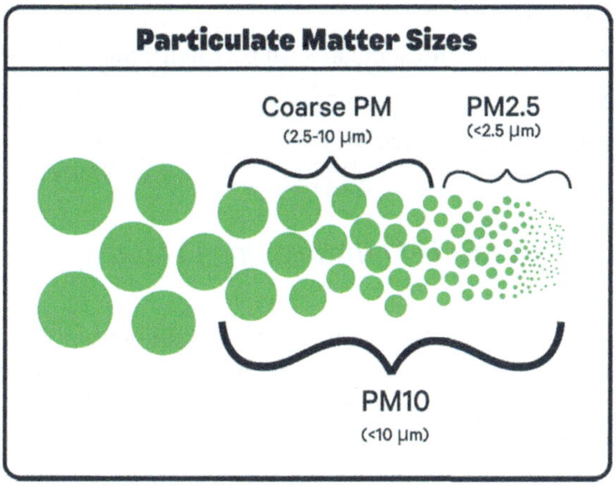

Fig. 1.17 PM10 and PM2.5

- 10 $\mu g/m^3$ annual mean
- 25 $\mu g/m^3$ 24-h mean.

Coarse particulate matter (PM10)

- 20 $\mu g/m^3$ annual mean
- 50 $\mu g/m^3$ 24-h mean.

5. **Carbon monoxide (CO)**:

Carbon monoxide is a colourless, odourless, tasteless and toxic air pollutant. The molecular structure is shown in Fig. 1.18.

Sources:

It is formed by the incomplete combustion of fuels containing carbon like gasoline, natural gas, oil, coal, wood, etc., in case of complete combustion it forms CO_2. The biggest source is from vehicular emissions [62].

Effects:

Outdoor higher levels of CO are generally not observed. Indoors, a high concentration of CO may mix with the haemoglobin to form carboxyhaemoglobin (COHb) which is a stable compound that hinders the normal flow of Oxygen in the human body and may lead to dizziness, nausea, vomiting, respiratory problems, chest pain called angina and even death in extreme cases.

Fig. 1.18 CO molecule

Recommended levels:

The recommended air quality standards set by Environmental Protection Agency (EPA) for CO is:

- 9 ppm averaged over an 8-h period.

1.4.2 Introduction to Air Pollution Open Data

Maharashtra Pollution Control Board (MPCB) is a statutory body entrusted with the responsibilities of reducing pollution and implementing the environmental laws in the state of Maharashtra. In keeping with the standards mentioned in the Air (Prevention and Control of Pollution) Act 1981, it monitors the air pollution levels in the state. MPCB has 12 regional offices (RO) spread over the entire state for monitoring the pollution levels. Here, the point of focus is on the RO-Pune.

Under the MPCB RO-Pune, there are eight places, out of which the five places of interest with their description are shown in Table 1.1 [57].

The dataset published by the Pune Municipal Corporation (PMC) under the MPCB-RO Pune consisted of air pollution data for the above-mentioned five places recorded between January 2018 and December 2019. The review of original air pollution open data is carried out using Python programming language. The steps used to describe the acquired data are mentioned below:

1. Importing the required python libraries 'numpy' and 'pandas' and reading the data file into the notebook:

```
import numpy as np
import pandas as pd
df = pd.read_excel('.../Desktop/Book1.xlsx')
```

Numpy stands for 'Numerical Python', and it is a core library for scientific computing using Python. It helps in numerical computation of high-dimensional

Table 1.1 Five locations where the reading of air pollution was taken

Sr. No.	Station code	Station name	Type	Location used in data collection
1	312	Bhosari	Industrial	MPCB-BSRI
2	379	Nal Stop	Rural and other	MPCB-NS
3	381	Swargate	Residential	MPCB-SWGT
4	708	Pimpri-Chinchwad	Residential	MPCB-PMPR
5		Karve Road	Residential	MPCB-KR

arrays and matrices. Pandas stand for 'Python Data Analysis Library' which is a flexible and power data manipulation tool in Python. df is the name of the data frame used to read the file 'Book1.xlsx' which is the excel file containing the already described the air pollution data.

2. First glimpse of the dataset can be achieved using following command:

```
df.head(10)
```

	Sr. No.	Date	SO_2 µg/m^3	NO_x µg/m^3	RSPM µg/m^3	SPM	CO_2 µg/m^3	AQI	Location
0	1	2018-02-01 00:00:00	49	87	263	375.0	NaN	213.0	MPCB-BSRI
1	2	2018-05-01 00:00:00	33	81	209	467.0	NaN	173.0	MPCB-BSRI
2	3	2018-09-01 00:00:00	34	77	201	305.0	NaN	167.0	MPCB-BSRI
3	4	2018-12-01 00:00:00	42	31	165	255.0	NaN	143.0	MPCB-BSRI
4	5	15-01-2013	33	79	196	297.0	NaN	164.0	MPCB-BSRI
5	6	19-01-2018	21	63	166	313.0	NaN	144.0	MPCB-BSRI
6	7	23-01-2018	33	87	183	3130	NaN	155.0	MPCB-BSRI
7	3	27-01-2018	44	35	152	308.0	NaN	135.0	MPCB-BSRI
8	9	30-01-2018	34	93	187	307.0	NaN	153.0	MPCB-BSRI
9	10	2018-02-02 00:00:00	31	93	209	368.0	NaN	173.0	MPCB-BSRI

The above command displayed the first ten rows of the dataset.

1.4.3 Data Profiling of Air Pollution Open Data

Data profiling is a mandatory step before proceeding to the next step of data cleaning and wrangling. It helps understand the data structure of the dataset, data values and null values present, outliers and junk values, and issues related to data quality (if any), etc. A complete review of each feature present in the dataset can help decide if the identified data source is worthy of efforts in subsequent steps.

The air pollution dataset has 823 records from 2018 and 181 records from 2019. The overview of the dataset is provided in this sub-section. There were a total of 9 columns/attribute/features in the dataset that described as follows:

1. *Serial number*: The serial number assigned to each record of the dataset. There are total 1004 records.
2. *Date*: The date on which the values were recorded.

3. *SO_2* μg/m^3: The concentration of sulphur dioxide (SO_2) measured in μg/m^3 in the particular area on a particular date.
4. *NO_x* μg/m^3: The concentration of nitrogen oxides (NO_x) measured in μg/m^3 in the particular area on a particular date.
5. *RSPM* μg/m^3: The concentration of Respirable Suspended Particulate Matter (RSPM) measured in μg/m^3 in the particular area on a particular date.
6. *SPM*: The concentration of Suspended Particulate Matter (SPM) measured in μg/m^3 in the particular area on a particular date.
7. *CO_2* μg/m^3: The concentration of carbon dioxide (CO_2) measured in μg/m^3 in the particular area on a particular date.
8. *AQI*: The air quality level (AQI) measured in the particular area on a particular date.
9. *Location*: The location of the place where the above details were measured and recorded. The five locations where the readings were taken are—MPCB-BSRI (111 records), MPCB-KR (414 records), MPCB-NS (95 records), MPCB-PMPR (311 records) and MPCB-SWGT (73 records).

There are following observations regarding air pollution dataset:

- The CO_2 column consisted of all null (or missing) values.
- The SPM consisted of a total of 725 missing values.
- 11 values were missing for the AQI column.
- The 'Date' column contained time-stamp for some of the recorded values.

1.4.4 Data Cleaning and Wrangling for Analysis

Before going on to perform some exploratory data analysis and data visualization, it is imperative to clean and pre-process the collected data. The cleaning and pre-processing was done using Python 3 programming language. Below are the explanations of the code and how that is used to clean the data.

1. Brief description of the dataset

```
df.describe ()
```

	Sr. No.	SO_2 µg/m^3	NO_x µg/m^3	RSPM µg/m^3	SPM	CO_2 µg/m^3	AQI
count	1004.000000	1004.000000	1004.000000	1004.000000	279.000000	0.0	993.000000
mean	103.615538	29.516932	63.575697	102.168327	26.781362	NaN	105.420947
std	91.359989	16.443041	31.460609	54.081606	138.430172	NaN	39.929752
min	1.000000	4.000000	9.000000	5.000000	11.000000	NaN	30.000000
25%	29.000000	14.000000	38.000000	63.000000	155.000000	NaN	78.000000
50%	67.000000	27.000000	57.000000	98.000000	277.000000	NaN	108.000000
75%	172.000000	41.000000	85.000000	135.000000	352.500000	NaN	129.000000
max	336.000000	93.000000	198.000000	331.000000	640.000000	NaN	281.000000

The describe command gives the count of each of the columns, the mean, the standard deviation, and the five number summary as described later in the 'boxplot section'. It is seen that the CO_2 column has no recorded values, and the AQI has 11 missing values.

2. The time-stamp is removed by using the following command, and the resulting dataset is printed again:

```
df['Date'] = pd.to_datetime(df['Date'],errors =
'coerce').dt.floor('d')
df.head()
```

	Sr. No.	Date	SO_2 μg/m^3	NO_x μg/m^3	RSPM μg/m^3	SPM	CO_2 μg/m^3	AQI	Location
0	1	2018-02-01	49	37	263	375.0	NaN	213.0	MPCB-BSRI
1	2	2018-05-01	33	81	209	467.0	NaN	173.0	MPCB-BSRI
2	3	2018-09-01	34	77	201	305.0	NaN	167.0	MPCB-BSRI
3	4	2018-12-01	42	31	165	255.0	NaN	143.0	MPCB-BSRI
4	5	2018-01-15	38	79	195	297.0	NaN	164.0	MPCB-BSRI

Now, since the time-stamp problem is solved, the focus can be shifted towards the CO_2 and SPM columns.

3. The 'CO_2 μg/m^3' column is now dropped on the virtue of having no recorded values. Also, the 'Sr No' column is dropped, as we already have an index. The 'read-me' file that was published with the same dataset specifically stated to not consider the SPM values. Also, the 'SPM' column had a total of 725 missing values. For these reasons, the 'SPM' column is also dropped.

```
df.drop(['CO2 μg/m3','Sr No','SPM'], axis = 1,
inplace = True)
df.head()
```

	Date	SO_2 μg/m^3	NO_x μg/m^3	RSPM μg/m^3	AQI	Location
0	2013-02-01	49	87	263	213.0	MPCB-BSRI
1	2013-05-01	33	81	209	173.0	MPCB-BSRI
2	2013-09-01	34	77	201	167.0	MPCB-BSRI
3	2013-12-01	42	81	165	143.0	MPCB-BSRI
4	2013-01-16	38	79	195	164.0	MPCB-BSRI

4. The next step involves the importing of the data visualization libraries in Python such as the 'matplotlib' and 'seaborn'.

```
import matplotlib.pyplot as plt
import seaborn as sns
```

5. Heat map is used as a visualization technique to check on the missing data. A heat map is a type of a graphical representation of data which uses a system of colour-coding to represent different values as shown in Fig. 1.19.

```
plt.figure(figsize = (10,8))
sns.heatmap(df.isnull())
```

The white lines in this above plotted heat map show the missing values. It is observed that only the AQI column has some missing values. Next, we need to find out where are the missing values for the AQI column.

6. The next step is to find out the column which has the missing values for the AQI.

```
df[df['AQI'].isnull()]
```

```
<matplotlib.axes._subplots.AxesSubplot at 0x1d703099f98>
```

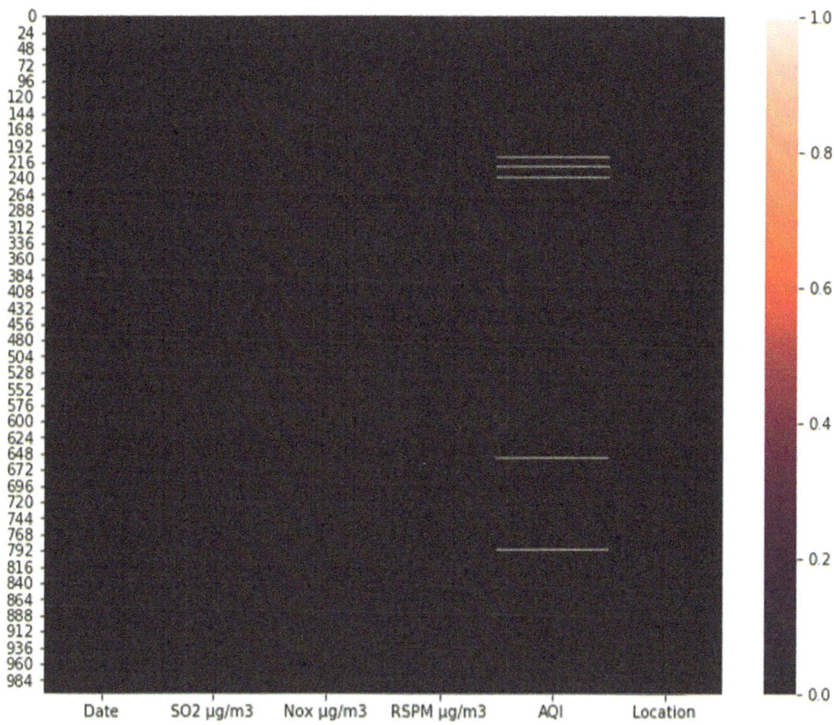

Fig. 1.19 Heat map showing the missing values for the columns

	Date	SO_2 µg/m^3	NO_x µg/m^3	RSPM µg/m^3	AQI	Location
123	2018-01-13	23	45	100	NaN	MPCB-KR
210	2013-04-14	18	43	100	NaN	MPCB-KR
211	2013-04-16	18	46	100	NaN	MPCB-KR
224	2018-01-05	18	47	100	NaN	MPCB-KR
229	2013-06-05	26	54	100	NaN	MPCB-KR
240	2018-05-18	13	33	100	NaN	MPCB-KR
260	2013-08-06	12	33	50	NaN	MPCB-KR
230	2018-06-28	12	31	50	NaN	MPCB-KR
555	2018-12-02	26	30	76	NaN	MPCB-PMPR
792	2018-08-30	15	40	34	NaN	MPCB-PMPR
846	2018-11-17	53	30	34	NaN	MPCB-PMPR

The output clearly shows 11 values of AQI are missing, out of which 8 belongs to MPCB-KR and 3 belongs to MPCB-PMPR.

7. Next step is to check for any suitable way to replace the missing values. Therefore, to find out that, a boxplot is drawn for all the five locations. A boxplot is a statistical plot which can describe the distribution of data based on the following number summary namely:

   - *Minimum*: The minimum value of the selected variable in the dataset.
   - *First quartile (Q1)*: It is the median of the lower half of the dataset. 25% of the numbers of the entire dataset lie below the value, and 75% lie above the value.
   - *Median*: It is the middle data in the dataset.
   - *Third quartile (Q3)*: It is the median of the upper half of the dataset. 25% of the numbers of the entire dataset lie above Q3, and 75% lie below it.
   - *Maximum*: The maximum value of the selected variable in the dataset.

(Note: Q3–Q1 is the interquartile range, which is the middle 50% of the values).

A box plot draws a box connecting the Q1 and Q3 with the median in the middle as demonstrated in Fig. 1.20. It also gives useful information about the outliers and their possible values.

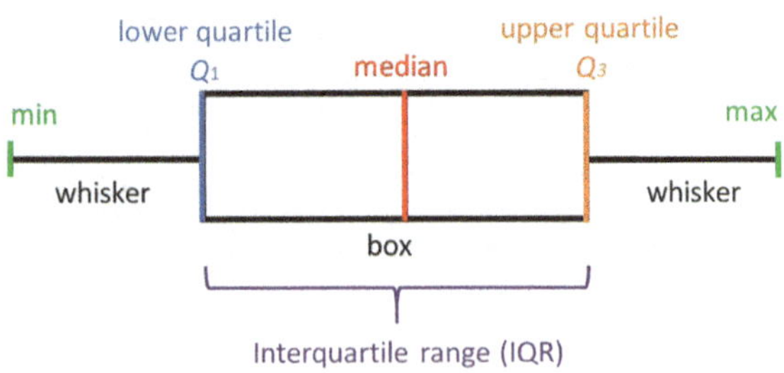

Fig. 1.20 Box plot representation

```
sns.set_style('darkgrid')
plt.figure(figsize = (10,8))
sns.boxplot(x = "Location", y = "AQI", data = df,
palette = 'rainbow')
```

Moving ahead, a box plot is plotted to compare the AQI values for the five locations (MPCB-BSRI, MPCB-KRR, MPCB-NS, MPCB-PMPR, MPCB-SWGT), and the same is shown in Fig. 1.21.

Now, from the above Boxplot, it is clearly seen that AQI median values of both MPCB-KR and MPCB-PMPR are very close to 100. Hence, the missing values in the 11 missing cases can be replaced with the median value, i.e. 100. It is important to note that the mean value is not considered here to fill in the missing values because mean is more sensitive to outliers, and median is a safer choice in that respect.

8. To replace the missing AQI values with 100, the 'fillna' command is used to fill in the missing value and the command 'inplace = True' is for making the change permanent in the data frame. By default, the value is set to False.

```
df['AQI'].fillna(100, inplace = True)
```

9. Finally, another heat map is drawn to confirm whether all the missing values have been removed, and the same is shown in Fig. 1.22.

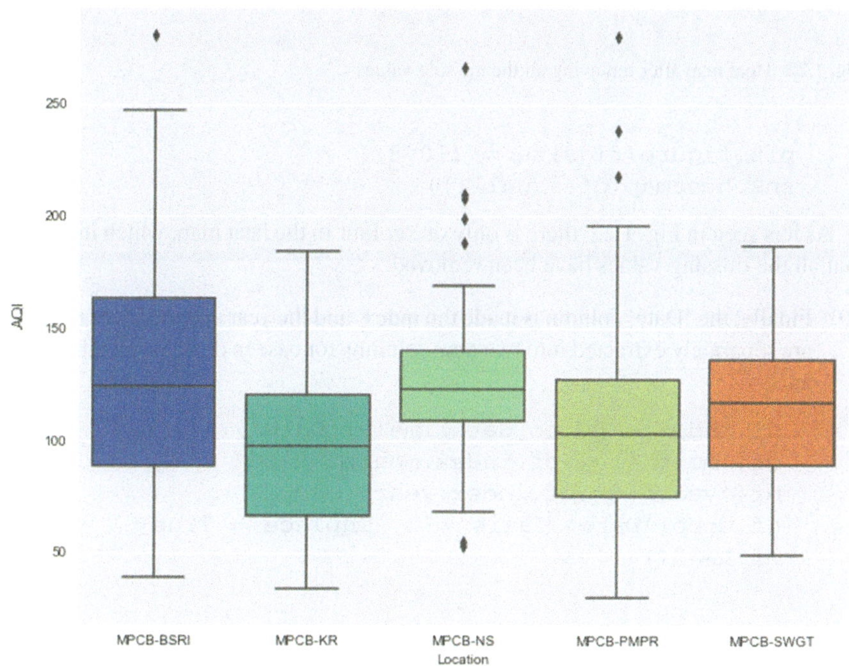

Fig. 1.21 Boxplot showing the AQI values for the five locations

Fig. 1.22 Heat map after removing all the missing values

```
plt.figure(figsize = (10,8))
sns.heatmap(df.isnull())
```

As it is seen in Fig. 1.22, there is only one colour in the heat map, which indicates that all the missing values have been removed.

10. Finally, the 'Date' column is made the index, and the year and month of the date are separately extracted into two new columns for ease in data visualization and analysis.

```
df.index = pd.to_datetime(df.Date)
df['month'] = df.index.month
df['year'] = df.index.year
df.drop('Date', axis = 1, inplace = True)
df.head()
```

Date	SO_2 μg/m^3	NO_x μg/m^3	RSPM μg/m^3	AQI	Location	Month	Year
2018-02-01	49	37	263	213.0	MPCB-BSRI	2	2018
2018-05-01	33	31	209	173.0	MPCB-BSRI	5	2018
2018-09-01	34	77	201	167.0	MPCB-BSRI	9	2018
2018-12-01	42	31	165	143.0	MPCB-BSRI	12	2018
2018-01-16	38	79	196	164.0	MPCB-BSRI	1	2018

The month and year have been extracted and added to two separate columns as shown above with the 'Date' column made as the index.

The data is now ready for exploratory data analysis.

1.5 Conclusion

In the view of inching towards smart cities and sustainable development, use of urban trees and forests in reducing air pollution levels is indispensable. In this chapter, the importance of the green cover and the air pollution characteristics has been explained in great detail. How the green cover can go a long way in curbing the air pollution levels in a particular place is also explained. Now, to make it a fact based study, both the tree census and air pollution data have been thoroughly cleaned and pre-processed, the steps have been explained in detail to make it perfectly ready for further analysis and drawing interesting insights in the subsequent chapters.

References

1. Cockell C, Koeberl C, Gilmour I (2006) Biological processes associated with impact events, 1 edn. Springer Science & Business Media, Berlin, pp 197–219
2. Algeo TJ, Scheckler SE (1998) Terrestrial-marine teleconnections in the Devonian: links between the evolution of land plants, weathering processes, and marine anoxic events. Philos Trans Roy Soc B Biol Sci 353(1365):113–130
3. Bond DPG, Wignall PB (2008) The role of sea-level change and marine anoxia in the Frasnian–Famennian (Late Devonian) mass extinction (PDF). Palaeogeogr Palaeoclimatol Palaeoecol 263(3–4):107–118
4. Gaston KJ (2000) Global patterns in biodiversity. Nature 405(6783):220–227
5. Field R, Hawkins BA, Cornell HV, Currie DJ, Diniz-Filho J, Alexandre F, Guégan J-F, Kaufman DM, Kerr JT, Mittelbach GG, Oberdorff T, O'Brien EM, Turner JRG (2009) Spatial species-richness gradients across scales: a meta-analysis. J Biogeogr 36(1):132–147
6. Gaston KJ, Spicer JI (2013) Biodiversity: an introduction. Wiley, Hoboken. ISBN 978-1-118-68491-7
7. Young A (2003) Global environmental outlook 3 (GEO-3): past, present and future perspectives. Geogr J 169:120
8. CBD monitoring and indicators: designing national-level monitoring programmes and indicators. Convention on Biological Diversity, Montreal (2003)

9. Ten Brink B, Tekelenburg T (2002) Biodiversity: how much is left? National Institute for Public Health and the Environment
10. Mora F (2019) The use of ecological integrity indicators within the natural capital index framework: the ecological and economic value of the remnant natural capital of México. J Nat Conserv 47:77–92
11. Czucz B, Horvath F, Molnar Z, Botta-Dukat Z (2008) The natural capital index of Hungary. Acta Bot Hung
12. Scholes RJ, Biggs R (2005) A biodiversity intactness index. Nature 434:45–49
13. UN World Summit on Sustainable Development: Johannesburg plan of implementation. United Nations, New York (2002)
14. United Nations Official Document. www.un.org
15. Transforming our world: the 2030 Agenda for Sustainable Development. United Nations—Sustainable Development Knowledge Platform
16. Mittermeier RA, Mittermeier CG (2005) Megadiversity: earth's biologically wealthiest nations. Cemex, Mexico
17. Maan JS, Chaudhry P (2019) People and protected areas: some issues from India. Anim Biodivers Conserv 42(1):79–90
18. https://in.one.un.org/
19. Nowak DJ, Crane DE (2002) Carbon storage and sequestration by urban trees in the United States. Environ Pollut 116:381–389
20. Cairns MA, Brown S, Helmer EH, Baumgardner GA (1997) Root biomass allocation in the world's upland forests. Oecologia 111:1–11
21. Moulton RJ, Richards KR (1990) Costs of sequestering carbon through tree planting and forest management in the United States. USDA Forest Service, General Technical Report WO-58, Washington, DC
22. Russo A, Escobedo FJ, Timilsina N, Schmitt AO, Varela S, Zerbe S (2014) Assessing urban tree carbon storage and sequestration in Bolzano, Italy. Int J Biodivers Sci Ecosyst Serv Manage 10(1):54–70
23. Nowak DJ, Crane DE, Stevens JC, Ibarra M (2002) Brooklyn's urban forest. Newtown Square (PA): Northeastern Research Station, United States Department of Agriculture, Forest Service, Borough of Brooklyn. General Technical Report NE-290
24. Zhao M, Kong Z, Escobedo FJ, Gao J (2010) Impacts of urban forests on offsetting carbon emissions from industrial energy use in Hangzhou, China. J Environ Manage 91:807–813
25. Escobedo F, Varela S, Zhao M, Wagner JE, Zipperer W (2010) Analyzing the efficacy of subtropical urban forests in offsetting carbon emissions from cities. Environ Sci Policy 13:362–372
26. Lawrence AB, Escobedo FJ, Staudhammer CL, Zipperer W (2012) Analyzing growth and mortality in a subtropical urban forest ecosystem. Landsc Urban Plan 104:85–94
27. Caragliu A, Del Bo C, Nijkamp P (2011) Smart cities in Europe. J Urban Technol 18(2):65–82
28. The Maharashtra (Urban Areas) Protection and Preservation of Trees Act (1975)
29. Census of India: Pune City Census (2011)
30. Kantakumar LN, Kumar S, Schneider K (2016) Spatiotemporal urban expansion in Pune metropolis. India using remote sensing. Habitat Int 51:11–22
31. Butsch C, Kumar S, Wagner PD, Kroll M, Kantakumar LN, Bharucha E, Schneider K, Kraas F (2017) Growing 'Smart'? Urbanization processes in the Pune urban agglomeration. Sustainability 9:2335
32. Pune towards Smart City: Vision Document, Pune Municipal Corporation (2018)
33. Hamel G (2017, July 21) The effects of cutting down trees on the ecosystem. Sciencing. https://sciencing.com/the-effects-of-cutting-down-trees-on-the-ecosystem-12000334.html
34. https://www.treepeople.org/tree-benefits
35. Trimble S (2019, April 2) Forest & plant canopy analysis—tools & methods. CID Bio-Science Inc. https://cid-inc.com/blog/forest-plant-canopy-analysis-tools-methods/
36. Nowak DJ (2002) The effects of urban trees on air quality. USDA Forest Service, pp 96–102

37. Nowak DJ (2017) Assessing the benefits and economic values of trees. In: Ferrini F, van den Bosch CCK, Fini A (eds) Handbook of urban forestry, Chap 11. Routledge, New York, NY, pp 152–163
38. McPherson EG (1994) Chicago's urban forest ecosystem: results of the Chicago Urban Forest Climate Project, vol 186. US Department of Agriculture, Forest Service, Northeastern Forest Experiment Station
39. Sofia Juliet R (2019, June 22) The nitty-gritty of a tree census. The Hindu. https://www.thehindu.com/news/cities/the-nitty-gritty-of-a-tree-census/article28110162.ece
40. Yang J, McBride J, Zhou J, Sun Z (2005) The urban forest in Beijing and its role in air pollution reduction. Urban For Urban Greening 3(2):65–78
41. Payne J (2016, May 4) What is the economic value of your city's trees? The Conversation. www.weforum.org/agenda/2016/05/what-is-the-economic-value-of-your-citys-trees/
42. Rahman MA, Ennos AR, What we know and don't know about the surface runoff reduction potential of urban trees
43. Ten Brink P, Mutafoglu K, Schweitzer J-P, Kettunen M, Twigger-Ross C, Baker J, Kuipers Y, Emonts M, Tyrväinen L, Hujala T, Ojala A (2016) The health and social benefits of nature and biodiversity protection. A report for the European Commission (ENV.B.3/ETU/2014/0039), Institute for European Environmental Policy, London/Brussels
44. Scharenbroch BC (2012) Urban trees for carbon sequestration. In: Lal R, Augustin B (eds) Carbon sequestration in urban ecosystems. Springer, Dordrecht
45. https://en.wikipedia.org/wiki/Carbon_sequestration/
46. Pune Tree Census August 2019. Pune Municipal Corporation Open Data Portal. treecensus.punecorporation.org
47. Agarwal S, How to measure the diameter, height and volume of a tree. Air Pollution. www.environmentalpollution.in/forestry/how-to-measure-a-tree/how-to-measure-the-diameter-height-and-volume-of-a-tree-forestry/4732/
48. Stecker T (2014, January 17) E & E: old trees store more carbon more quickly, than younger trees. E&E ClimateWire. www.pacificforest.org/ee-old-trees-store-more-carbon-more-quickly-than-younger-trees
49. Phase 3 Field Guide—Crowns: Measurements and Sampling (2005, October) Forest Inventory and Analysis (FIA) Program of the U.S. Forest Service. www.fia.fs.fed.us/library/field-guides-methods-proc/docs/2006/p3_3-0_sec12_10_2005.pdf
50. Branch Collar (2018 November 10). www.thedailygarden.us/garden-word-of-the-day/branch-collar
51. treeinspection.com/advice/how-to-spot-a-dangerous-tree?showal=1
52. https://en.wikipedia.org/wiki/Botanical_name
53. https://en.wikipedia.org/wiki/Common_name
54. https://en.wikipedia.org/wiki/Phenology
55. Pune Municipal Corporation Open Data Portal: "Air Quality Parameters—MPCB" (2018 March, 2019). opendata.punecorporation.org/Citizen/CitizenDatasets/Index?categoryId=17
56. https://www.airnow.gov/aqi/aqi-basics/
57. http://www.mpcb.gov.in/sites/default/files/focus-area-reports-documents/Air_Quality_Report_2017-18_29052019.pdf
58. https://www.pca.state.mn.us/air/sulfur-dioxide
59. https://www.downtoearth.org.in/news/air/india-emits-the-most-sulphur-dioxide-in-the-world-66230
60. https://www.who.int/news-room/fact-sheets/detail/ambient-(outdoor)-air-quality-and-health
61. http://www.npi.gov.au/resource/particulate-matter-pm10-and-pm25
62. https://www.teriin.org/sites/default/files/2018-08/AQM-SA_0.pdf

Chapter 2
Exploring Air Pollution and Green Cover Dataset—A Quantitative Approach

2.1 Data Exploration: Tree Census

'Money actually does grow on trees'! from the perspective of urban planning, in contrast to expansive forest landscapes in a city, small clusters or even individual trees can provide measurable economic, environmental, social and health benefits for urban populations. According to the India State of Forest Report 2019, released by the Union ministry of environment and forests, the forest cover in Pune District of Maharashtra State in India was 1708.00 km^2 in 2017 and 1710.86 km^2 in 2019, resulting in the rise of 2.86 km^2 over two years. The forest cover of Pune District is nearly 11% of the total geographical area. Besides, urban canopies in the city directly contribute to meeting a regulatory regarding clean air requirement. Thus, Pune has both a significant forest area and an urban forest cover, which has a great potential to increase its overall green cover.

After describing and cleaning the tree census data in Chap. 1, the analysis now moves on to exploring the data. While cleaning, 119 inconsistent records from the tree census dataset were dropped, which resulted in 40,09,504 tree records in the pandas data frame. Figure 2.12 of the last chapter, a blue coloured scatter plot obtained using the information on northings and eastings for all the trees, gave an idea of the areas of Pune that were covered in the tree census project and the areas that were left out.

This section explores each and every attribute of the tree census dataset with the help of various charts and plots. Graphical presentation helps to comprehend the data effectively.

N. Sharma et al., *Open Data for Sustainable Community*,
Advances in Sustainability Science and Technology,
https://doi.org/10.1007/978-981-33-4312-2_2

2.1.1 Hexbin Plot Representing Tree Density

A Hexbin plot is used to visualize the concentration of trees throughout the city, which takes a list of X and Y values as input. It looks similar to a scatter plot, with the entire graphing space divided into hexagons, with all points grouped into their respective hexagonal regions and a colour gradient indicating the density. More data points a hexagon covers the darker is its colour. A scatter plots displays all the data points even when they are overlapping. Hexbin plots are said to be most suitable for spatial data, i.e. the data containing latitudes and longitudes, northings and eastings, etc.

The hexbin plot in Fig. 2.1 shows the concentration of trees in different regions in Pune. More the number of trees covered by the hexagon darker is its colour. This plot helps in identifying the areas inside the city that have less green cover. These are the areas where the new trees should be planted keeping in mind all the environmental

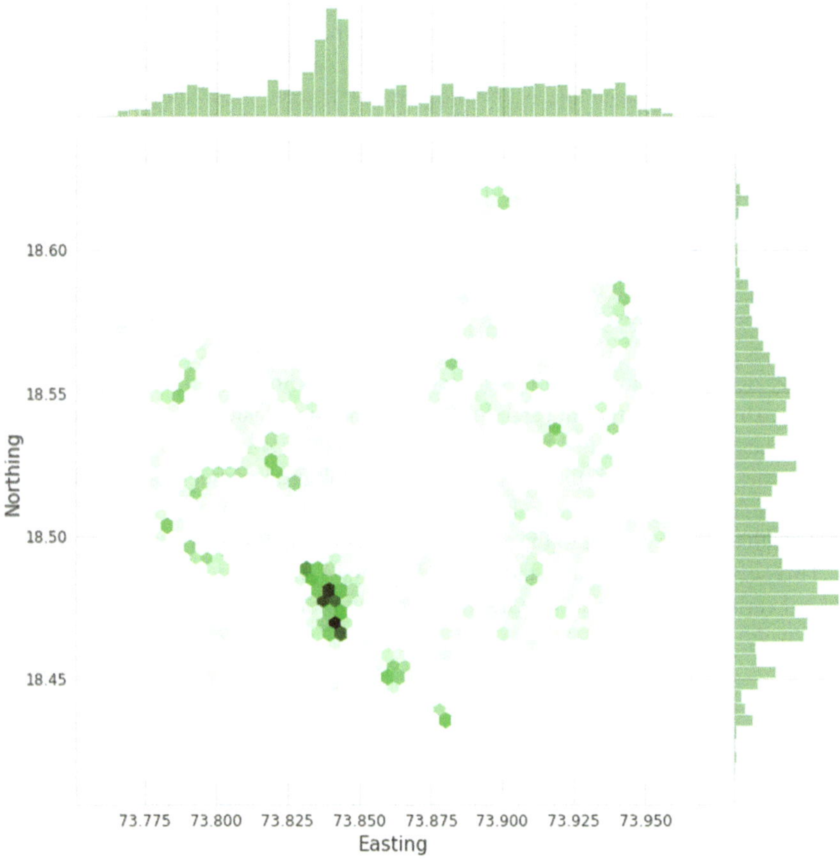

Fig. 2.1 Hexbin plot showing tree concentration in Pune

factors including the location. The following Python code generates the required hexbin plot:

```
[]
g = (sns.jointplot(x='easting', y='northing',
data=tree, kind='hex', color=
'green').set_axis_labels('Easting', 'Northing'))
g.fig.set_size_inches(10,10);
```

2.1.2 Pie Chart Representing the Categories of Tree Condition

A pie chart is a circular graph which represents a whole, and its slices represent portions of the whole that belong to a particular category. The four categories of tree condition as per tree census dataset are 'Healthy', 'Average', 'Poor', 'Dead'. After removing the 119 inconsistencies, the pie chart of the four categories of tree condition from the remaining 40,09,504 records is plotted using the following lines of code and presented in Fig. 2.2:

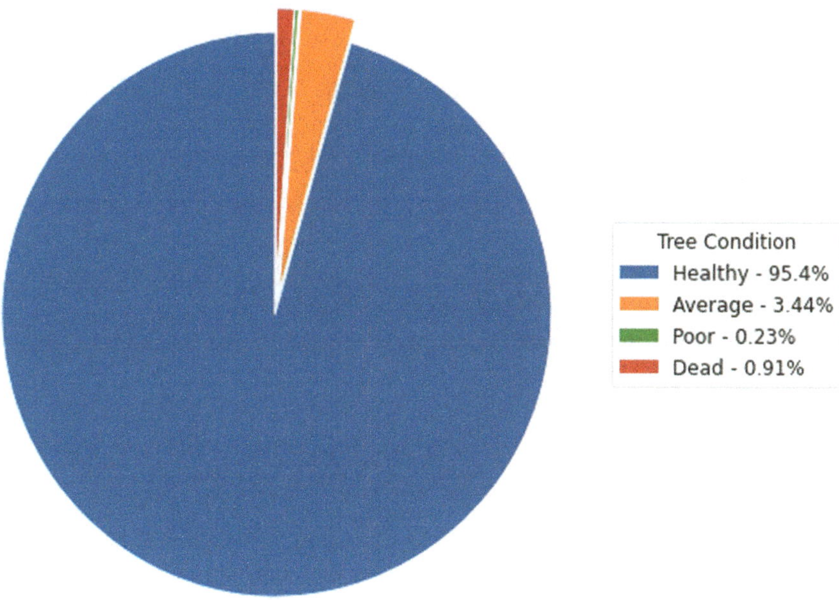

Fig. 2.2 Pie chart representing the tree conditions

```
[]
import matplotlib as mpl
mpl.rcParams['font.size'] = 12.0
# Pie chart, where the slices will be ordered and plotted counter-
clockwise:
labels = 'Healthy - 95.4%', 'Average - 3.44%', 'Poor - 0.23%', 'Dead -
 0.91%'
label = 'Healthy', 'Average', 'Poor', 'Dead'
sizes = [3834030, 125692, 8892, 40890]
explode = (0, 0.1, 0.1, 0.1)  # only "explode" the 2nd slice (i.e. 'Hogs')

fig1, ax1 = plt.subplots(figsize=(8,8))
pie = ax1.pie(sizes, explode=explode, labels=label,
        startangle=90, radius=1.2, rotatelabels=True)
ax1.axis('equal')  # Equal aspect ratio ensures that pie is drawn as a circ
le.

ax1.legend(pie[0],labels,title="Tree Condition",
          loc="center left",
          bbox_to_anchor=(1, 0, 0.5, 1))

for pie_wedge in pie[0]:
    pie_wedge.set_edgecolor('white')

plt.tight_layout()
plt.show()
```

It is observed that 95.4% of the trees are healthy, 3.44% were reported to be in average condition, whereas 0.23% in poor condition. Approximately, 0.91% of the trees were found to be dead. So, 0.91% of 40,09,504 trees is approximately 36,486 dead trees!

2.1.3 *Violin Plots to Study Health-Category-Wise Distribution of Girth*

A violin plot shows the distribution of a variable (here tree girth in metres). The violin plot is wider for the range of values of a variable that occurs most frequently. Violin plots are best used when comparing two or more distributions.

Figure 2.2 in Sect. 2.1.2 presents an over-optimistic picture of the health condition of the green cover in the city by merely looking at the percentage of healthy trees (95.4%). However, health conditions can also be determined using girth of the tree as it gives a fair idea about the age of a tree. Figure 2.3 compares the distribution of the girth of trees based upon the four health conditions through their violin plots obtained using the following code:

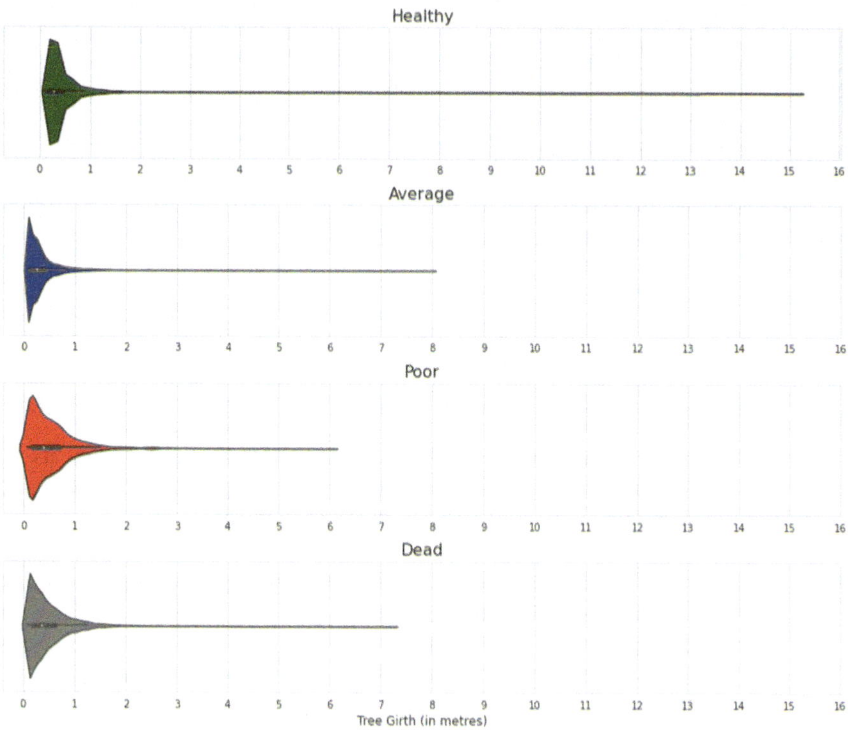

Fig. 2.3 Violin plots to compare the distribution of tree girths (in metres) based on the tree condition

```
[]
fig = plt.figure(figsize=(12,10))

ax = fig.add_subplot(411)
sns.violinplot(tree[tree['condition'] == 'Healthy']['girth_m'], kde=False,
color='g')
plt.title('Healthy', fontsize=16)
plt.xticks([i for i in range(17)])
plt.xlabel('')

ax = fig.add_subplot(412)
sns.violinplot(tree[tree['condition'] == 'Average']['girth_m'], kde=False,
color='b')
plt.title('Average', fontsize=16)
plt.xlabel('')
plt.xticks([i for i in range(17)])
```

```
ax = fig.add_subplot(413)
sns.violinplot(tree[tree['condition'] == 'Poor']['girth_m'], kde=False, col
or='r')
plt.title('Poor', fontsize=16)
plt.xlabel('')
plt.xticks([i for i in range(17)])

ax = fig.add_subplot(414)
sns.violinplot(tree[tree['condition'] == 'Dead']['girth_m'], kde=False, col
or='gray')
plt.title('Dead', fontsize=16)
plt.xlabel('Tree Girth (in metres)', fontsize=12)
plt.xticks([i for i in range(17)]);
```

From these violin plots, three inferences can be drawn:

1. Very large concentrations of healthy trees have girth below 2 m; i.e. they are all young trees or the newly planted trees. So it is very much possible that the health of many trees in Pune City might be deteriorating with age.
2. The concentration of these trees, i.e. the young ones with girth less than 2 m, is higher in the other three categories of tree condition which are average, poor and dead. This fact points to the possibility that the younger trees are at a higher risk of not surviving harsh environmental and climatic conditions if not taken care of. Therefore, the trees which are newly planted should be given special care.
3. It is also clear that the maximum girth of the healthy trees is almost double that of the maximum girth of trees in all the other three categories. This shows that unlike humans, the older trees tend to be healthier and more active, thus supporting more life. This happens when either the tree is taken proper care of throughout its entire life or if it is left away from human intervention. It was found on further investigation that Pune has 82 trees that have girth greater than or equal to 8 m, out of which 44 were Banyan trees.

2.1.4 Scatter Plot to Spot Clusters of Poor Quality and Dead Trees

Figure 2.4 is the scatter plot of 'northing' versus 'easting' with only those data points that correspond to the trees that are either in 'Poor' condition or are 'Dead'. The following code generates the required scatter plot:

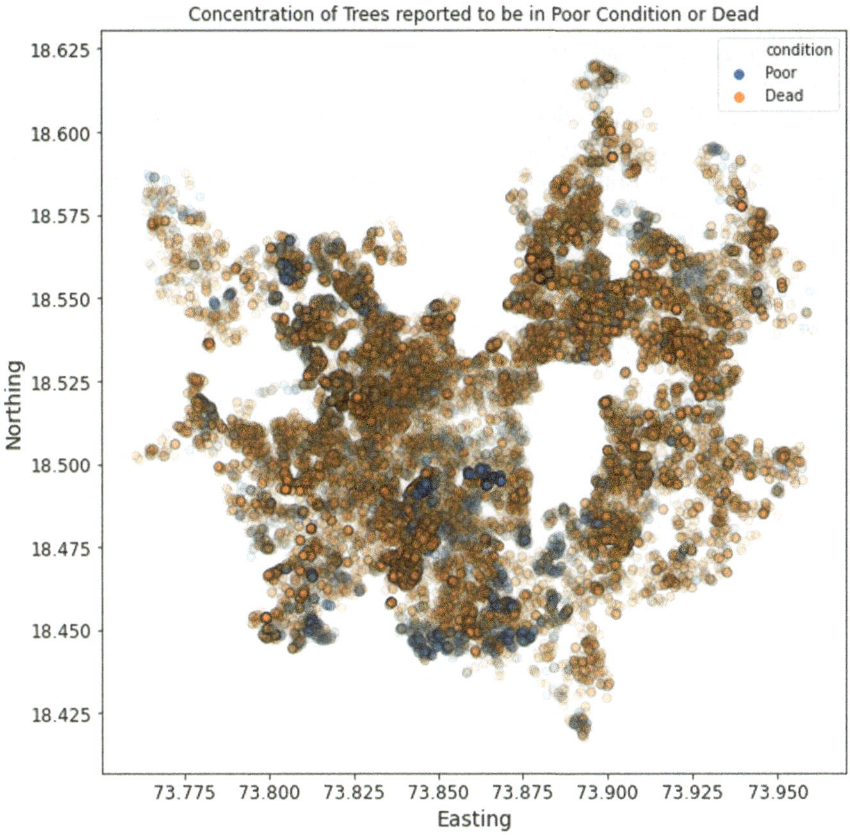

Fig. 2.4 Scatter plot showing the distribution of dead trees and those in poor condition

```
[]
plt.figure(figsize=(9,9))
sns.scatterplot( x='easting', y='northing', data = tree,edgecolor='black',
plt.figure(figsize=(9,9))
sns.scatterplot( x='easting', y='north-
ing', data = tree,edgecolor='black',alpha=0.1, hue=tree[(tree['condi-
tion'] != 'Healthy') & (tree['condition'] != 'Average')]['condition'])
plt.xlabel('Easting')
plt.ylabel('Northing');
alpha=0.1, hue=tree[(tree['condition'] != 'Healthy')&(tree['condi-
tion'] != 'Average')]['condition'])
plt.xlabel('Easting')
plt.ylabel('Northing');
```

Orange colour represents 'Dead', and blue represents 'Poor' health condition of a tree. The darker areas of either colour in this plot indicate the regions of high concentration of data points corresponding to that category; i.e. the dark orange region shows that there is a very high concentration of 'Dead' trees in that region. Clusters/groups of 'Dead' trees are clearly visible, which is an evidence that the diseases tend to spread from one tree to another. Also, other factors like the rarity of these trees along with other environmental factors should be studied in depth especially for the areas where larger clusters of such trees are found; in order to learn about what other environmental factors might have had a significant impact on the health of these trees.

2.1.5 *Mode of All Categorical Variables*

Mode of the observations or the distribution of values of a variable is the value which has the highest frequency of occurrence (or the number of occurrences) among all its values. The following code gives the mode of all the important categorical features that were recorded in the tree census:

```
[In] tree.mode().iloc[0]

[Out]
```

```
condition                                    Healthy
ownership                                 Government
society_name                        Bhamburda Forest
road_name                           Taljai Hill Road
botanical_name     Gliricidia sepium (Jacq.) Walp.
common_name                               Giripushpa
local_name                                Giripushpa
economic_i                               Timber wood
phenology                                   Seasonal
flowering                                 March-June
ward                                              53
is_rare                                        false
```

It was found that the mode of the society name, i.e. the society name to which maximum number of tree records corresponds to was found to be the Bhamburda Forest region. Also the Taljai hill road, among all the roads where the tree data was collected from, was found to have the biggest share of trees. The mode of local name and common name yielded that the most frequently occurring tree in Pune is the Giripushpa. The largest share of the trees covered in this census yield timber wood. Among all the 77 wards covered in the census, ward number 53 was found to have the largest number of trees, i.e. around 7,71,187 trees, which is 5,46,313 trees more than the ward number 22 with the next highest number of trees, i.e. 2,24,874 trees.

2.1.6 Top 10 Wards with Highest Number of Trees

This section presents the wards in the Pune City that has highest number of trees. This information helps to understand the actual green cover of the city and also the wards that are sparse.

```
[In] tree['ward'].value_counts().head(10)

[Out]
53    771187
20    224874
9     203800
1     186080
2     165991
75    159436
45    158082
29    138512
11    115419
30     97158
Name: ward, dtype: int64
```

2.1.7 In-Depth Analysis of Tree Condition and Ownership Type

2.1.7.1 Count Plot of Total Number of Trees by Ownership Type

A count plot shows the categories of a qualitative or integral variable on the x-axis and the number of observations corresponding to each category on the y-axis in the form of bars. The analysis of the trees with respect to the ownership types shows that largest share of healthy trees are owned by the government entities, followed by the private entities as shown in Fig. 2.5.

```
[]
plt.figure(figsize=(23,6))
sns.countplot(tree['ownership'], color='g', alpha=0.7,
              order=['Govt', 'Private', 'Public', 'Garden', 'On Road', 'On
Foot Path', 'Semi Government', 'Industrial', 'Avenues', 'On Divider', 'On B
ridge', 'In Well'])
plt.title('Total Count of Trees by Ownership Type', fontsize=20)
plt.xticks(fontsize=18, rotation=45)
plt.yticks(fontsize=18)
plt.xlabel('Ownership Types', fontsize=18)
plt.ylabel('Count', fontsize=18);
```

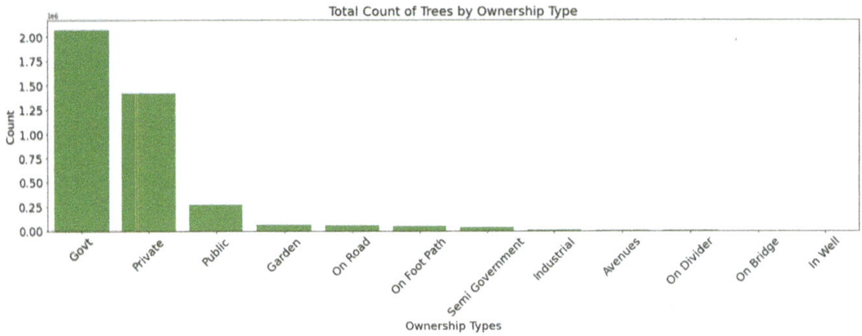

Fig. 2.5 Count plot of the total number of trees corresponding to each ownership type (y-axis scale in 10^6 trees)

2.1.7.2 Count Plot of the Dead Trees by Ownership Type

Similar analysis is run for only the dead trees owned by each of these ownership types. On running the following lines of code, we get the count plot of the dead trees for each ownership type as presented in Fig. 2.6:

```
[]
plt.figure(figsize=(23,6))
sns.countplot(tree[tree['condition'] ==
'Dead']['ownership'], color='r',
alpha=0.7,order=['Govt', 'Private', 'Public', 'Garden',
 'On Road', 'On Foot Path', 'Semi Government',
'Industrial', 'Avenues', 'On Divider',
'On Bridge', 'In Well'])
plt.title('Number of Dead Trees by Ownership Type',
 fontsize=20)
plt.xticks(fontsize=18, rotation=45)
```

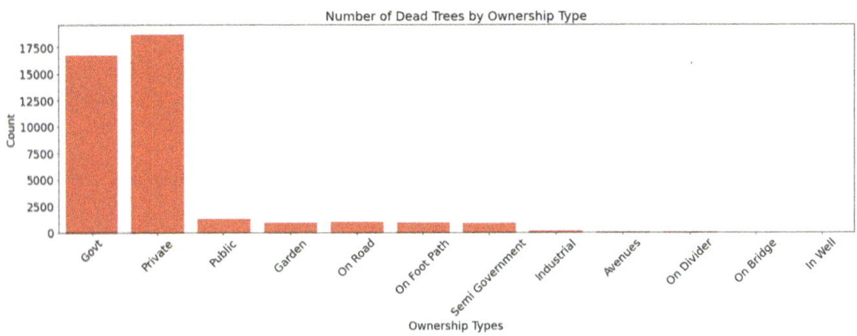

Fig. 2.6 Count plot of 'dead' trees corresponding to each 'Ownership' type

```
plt.yticks(fontsize=18)
plt.xlabel('Ownership Types', fontsize=18)
plt.ylabel('Count', fontsize=18);
```

The plot in Fig. 2.6 shows that the largest share of trees that have been reported dead is owned by the private entities despite the fact that they own lesser number of trees as compared to the government entities as shown in Fig. 2.5. Thus, from the count plots of total trees and dead trees, we conclude that the health of the trees owned by the private entities and those growing on roads, on bridges, etc., is compromised more than that of the trees owned by the Government. The private owners must take better care of the trees growing on their properties, and also the trees that are planted on bridges and on the road sides should be planted keeping in mind whether they can survive among all the pollution emitted from the vehicles or not.

2.1.7.3 Counts, Absolute and Cumulative Percentages of Trees for Different Tree Conditions by Ownership Types

To analyse and understand the ownership type under which the trees are at a greater risk, the following code was run to obtain the data frame with the percentage of dead trees (of the total number of dead trees counted in the tree census) owned by each ownership type (named as 'Perc_dead'), and the cumulative of these percentages (column named as 'cumulative_perc').

```
[In]
df3 = pd.DataFrame(tree[tree['condition'] ==
'Dead']['ownership'].value_cou
nts() / float(tree[tree['condition'] ==
'Dead'].shape[0]) * 100)
df3.columns = ['Perc_dead']
df3['cumulative_perc'] = df3['Perc_dead'].cumsum()
df3
[Out]
```

	Perc_deadi	Cumulative_perc
Private	45.717779	45.717779
Govt	41.046711	86.76449
Public	3.230619	89.995109
On road	2.404011	92.39912
On foot Path	2.323306	94.722426
Garden	2.279286	97.001712
Semi-government	2.169235	99.170946
Industrial	0.420641	99.591587

(continued)

(continued)

	Perc_deadi	Cumulative_perc
On wall	0.183419	99.775006
Avenues	0.114943	99.889949
On divider	0.100269	99.990218
In well	0.007337	99.997554
On bridge	0.002446	100

The above data frame shows that the private and government entities own around 86.8% of all the dead trees counted in the tree census. This is equivalent to saying that approximately 87% of the dead trees are owned by only 15.4% of all the ownership types (i.e. only 2 out of the 13 ownership types). This can be interpreted with the help of the Pareto Principle. The Pareto Principle states that '80% of the consequences come from 20% of causes' [1]. Though this is not sufficient to know under which ownership type, the trees deteriorate in their health the most.

For this further analysis was carried out to study what count and percentage of trees owned by each ownership types are at a greater risk. The code below gives us the following data frame. This data frame consists of 7 columns whose description is given below:

1. **Total_tree_count** Total number of trees owned by each ownership type
2. **Perc_total** Percentage of total number of trees owned by the corresponding ownership type
3. **Live count** Total number of live trees (i.e. the sum of all the trees reported as healthy, average or poor)
4. **Dead count** Total number of dead trees owned by the corresponding ownership type
5. **Perc_live** Percentage of live trees out of the total number of trees owned by that ownership type
6. **Perc_dead** Percentage of dead trees out of the total number of trees owned by that ownership type
7. **Cumulative_perc_total** Cumulative percentage of the total trees owned by different ownership types

```
[In]
ser1 = tree['ownership'].value_counts()
ser2 = tree['ownership'].value_counts() / float(tree.shape[0]) * 100
ser3 = tree[tree['condition'].isin(['Healthy', 'Average', 'Poor'])]['owners
hip'].value_counts()
ser4 = tree[tree['condition'] == 'Dead']['ownership'].value_counts()

list1 = []
list2 = []
for own in ['Govt', 'Private', 'Public', 'Garden', 'On Road', 'On Foot Path
', 'Semi Government', 'Industrial', 'On Divider', 'Avenues', 'On Wall', 'In
 Well', 'On Bridge']:
  a = len(tree[tree['condition'].isin(['Healthy', 'Average', 'Poor']) &
(tree['ownership'] == own)]) / float(len(tree[tree['ownership'] == own])) *
 100
  b = len(tree[(tree['ownership'] == own) & (tree['condition'] == 'Dead')])
 / float(len(tree[tree['ownership'] == own])) * 100
  list1.append(a)
  list2.append(b)
ser5 = pd.Series(list1)
ser6 = pd.Series(list2)

ser = pd.Series(['Govt', 'Private', 'Public', 'Garden', 'On Road', 'On Foot
 Path', 'Semi Government', 'Industrial', 'On Divider', 'Avenues', 'On Wall'
, 'In Well', 'On Bridge'])
df1 = pd.concat([ser5, ser6, ser], axis=1)
df1.columns = ['Perc_live', 'Perc_dead', 'Ownership']
df1.set_index('Ownership', inplace=True)

df2 = pd.concat([ser1, ser2, ser3, ser4], axis=1)
df2.columns = ['Total Tree Count', 'Percentage from Total', 'Live Count', '
Dead Count']

d = pd.concat([df2, df1], axis=1)
d = pd.concat([df2, df1], axis=1)
d['cumulative_perc_total'] = d['Perc_total'].cumsum()
d
[Out]
```

Ownership	Total_Tree_Count	Perc_totall	Live count	Dead count	perc_live	Perc_dead	Cumulative_perc_total
Govt	2073308	51.709837	2056524	16784	99.190472	0.809528	51.709837
Private	1422231	35.471495	1403537	18694	98.685586	1.314414	87.181332
Public	275070	6.860450	273749	1321	99.519759	0.480241	94.041782
Garden	62881	1.568299	61949	932	98.517835	1.482165	95.610080
On Road	56406	1.406807	55423	983	98.257278	1.742722	97.016888
On Foot Path	54957	1.370668	54007	950	98.271376	1.728624	98.387556
Semi-Government	40402	1.007656	39515	887	97.004564	2.195436	99.395212
Industrial	11451	0.285596	11279	172	98.497948	1.502052	99.680808
On Divider	6101	0.1521 63	6060	41	99.327979	0.672021	99.832972
Avenues	3714	0.092630	3667	47	98.734518	1.265482	99.925602
On Wall	2765	0.068961	2690	75	97.287523	2.712477	99.994563
In Well	187	0.004664	184	3	98.395722	1.604278	99.999227
On Bridge	31	0.000773	30	1	96.774194	3.225806	100.000000

Key findings and conclusions from the above data frame:

1. **Perc_total**
   51.71% trees are owned by the government; i.e. more than half the trees counted in the tree census are on government property. Private entities own 35.47% trees. This implies that around 87.2% of the trees are owned by these two ownership types alone.
2. **Perc_live**
   Percentage of live trees owned by all ownership types is greater than 97%. This is a good sign.
3. **Perc_dead**

   a. As per the analysis, government and private entities together own the largest percentage of dead trees. However, when the percentage of dead trees owned by all the ownership types is compared, it shows that only 0.8% of government owned trees and 1.3% of those owned by the private entities, are found to be dead.
   b. The trees growing on bridges are at the highest risk with the highest percentage (approximately 3.28%) of dead trees. These are followed by the trees growing on walls, roads, footpaths, etc. The tree owned by semi-government entities also seem to be at a very high risk.

Bar Plot of Percentage of Dead Trees by Ownership Type

A bar plot is a plot which represents the category-wise values of a particular quantity on the y-axis and the corresponding categories on the x-axis. In order to study, the percentage of dead trees of the total number of trees owned by the particular ownership type, the bar plot as shown in Fig. 2.7, was obtained by running the following lines of code.

```
[]
d.plot.bar(x='Ownership', y='Perc_dead', rot=45,
figsize=(10,5), color='r')
plt.ylabel('Percentage of Dead Trees');
```

2.1.8 *Box Plot of Tree Girth and Canopy by Tree Condition and Rarity*

It is known that trees need favourable environmental conditions for their growth and survival. Now, to compare the growth of the rare species of trees with the indigenous ones, the boxplot of the tree girth as shown in Fig. 2.8 was obtained using the following lines of code:

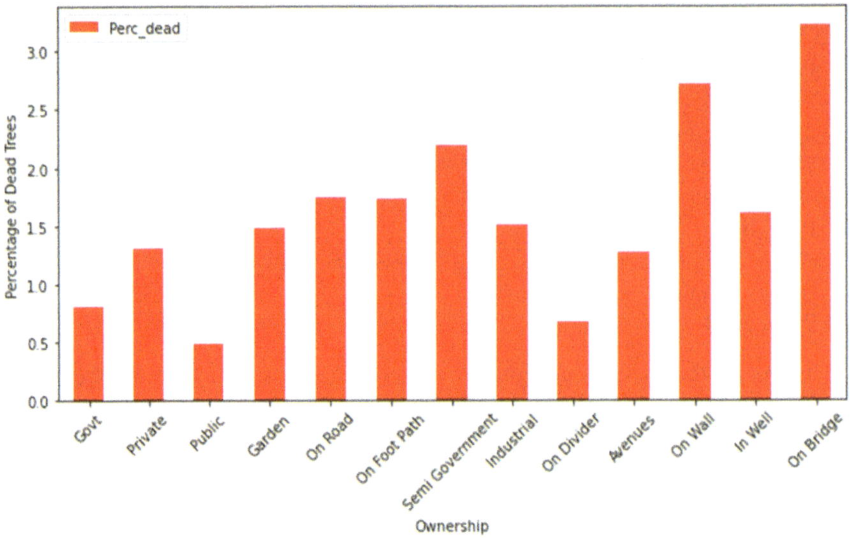

Fig. 2.7 Percentage of dead trees out of the total number of trees owned by each ownership type

```
[]
plt.figure(figsize=(10,10))
sns.boxplot(y='girth_m', x='condition', data=tree,
hue=tree['is_rare'],
palette='YlGnBu_r')
plt.xlabel('Condition', fontsize=14)
plt.ylabel('Tree Girth (in meters)', fontsize=14);
```

The above boxplot shows that the median girth of the rare trees in all the four categories of tree condition is less than that of the indigenous species. The following lines of code were written to generate the box plot for canopy diameter as depicted in Fig. 2.9:

```
[]
plt.figure(figsize=(10,10))
sns.boxplot(y='canopy_dia_m', x='condition', data=tree,
hue=tree['is_rare'], palette='YlGnBu_r')
plt.xlabel('Condition', fontsize=14)
plt.ylabel('Tree Canopy Diameter (in meters)',
fontsize=14);
```

The median canopy diameter of rare trees is lower than that of the indigenous species. The last two plots show that maybe the rare tree species in Pune are not able to grow to their full potential not having all the climatic conditions favourable to their growth.

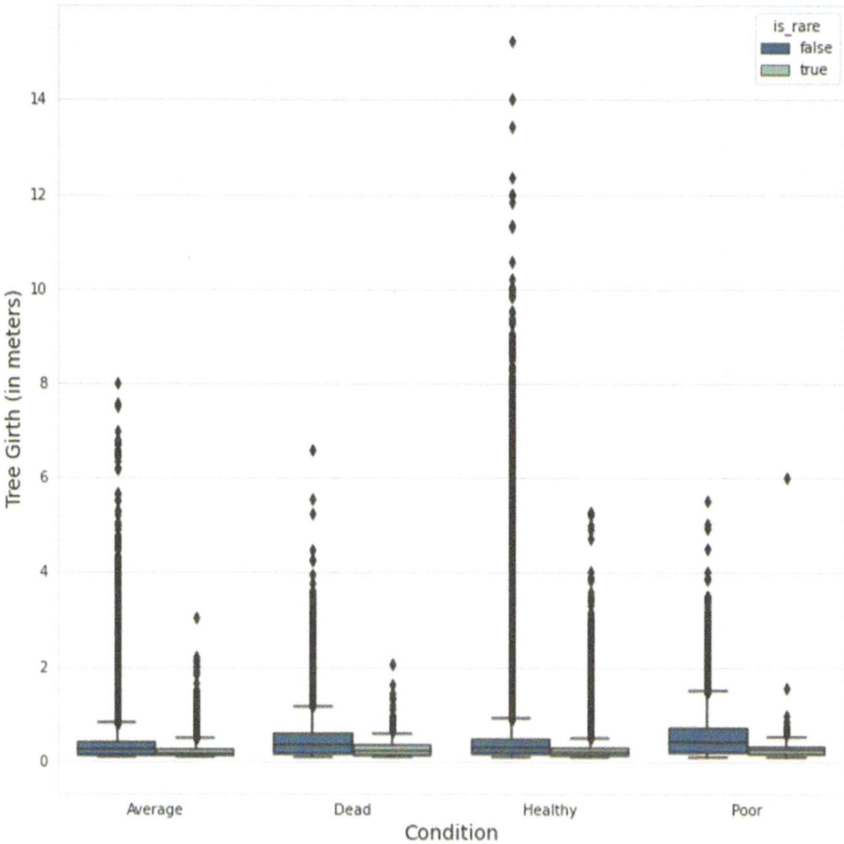

Fig. 2.8 Box plot of the tree girth (in metres) based on rarity of the trees and their health condition

2.1.9 Count Plot of Trees by Their Yield Type

The following count plot in Fig. 2.10 shows the number of trees by the type of their yield and hence, provides an idea about their economic value and usage. Following lines of code generate the required count plot:

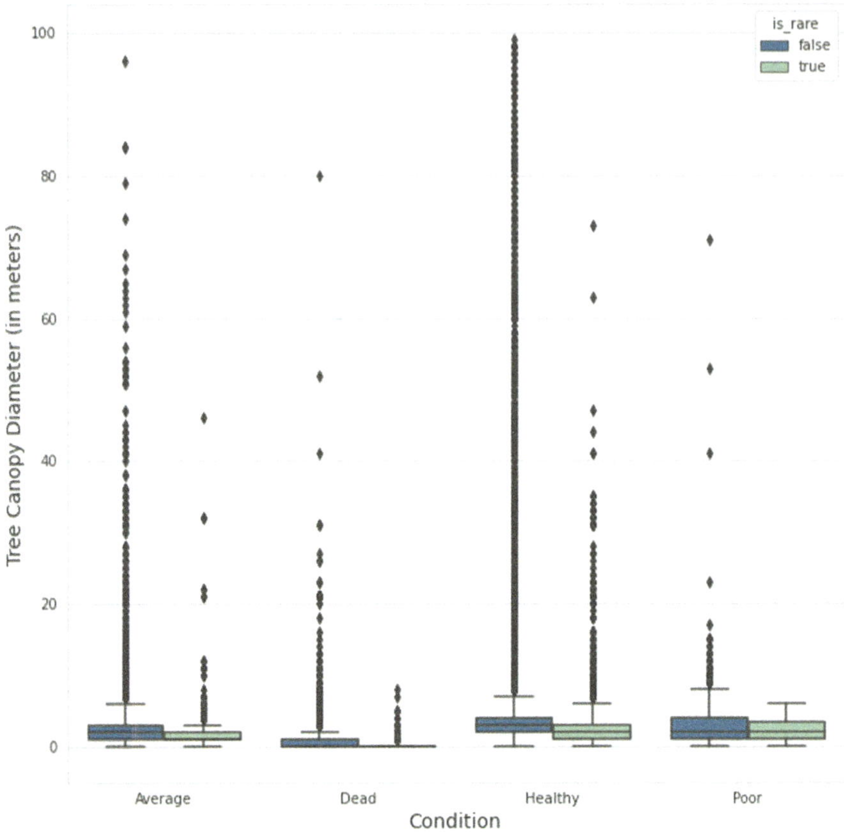

Fig. 2.9 Box plot of the tree girth (in metres) based on rarity of the trees and their health condition

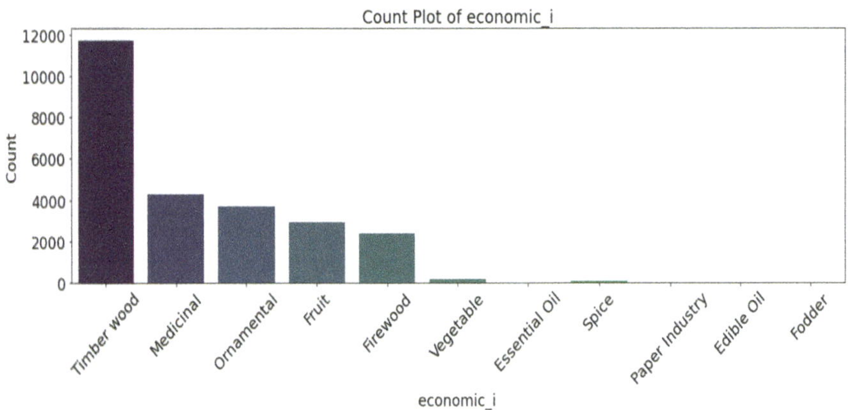

Fig. 2.10 Count plot of various categories of 'economic_i' feature

```
[]
plt.figure(figsize=(18,5))
sns.countplot(x='economic_i', data=tree[tree['condition'] == 'Dead'], palet
te='viridis',
              order=['Timber wood', 'Medicinal', 'Ornamental', 'Fruit', 'Fi
rewood',  'Vegetable', 'Essential Oil', 'Spice', 'Paper Industry', 'Edible
Oil', 'Fodder'])
plt.title('Number of Trees corresponding to different Economic Values')
plt.title('Count Plot of economic_i', fontsize=20)
plt.xticks(fontsize=18, rotation=45)
plt.yticks(fontsize=18)
plt.xlabel('economic_i', fontsize=18)
plt.ylabel('Count', fontsize=18);
```

From the above count plot, we can infer that most of the trees in Pune are grown for the timber wood, followed by the trees that have some kind of medicinal value. Then, come the trees that are grown for their ornamental values, followed by those which give fruits and then the ones from which firewood is obtained. The trees that yield vegetables and spices, essential oils and edible oils, fodder and raw material for the paper industry seem to be negligible in number as compared to the others.

2.1.10 Counts of Top 10 Most Commonly Occurring Trees in Pune Which Yield Timber Wood

The following code gives the number of unique tree species that are grown for their timber yield:

```
[In]
# Types of trees that yield Timber wood
tree[tree['economic_i'] == 'Timber wood']['local_name'].nunique()

[Out]
          79
```

Thus, there are 79 different tree species in Pune that yield timber wood, out of which the top 10 most frequently occurring tree types are obtained by running the following lines of code:

```
[In]
tree[tree['economic_i'] == 'Timber wood']['local_name'].value_counts().
head(10)
```

```
[Out]
Subabul              732010
Patangi              288509
Babool               112365
Villayati Babul      110034
Silver Oak            29949
Villayati Chinch      26715
Rain Tree             20369
Sag                   16665
Vavla                 15998
Mothi Mahogani        14039
Name: local_name, dtype: int64
```

The top 10 most frequently occurring tree types that yield timber wood are Subabul, Patangi, Babool, Villayati Babul, Silver Oak, Villayati Chinch, Rain Tree, Sag, Vava and Mothi Mahogani.

2.1.11 Counts of Balanced and Unbalanced Trees

As discussed in Chap. 1, the unbalanced trees need special monitoring if they are growing at a high risk location, such as inside a parking lot, inside children's play area, on a roadside or a highway. The trees that are unbalanced have a potential to fall and may cause serious damage to life and/or property. The reason behind the tree starting to lean over could be gradual weakening of the trunk and branches of the tree. Out of all the trees counted in the census, 99.36% are balanced and rest were found to be unbalanced. The unbalanced trees should be located by the civic body and be given special attention. If they are at very high risk spot, then they must be cut down.

```
[In]
tree['balanced'].value_counts() / len(tree) * 100
```

```
[Out]
True      99.362259
False      0.637741
Name: balanced, dtype: float64
```

The unbalanced trees and their location in the city should be carefully examined. It should be determined whether the tree has a potential to fall by looking for any cracks on the trunk or at the join of the trunk and its branches. These trees should be checked regularly.

2.1.12 Count of Trees with Respect to the Reported Signs of Stress/Damage on the Tree

The 'remarks' column of tree census dataset gives the information about the signs of physical damage and stress on the tree that were recorded during the census. The following code gives the percentage of the trees with different categories of remarks, i.e. the signs of visible damage on the tree, such as an infection or the stress that the tree is under.

```
[In] tree['remarks'].value_counts() / len(tree) * 100

[Out]        Mechanically cut     54.650549
             Diseased              0.216055
             Uprooted              0.190866
             On The Wall           0.000224
             Dangerous             0.000025
             Name: remarks, dtype: float64
```

Thus, it is found that more than 54.65% of the trees show signs of a mechanical cut.

2.1.13 Count Plot of Trees by the Rarity of Their Occurrence

The rare trees as seen earlier in this chapter showed signs of plausible problems in growth, making them less likely to sustain for long in the conditions that are not suitable for their growth and survival. The following plot in Fig. 2.11 is the count

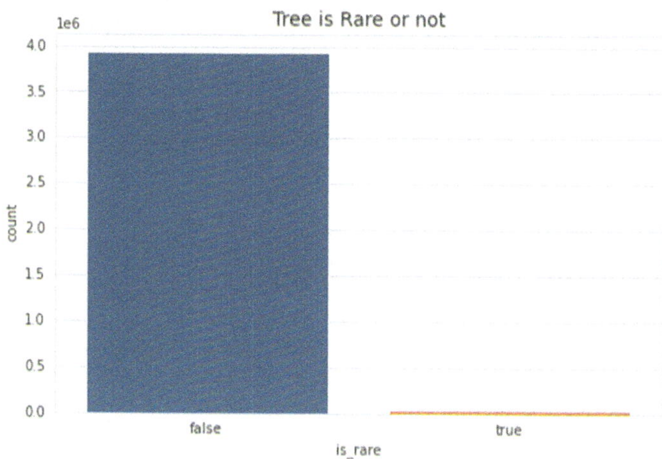

Fig. 2.11 Count plot to show the number of rare and indigenous trees

plot showing the number of trees in both these categories, i.e. rare (coded as 'true') and not rare (coded as 'false').

```
[]
fig = plt.figure(figsize=(8,5))
sns.countplot(x='is_rare', data=tree)
plt.title('Tree is Rare or not', size=14)
```

Figure 2.11 clearly shows that the trees that are not rare are indigenous to Pune and nearby areas and are far greater in number than rare trees. During tree plantation drives or on any occasion of planting a new tree, one should always research about what kind of a tree will grow best in that particular spot and can be maintained depending upon the climate, location and various other factors. All the time, effort and money that go into planting a particular tree or the investment of big sums of money in large-scale afforestation projects taken up by the government must be effective enough in the long run.

2.1.14 Count Plot of Trees by Their Phenology Category

Phenology of a tree is the study of the natural changes in the trees with the change in the season, such as flowering, appearing of new leaves, ripening of the fruit, shedding of leaves and the tree getting bare. The phenology of the tree species in Pune as recorded in the tree census, showed that only 7.62% of the total number of known observations in the 'phenology' column (i.e. excluding the missing values), has the value 'Seasonal'. This means that approximately 99.38% of all the tree records have been reported to have life cycle events that go on throughout the year.

Most of the trees in Pune which have been included in the tree census have a seasonal phenology as shown in Fig. 2.12.

```
[]
fig = plt.figure(figsize=(8,5))
sns.countplot(x='phenology', data=tree)
plt.title('Phenology of Tree', size=14);
```

2.1.15 Flowering Season of the Trees

The flowering season is a season in which a particular tree species begins to show flower buds, and subsequently blossom over the season and cover the tree. The following code gives the number of unique values in the 'flowering' column:

```
[In]
tree['flowering'].nunique()
```

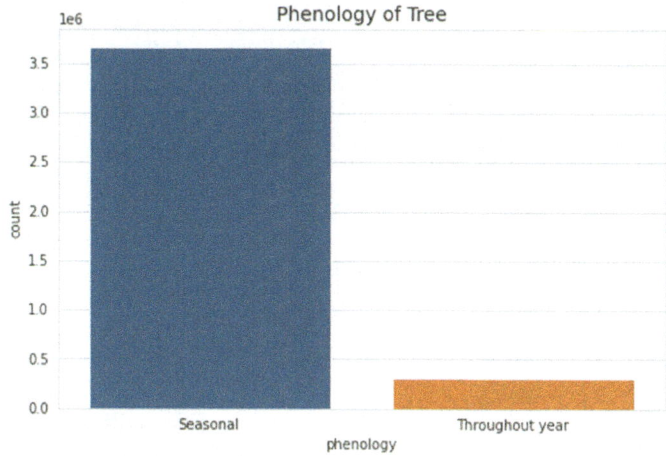

Fig. 2.12 Count plot of 'phenology' column

```
[Out]
98
```

2.1.16 *Pair Plot of All the Numerical Variables*

To inspect the correlations between the numerical variables of the tree census data, a pair plot is used. A pair plot gives a matrix of the pairwise scatter plots of all numerical variables in a data frame. This matrix is a symmetric matrix; i.e. if the rows and columns of this matrix are exchanged or as defined in linear algebra the transpose of the matrix is taken, then the new matrix obtained is identical to the old one. The diagonal of this matrix running from top-left to its bottom-right, also known as the principal diagonal of a matrix contains the distribution plots, or histograms of each of the numerical variable. A histogram gives the frequency distribution of a numerical variable. From these distribution plots, one can get a fair idea about the skewness of the distribution of these variables. The following code is run to generate a pair plot of all its numerical variables (Fig. 2.13).

```
[]
sns.pairplot(tree[['girth_m', 'canopy_dia_m',
'height_m', 'northing', 'easting']]);
[]
```

From the above pair plot, it can be inferred that none of the pairs show any significant correlation or pattern.

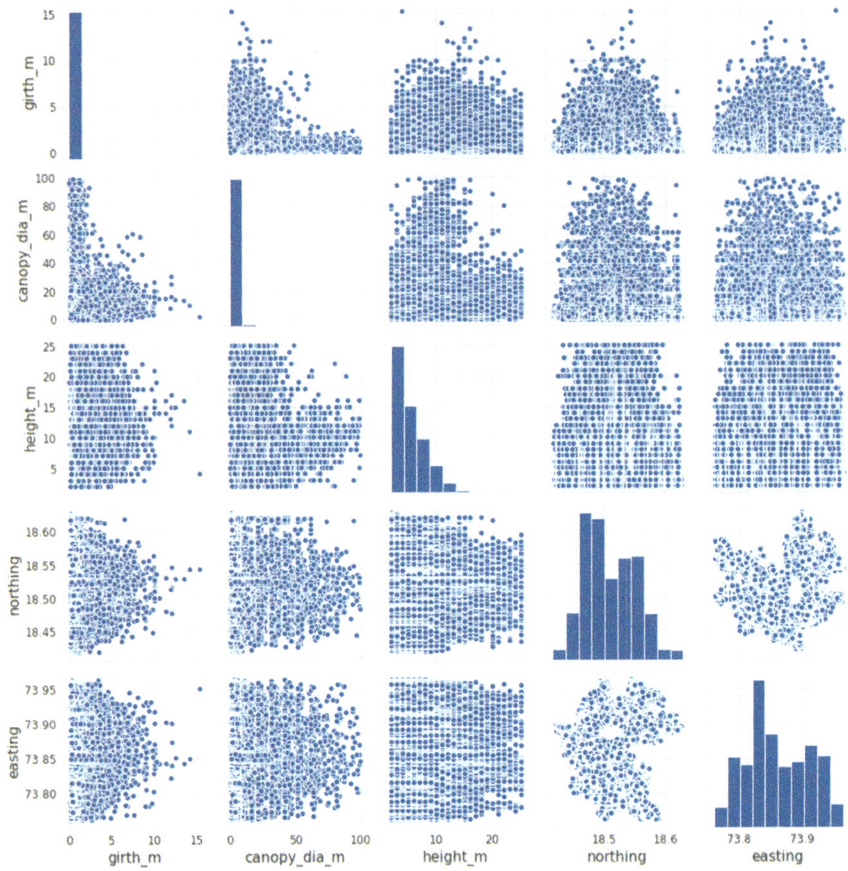

Fig. 2.13 Pair plot of the tree data frame

2.1.17 K-Means Clustering

Just like humans can make decisions on their own based on different situations, machine learning is a field of study that gives computers the capability to learn without being explicitly programmed. Broadly, machine learning is divided into four categories as shown in Fig. 2.14, which are:

Clustering is an unsupervised learning technique, which includes broad set of techniques for finding subgroups or clusters, in a dataset. It helps to determine the intrinsic grouping among the unlabelled data, or grouping the similar observations together, while observations in different groups are quite different from each other. There are different clustering algorithms for every answer of the question, 'What defines that two or more observations are similar to each other or are different'? Now this depends upon the domain and the type of data on which it is required to apply the clustering algorithm.

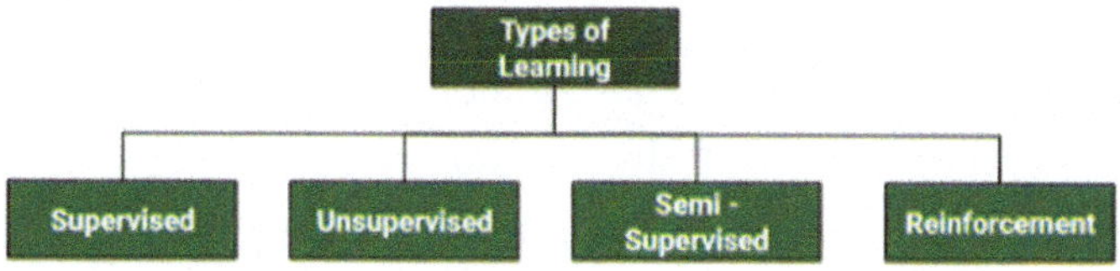

Fig. 2.14 Four basic types of machine learning

There are five classes of clustering, i.e. hierarchical, partitioning method (k-means, PAM, CLARA), density-based clustering, model-based clustering, fuzzy clustering [2]. Out of all these non-hierarchical methods, K-means clustering has been chosen here. The whole procedure for obtaining the (possible) clusters using tree girth and tree coordinates has been described in this section. Here, the three features Northing, Easting and Tree Girth (in metres) have been chosen for clustering with the following objective in mind:

To find possible clusters of trees with respect to their girth, i.e. a measure of their age.

The objective of K-means clustering is simple as it groups similar data points together and helps discover the underlying patterns. To meet the objective, the first input needed by the algorithm is an optimal number of clusters K, which is subsequently used to group the dataset into desired number of clusters.

The following procedure was used to execute this algorithm and obtain the clusters.

Step 1: To display the glimpse of dataset used for clustering

```
[] tree.head()
[]
```

	girth_cm	northing	easting
0	10	18.486828	73.895439
1	115	18.557149	73.816301
2	15	18.505884	73.792023
3	13	18.557148	73.816269
4	25	18.557150	73.816262

Step 2: To standardize the outliers in the dataset

It is very evident from the above output and also from the first boxplot of Fig. 2.10 from Chap. 1, that the 'girth_cm' column has outliers. Therefore, it is essential to convert the values of the above data frame to their z-scores, and to achieve this, there is a need to import the required 'stats' package.

```
[]
# Importing the required Python packages
from scipy import stats

# Obtain the values of features in the above
DataFrame to their z-scores
tra = stats.zscore(tree)
```

The final data frame after the conversion to z-scores is as shown below:

```
[]
tr_std = pd.DataFrame(tra, columns=['girth_z',
'northing_z', 'easting_z'])
tr_std.head()

[]
```

	girth_z	northing_z	easting_z
0	−0.765818	−0.553631	0.729385
1	2.218424	1.179829	−0.907250
2	−0.623712	−0.083881	−1.409345
3	−0.680554	1.179806	−0.907915
4	−0.339498	1.179853	−0.908061

Step 3: To calculate the cost function and map it with the corresponding number of clusters

The objective of the cost function is to minimize variance within the cluster. To achieve this, we plot sum of square errors (i.e. variance) on y-axis and the number of clusters on x-axis. The code below finds the weighted sum of square (WSS) scores for all the values for arbitrary number of clusters (k) from 1 to $k_{\max}$.

```
[]
from sklearn.cluster import KMeans

# function returns WSS score for k values from 1 to kmax

def calculate_WSS(points, kmax):
  sse = []
  for k in range(1, kmax+1):
    kmeans = KMeans(n_clusters = k).fit(points)
    centroids = kmeans.cluster_centers_
    pred_clusters = kmeans.predict(points)
    curr_sse = 0
```

```
    # calculate square of Euclidean distance of each point from its   clust
er center and add to current WSS
    for i in range(len(points)):
      curr_center = centroids[pred_clusters[i]]
      curr_sse += (points[i, 0] - curr_center[0]) ** 2 + (points[i, 1] -
 curr_center[1]) ** 2

    sse.append(curr_sse)
  return sse
```

```
[] calculate_WSS(tra, 10)
[]      [8019007.999999495,       6991177.476223489,       4843759.231755912,
3171856.6033142745,       2895224.0065653613,       2353303.3628638308,
2041389.0016338741,        1839798.909059414,       1666788.4833885757,
1525814.4688175553]
```

Now, since the weighted sum of square (WSS) values have been generated, the same is used to plot cost and number of clusters to find the optimal value of K. The following code is used to obtain the required plot as shown in Fig. 2.15.

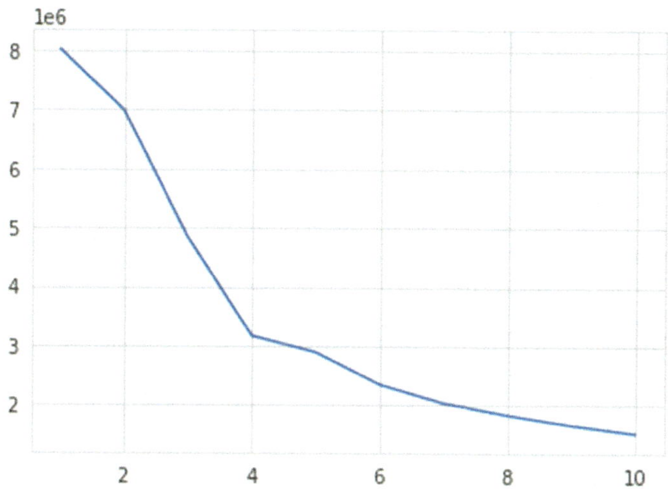

Fig. 2.15 Plot of the cost function. x-axis = Number of clusters k, y-axis = The cost corresponding to k clusters

```
[]
sse = pd.Series([8019007.999999495,
 6991177.476223489,
 4843759.231755912,
 3171856.6033142745,
 2895224.0065653613,
 2353303.3628638308,
 2041389.0016338741,
 1839798.909059414,
 1666788.4833885757,
 1525814.4688175553], index=[i for i in range(1, 11)])

sns.set_style('whitegrid')
plt.plot(sse)

[]
```

The above plot of the cost function, i.e. the weighted sum of squares (WSS), clearly shows an elbow point at $k = 4$. The 'elbow point' is a bend which represents the optimal number of clusters (K) to be generated.

Step 4: To generate the clusters

The value of K is supplied as an argument to the K-means algorithm. The following lines of code generate the clusters:

```
[In]
kmeans = KMeans(n_clusters=4, random_state=0)
.fit(tr_std)
labels = kmeans.labels_
tr['clusters'] = labels
centroids = kmeans.cluster_centers_
pred_clusters = kmeans.predict(tr_std)
tr.head()

[Out]
```

	northing	easting	girth_CB	clusters
0	18.486828	73.895439	10	2
1	18.557149	73.816301	115	1
2	18.505884	73.792023	15	0
3	18.557148	73.816269	13	0
4	18.557150	73.816262	25	0

The labels of the four clusters thus obtained are assigned to original data frame with the three features. The following code gives the plot of the four clusters in two dimensions. Figure 2.16 presents the plot.

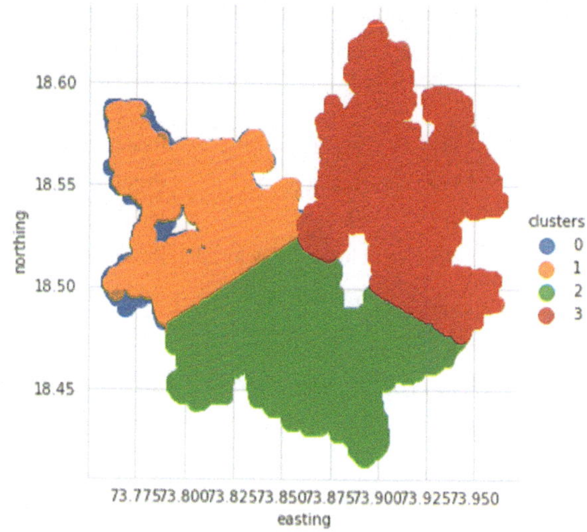

Fig. 2.16 Plot showing the four clusters

```
[]
sns.lmplot('easting', 'northing',
           data=tr,
           fit_reg=False,
           hue='clusters',
           scatter_kws={'marker': 'D',
                        's': 100})
plt.xlabel('easting')
plt.ylabel('northing')
```

2.2 Data Exploration: Air Pollution Data

In the last chapter, the consequences of air pollution, its sources and effects of the five major pollutants as well as their recommended levels were discussed. Also, the data was described, pre-processed, cleaned and made ready for the statistical analysis. In this section, the same dataset obtained from Maharashtra Pollution Control Board – Regional Office, Pune (MPCB-RO), is used for exploratory data analysis and to find some interesting insights. But, before that, let us first look at the basic description of the five locations as shown in Table 2.1, with respect to the data recorded.

Table 2.1 Five locations where air pollution parameters were recorded

S. No.	Station code	Station name	Type	Location used in data collection	Number of recorded observations
1	381	Swargate	Residential	MPCB-SWGT	73
2	379	Nal Stop	Rural & other	MPCB-NS	95
3	312	Bhosari	Industrial	MPCB-BSRI	111
4	708	Pimpri-Chinchwad	Residential	MPCB-PMPR	311
5		Karve Road	Residential	MPCB-KR	414

2.2.1 *Descriptive Statistics of Pollutants for the Five Locations*

In this section, the distribution of data in the dataset has been presented with the help of mean, median, standard deviation, quartiles and the extreme values of the pollutants for five locations.

a. **MPCB-SWGT (Swargate)**

```
df[(df['Location']=='MPCB-SWGT')].describe()
```

(Note: df is the Python data frame which was pre-processed in the previous chapter).

	SO_2 μg/m^3	No_x μg/m^3	RSPM μg/m^3	AQI	Month	Year
Count	73.000000	73.000000	73.000000	73.000000	73.000000	73.000000
Mean	35.849315	89.438356	96.246575	113.671233	5.794521	2018.095890
Std	12.165150	32.437599	51.195643	34.620534	3.748922	0.296479
Min	9.000000	32.000000	13.000000	49.000000	1.000000	2018.000000
25%	27.000000	64.000000	50.000000	89.000000	2.000000	2018.000000
50%	36.000000	94.000000	95.000000	117.000000	6.000000	2018.000000
75%	44.000000	112.000000	“35.000000	136.000000	9.000000	2018.000000
Max	73.000000	167.000000	210.000000	187.000000	12.000000	2019.000000

(Note: RSPM μg/m^3 is actually the value of PM10 in μg/m^3, as mentioned in the 'readme' file published by the Pune Municipal Corporation.)

Few important observations:

1. A total of 73 values had been recorded for the location Swargate for 2018 and 2019.
2. The mean value of air quality index (AQI) is 113.67, while those of SO_2, NO_x and PM10 are 35.85, 89.44 and 96.25 μg/m^3, respectively.
3. The maximum and minimum values are:

i. AQI: 49 and 187
ii. SO_2: 9 and 73 μg/m^3
iii. NO_x: 32 and 167 μg/m^3
iv. PM10: 13 and 210 μg/m^3.

b. **MPCB-NS (Nal Stop)**

```
[] df[(df['Location']=='MPCB-NS')].describe()
```

	SO_2 μg/m^3	No_x μg/m^3	RSPM μg/m^3	AQI	Month	Year
Count	95.000000	95.000000	95.000000	95.000000	95.000000	95.000000
Mean	40.273634	80.5-5739	125.326316	124.400000	6.221053	201 8.210526
Std	13.790563	24.536063	57.497908	33.998874	3.815364	0.409345
Min	8. 000000	14.000000	16.000000	53.000000	1.000000	2018.000000
25%	31.500000	63.000000	80.500000	108.500000	2.500000	2018.000000
50%	40.000000	82.000000	133.000000	123.000000	6.000000	2018.000000
75%	49.500000	97.000000	157.500000	40.000000	9.500000	2018.000000
Max	74.000000	147.000000	316.000000	266.000000	12.000000	2019.000000

Few important observations:

1. A total of 95 values had been recorded for the location Nal Stop for 2018 and 2019.
2. The mean value of AQI is 124.4, while those of SO_2, NO_x and PM10 are 40.27, 80.52 and 125.33 μg/m^3, respectively.
3. The maximum and minimum values are:

i. AQI: 53 and 266
ii. SO_2: 8 and 74 μg/m^3
iii. NO_x: 14 and 147 μg/m^3
iv. PM10: 16 and 316 μg/m^3.

c. **MPCB-BSRI (Bhosari)**

```
[] df[(df['Location']=='MPCB-BSRI')].describe()
```

	SO_2 μg/m^3	No_x μg/m^3	RSPM μg/m^3	AQI	Month	Year
Count	111.000000	111.000000	111.000000	111.000000	111.000000	111.000000
Mean	41.477477	73.549550	133.972973	120.009009	6.117117	2018.207207
Std	15.520571	26.861848	77.5026S4	51.192950	3.642523	0.407143

(continued)

(continued)

	SO_2 µg/m^3	No_x µg/m^3	RSPM µg/m^3	AQI	Month	Year
Mi n	10.000000	31.000000	5.000000	39.000000	1.000000	2018.000000
25%	33.000000	52.000000	59.500000	88.500000	3.000000	2018.000000
50%	41.000000	73.000000	134.000000	124.000000	6.000000	2018.000000
75%	48.500000	90.500000	189.500000	163.500000	9.000000	2018.000000
Max	93.000000	178.000000	331.000000	231.000000	12.000000	2019.000000

Few important observations:

1. A total of 111 values had been recorded for the location Bhosari for 2018 and 2019.
2. The mean value of AQI is 128.01, while those of SO_2, NO_x and PM10 are 41.48, 73.55 and 133.97 µg/m^3, respectively.
3. The maximum and minimum values are:

   i. AQI: 39 and 281
   ii. SO_2: 10 and 93 µg/m^3
   iii. NO_x: 31 and 178 µg/m^3
   iv. PM10: 5 and 331 µg/m^3.

d. **MPCB-PMPR (Pimpri-Chinchwad)**

```
[] df[(df['Location']=='MPCB-PMPR')].describe()
```

	SO_2 µg/m^3	No_x µg/m^3	RSPM µg/m^3	AQI	month	year
Count	311.000000	311.000000	311.000000	311.000000	311.000000	311.000000
Mean	38.302251	72.877814	91.405145	103.617363	5.961415	2018.170418
Std	14.659313	30.895106	50.770863	37.639234	3.581105	0.376606
Min	9.000000	22.000000	9.000000	30.000000	1.000000	2018.000000
25%	28.000000	43.500000	52.500000	75.000000	3.000000	2018.000000
50%	37.000000	69.000000	89.000000	103.000000	6.000000	2018.000000
75%	47.000000	91.000000	124.000000	126.500000	9.000000	2018.000000
Max	76.000000	193.000000	330.000000	280.000000	12.000000	2019.000000

Few important observations:

1. A total of 311 values had been recorded for the location Pimpri-Chinchwad for 2018 and 2019.
2. The mean value of AQI is 103.62, while those of SO_2, NO_x and PM10 are 38.30, 72.88 and 91.41 µg/m^3, respectively.
3. The maximum and minimum values are:

i. AQI: 30 and 280
ii. SO_2: 9 and 76 $\mu g/m^3$
iii. NO_x: 22 and 198 $\mu g/m^3$
iv. PM10: 9 and 330 $\mu g/m^3$.

e. **MPCB-KR (Karve Road)**

```
[] df[(df['Location']=='MPCB-KR')].describe()
```

	SO_2 $\mu g/m^3$	No_x $\mu g/m^3$	RSPM $\mu g/m^3$	AQI	Month	Year
Count	414.000000	414.000000	414.000000	414.000000	414.000000	414.000000
Mean	16.125604	45.466134	97.456522	94.765700	5.852657	2018.138406
Std	7.214035	23.857579	42.990418	35.296471	3.523705	0.391509
Min	4.000000	9.000000	12.000000	34.000000	1.000000	2018.000000
25%	11.000000	28.000000	64.000000	68.250000	3.000000	2018.000000
50%	14.000000	37.000000	96.000000	98.000000	6.000000	2018.000000
75%	21.000000	52.750000	127.000000	119.750000	9.000000	2018.000000
Max	50.000000	164.000000	219.000000	184.000000	12.000000	2019.000000

Few important observations:

1. A total of 414 values had been recorded for the location Karve Road for 2018 and 2019.
2. The mean value of AQI is 94.77, while those of SO_2, NO_x and PM10 are—16.13, 45.47 and 97.46 $\mu g/m^3$, respectively.
3. The maximum and minimum values are:

i. AQI: 34 and 184
ii. SO_2: 4 and 50 $\mu g/m^3$
iii. NO_x: 9 and 164 $\mu g/m^3$
iv. PM10: 12 and 219 $\mu g/m^3$.

So, from the detailed data description, it is clearly evident that industrial area Bhosari has the highest AQI making it the most polluted among the five locations.

2.2.2 *Visualizing Air Quality Index (AQI)*

Now, that the basic data description for five locations is completed, a comparative study with respect to the AQI and the three pollutants (SO_2, NO*x* and PM10) can be done. A box-plot is used to extract information from the dataset regarding minimum, maximum, median and interquartile ranges, suitable for comparing the five locations graphically and to ease our understanding of the dataset. Besides, a swarm plot is also drawn to visualize the data from different perspective. A swarm plot plots the exact locations of all the data points in a two-dimensional plane, providing a good overview of dispersion of the data points in a graph.

2.2.2.1 Boxplot of the AQI at the Five Locations

A box plot is drawn by plotting the five locations against their corresponding air quality index as shown in Fig. 2.17. (To know more about box plot, refer to Chap. 1)

```
[]
sns.set_style('darkgrid')
plt.figure(figsize=(10,8))
sns.boxplot(x="Location", y="AQI", data=df,
palette='rainbow')
```

Key findings and conclusions from the above boxplot:

1. The median AQI for Bhosari is the highest, i.e. 124, followed by Nal Stop 123, Swargate 117, Pimpri-Chinchwad 103 and the lowest being 98 for Karve Road. Bhosari has the highest mean value of 128.01 followed by Nal Stop at 124.04.

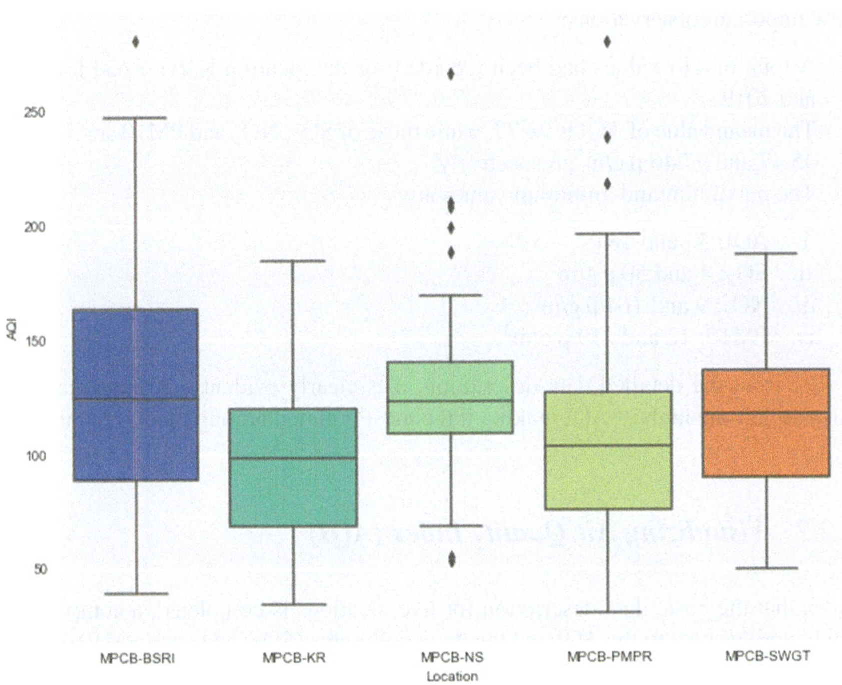

Fig. 2.17 Boxplot of AQI at the five locations

So, it can be safely concluded that among the five locations, Karve Road had the best air quality index and Bhosari had the worst air quality index. Bhosari, being the only industrial area among the five locations, is expected to have the worst AQI, and our findings confirm this fact.

2. The values of outliers for Pimpri-Chinchwad are more than the other locations. Also, the values of AQI for Pimpri-Chinchwad have the largest interquartile range (25–75%) of the data points. This suggests that the variation in the data collected was highest for this location. (standard deviation of AQI was highest at 51.2).

2.2.2.2 Swarm Plot of AQI at the Five Locations

A swarm plot is used when all the points along with their corresponding distribution need to be on a single plot. It serves as a good alternative to box and violin plots. Following code is written to create a swarm plot which is presented in Fig. 2.18.

```
[] sns.swarmplot(x="Location", y="AQI", data=df)
```

In the above swarm plot, each and every data point is visualized in a two-dimensional space. Without going into intricate statistical analysis, it is observed that most of the AQI values lie between 50 and 150, as the concentration of the

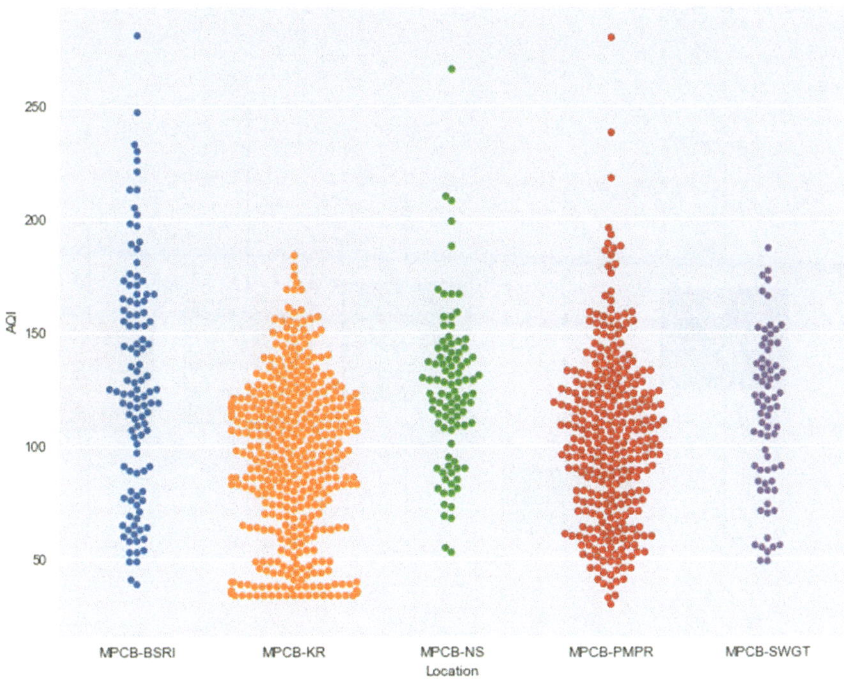

Fig. 2.18 Swarm plot of the AQI for the five locations

points in this range is the maximum. Also, it is noticed that the number of data points recorded are high in case of Karve Road and Pimpri-Chinchwad as compared to the other three locations.

2.2.3 Visualizing Individual Pollutant Levels

In this section, an attempt has been made to study the concentration of each pollutant in different locations.

2.2.3.1 Boxplot of SO_2 μg/m^3 at the Five Locations

The following code generates the boxplot of SO_2 concentration for the five locations as shown in Fig. 2.19.

```
[] sns.boxplot(x="Location", y="SO2 μg/m3", data=df,
palette='rainbow')
```

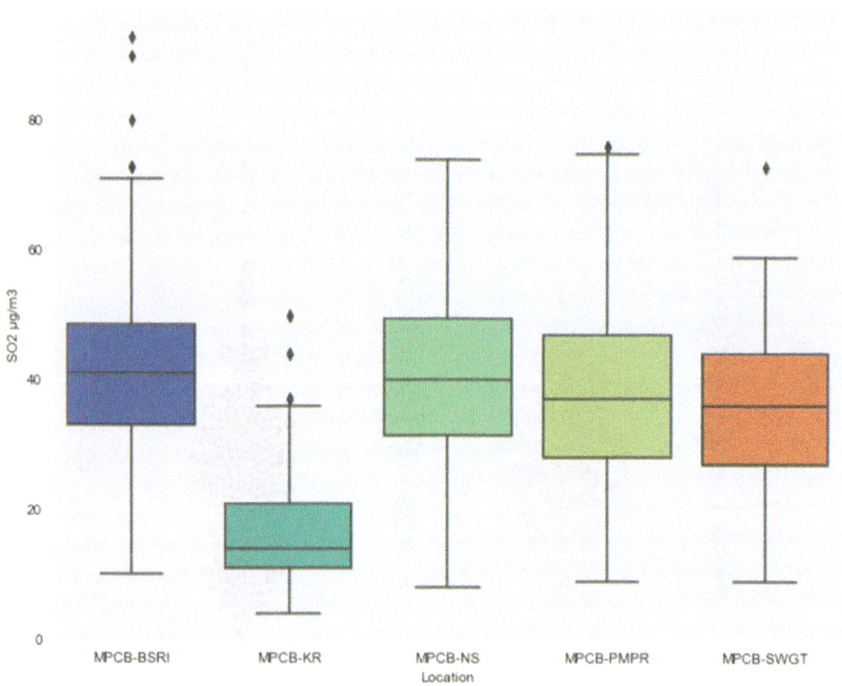

Fig. 2.19 Boxplot of SO_2 concentration for the five locations

Key findings and conclusions from the above boxplot:

1. The median value for the pollutant SO_2 is highest for Bhosari, i.e. 41 $\mu g/m^3$, followed by Nal Stop 40 $\mu g/m^3$, Pimpri-Chinchwad 37 $\mu g/m^3$, Swargate 36 $\mu g/m^3$ and the lowest being 14 $\mu g/m^3$ for Karve Road. Bhosari has the highest mean SO_2 value of 41.48 $\mu g/m^3$ followed by Nal Stop at 40.28 $\mu g/m^3$. So, it can be concluded from Fig. 2.19 that among the five locations, Karve Road has the lowest SO_2 concentration and Bhosari has the highest SO_2 concentration. Also, Bhosari, being the only industrial area among the five locations, was expected to have the highest mean SO_2 concentration, and our findings confirm this fact.
2. The values of outliers for Bhosari are more than the others.
3. Even though being residential areas, Nal Stop and Pimpri-Chinchwad showed almost similar SO_2 levels as compared to the industrial area Bhosari.

2.2.3.2 Boxplot of NO_x $\mu g/m^3$ at the Five Locations

The following code generates the boxplot of NO_x concentration for the five locations as shown in Fig. 2.20.

```
[] sns.boxplot(x="Location", y="Nox µg/m3", data=df,
palette='rainbow')
```

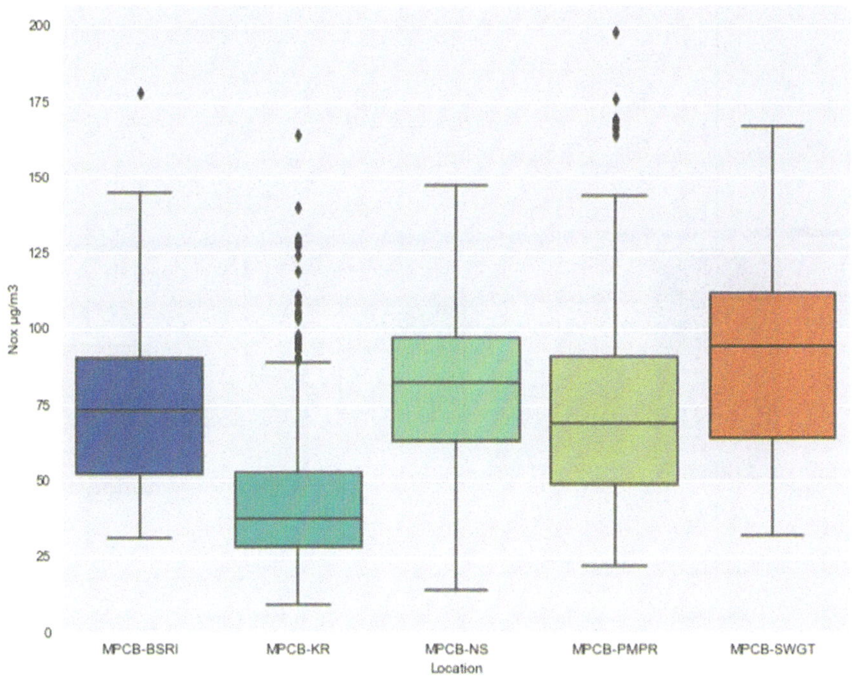

Fig. 2.20 Boxplot of NOx concentration for the five locations

Key findings and conclusions from the above boxplot:

1. The median NO_x for Swargate is the highest, i.e. 94 μg/m^3, followed by Nal Stop 82 μg/m^3, Bhosari 73 μg/m3, Pimpri-Chinchwad 69 μg/m^3 and the lowest being 37 μg/m^3 for Karve Road. Swargate has the highest mean NO_x value of 89.43 μg/m^3 followed by Nal Stop at 80.52 μg/m^3. So, it can be concluded from Fig. 2.20 that among the five locations, Karve Road has the lowest NO_x concentration with a mean of 45.46 μg/m^3 and Swargate has the highest average NO_x concentration.
2. Also, Swargate shows the highest standard deviation in its recorded values of NO_x in μg/m^3; i.e. the values are more widely dispersed as compared to the other four locations.

2.2.3.3 Boxplot of PM10 (RSPM) μg/m^3 at the Five Locations

The following code generates the boxplot of PM10 concentration for the five locations as shown in Fig. 2.21.

```
[] sns.boxplot(x="Location", y="RSPM μg/m3", data=df,
palette='rainbow')
```

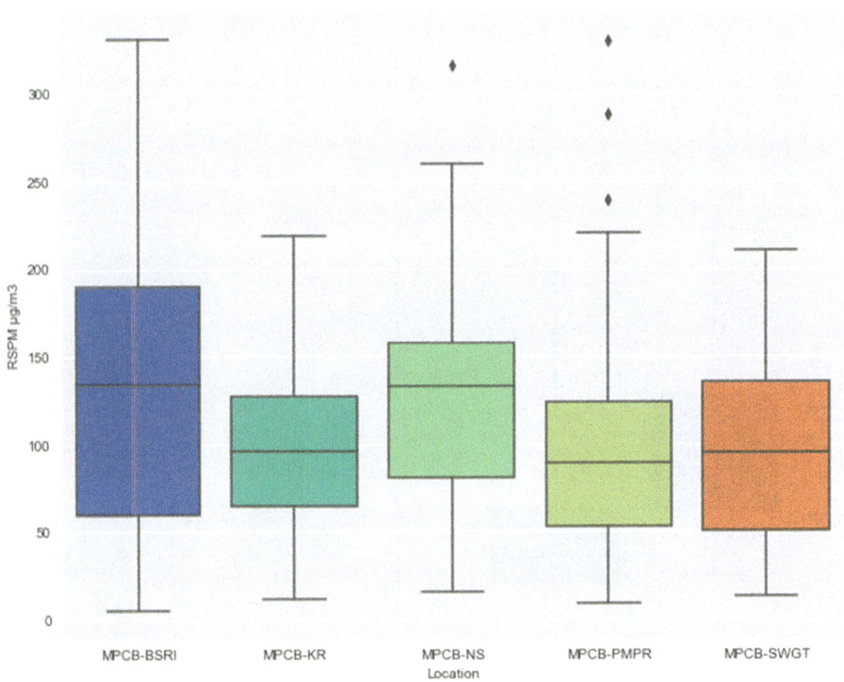

Fig. 2.21 Boxplot of PM10 concentration for the five locations

Key findings and conclusions from the above boxplot:

1. The median PM10 for Bhosari is the highest, i.e. 134 μg/m^3, followed by Nal Stop 133 μg/m^3, Karve Road 96 μg/m^3, Swargate 95 μg/m^3 and the lowest being 89 μg/m^3 for Pimpri-Chinchwad. Bhosari has the highest mean value of 133.98 μg/m^3 followed by Nal Stop at 125.33 μg/m^3. Bhosari, being an industrial area, is expected to have the highest concentration of PM10 and the findings here confirm that.
2. Bhosari has the highest standard deviation in its recorded values, which shows that it has enough fluctuations.
3. Considering the outliers, Swargate and Karve Road has lesser extreme values in comparison with others, and an all over lesser concentration of PM10.

2.2.4 *Interrelationships Between AQI, SO_2 and NOx (in μg/m^3) Concentration*

It would be interesting to study the variation of AQI with respect to the individual pollutants at all the five locations. For this, joint plots are used, which depicts the pairwise relationships in a dataset. It displays the relationship between two variables along with their individual distribution. Python allows us to draw different kinds of joint plots in the form of regression, scatter plot and so on.

2.2.4.1 Variation of AQI with Respect to SO_2 (in μg/m^3)

The joint plot is created by plotting AQI on the y-axis (dependent variable) and the SO_2 values (in μg/m^3) on the x-axis (independent variable) The code for the same is given below, and the output is presented in Figs. 2.22, 2.23, 2.24, 2.25 and 2.26.

```
[]
sns.jointplot(x='SO2     µg/m3',y='AQI',data     =df[df['Location']=='MPCB-
PMPR'],kind='reg')
sns.jointplot(x='SO2     µg/m3',y='AQI',data     =df[df['Location']=='MPCB-
KR'],kind='reg')
sns.jointplot(x='SO2     µg/m3',y='AQI',data     =df[df['Location']=='MPCB-
BSRI'],kind='reg')
sns.jointplot(x='SO2     µg/m3',y='AQI',data     =df[df['Location']=='MPCB-
NS'],kind='reg')
sns.jointplot(x='SO2     µg/m3',y='AQI',data     =df[df['Location']=='MPCB-
SWGT'],kind='reg')
```

Note: The kind = 'reg' refers to drawing a regression line using the plotted points. It is observed that the distributions of both the variables are available on their corresponding sides. These types of plots demonstrate the relationship between the two variables along with their individual distributions; hence, it is very useful.

Key findings and conclusions from these joint plots:

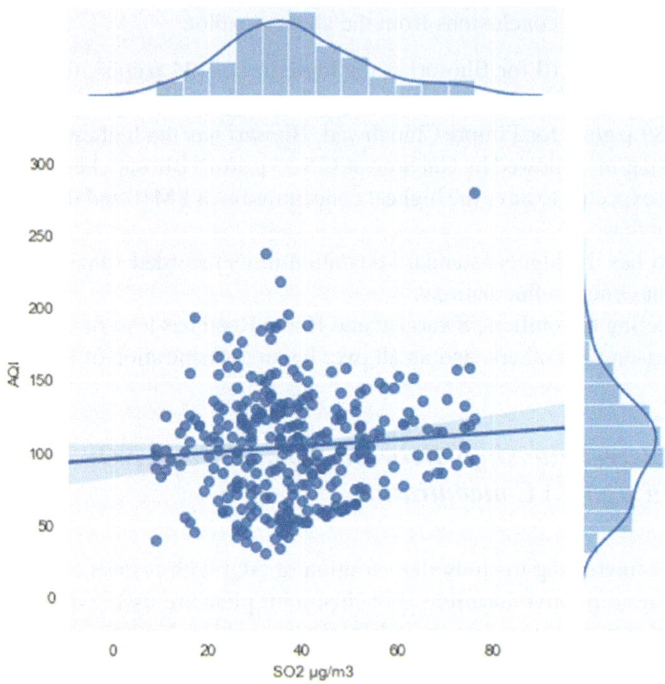

Fig. 2.22 MPCB-PMPR

1. For Karve Road, the increase in AQI with the increase in SO_2 levels (in $\mu g/m^3$) shows a linear relationship. This linear increase is more as compared to the other locations. This is evident from the regression line drawn.
2. For Bhosari and Swargate, the AQI levels almost remained constant (or increased negligibly) with increase in SO_2 levels (in $\mu g/m^3$).
3. For Pimpri-Chinchwad, the AQI increased with the SO_2 levels (in $\mu g/m^3$), but the rate of increase is much less as compared to Karve Road.
4. For Nal Stop, the AQI level almost remained constant (a slight decrease thereafter) with the increase in SO_2 levels (in $\mu g/m^3$).

2.2.4.2 Variation of AQI with Respect to NO_x (in $\mu g/m^3$)

The joint plot is created by plotting AQI on the y-axis (dependent variable) and the NOx values (in $\mu g/m^3$) on the x-axis (independent variable) The code for the same is given below, and the output is presented in Fig. 2.27, 2.28, 2.29, 2.30 and 2.31.

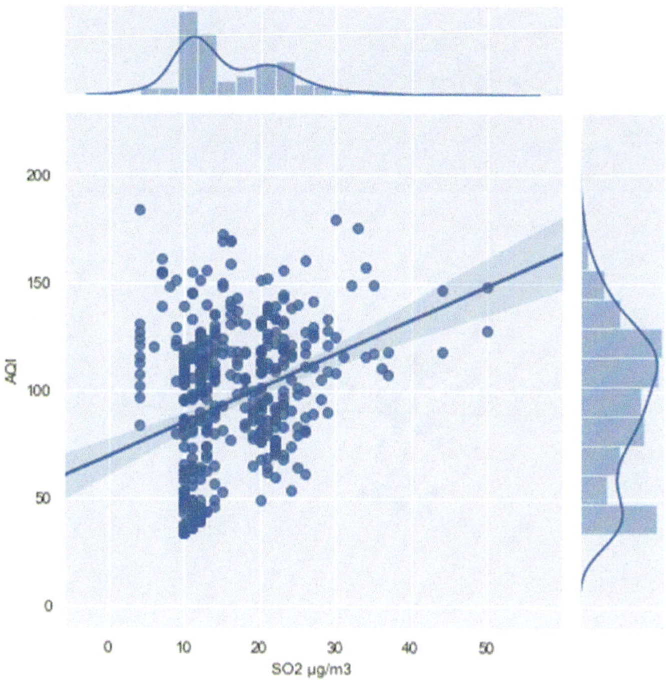

Fig. 2.23 MPCB-KR

```
[]
sns.jointplot(x='Nox     µg/m3',y='AQI',data     =df[df['Location']=='MPCB-
PMPR'],kind='reg', color='Red')
sns.jointplot(x='Nox     µg/m3',y='AQI',data     =df[df['Location']=='MPCB-
KR'],kind='reg', color='Red')
sns.jointplot(x='Nox     µg/m3',y='AQI',data     =df[df['Location']=='MPCB-
BSRI'],kind='reg', color='Red')
sns.jointplot(x='Nox     µg/m3',y='AQI',data     =df[df['Location']=='MPCB-
NS'],kind='reg', color='Red',xlim=(0,175)
sns.jointplot(x='Nox     µg/m3',y='AQI',data     =df[df['Location']=='MPCB-
SWGT'],kind='reg', color='Red')
```

Key findings and conclusions from the above joint plots:

1. It is observed that in all the five locations, there is a linear relationship between AQI and NO_x (in $\mu g/m^3$). Hence, the AQI increases, with the increase in NO_x.
2. For Karve Road, a high concentration of values is observed between the coordinates (25, 35) and (75, 150).
3. The recorded values show high scattering for Bhosari and Nal Stop.
4. The rate of increase of AQI with respect to NO_x is observed to be the maximum at Bhosari as compared to the other locations.
5. The slope of increase is the minimum at Nal Stop as the highest recorded value of NOx is 147 $\mu g/m^3$.
6. The linear relationship has been mostly perfect in the case of Swargate.

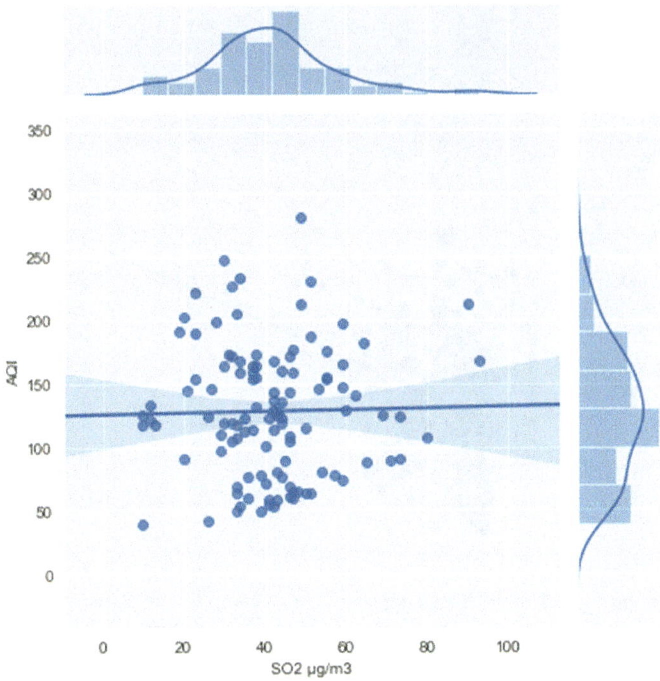

Fig. 2.24 MPCB-BSRI

2.2.4.3 Variation of AQI with Respect to PM10 (RSPM) (in µg/m^3)

The joint plot is generated by plotting AQI on the y-axis (dependent variable) and the PM10 (RSPM) values (in µg/m^3) on the x-axis (independent variable) The code for the same is given below, and the output is presented in Figs. 2.32, 2.33, 2.34, 2.35 and 2.36.

```
[]
sns.jointplot(x='RSPM     µg/m3',y='AQI',data     =df[df['Location']=='MPCB-
PMPR'],kind='reg', color='Green', xlim=(0,350), ylim=(0,250))
sns.jointplot(x='RSPM     µg/m3',y='AQI',data     =df[df['Location']=='MPCB-
KR'],kind='reg', color='Green', xlim=(0,250), ylim=(0,250))
sns.jointplot(x='RSPM     µg/m3',y='AQI',data     =df[df['Location']=='MPCB-
BSRI'],kind='reg', color='Green', xlim=(0,350), ylim=(0,250))
sns.jointplot(x='RSPM     µg/m3',y='AQI',data     =df[df['Location']=='MPCB-
NS'],kind='reg', color='Green', xlim=(0,250), ylim=(0,250))
sns.jointplot(x='RSPM     µg/m3',y='AQI',data     =df[df['Location']=='MPCB-
SWGT'],kind='reg', color='Green', xlim=(0,250), ylim=(0,250))
```

Key findings and conclusions from these joint plots:

1. There is almost a linear relationship between AQI and PM10 µg/m^3 levels. AQI increases with the increase in PM10 µg/m^3 concentration.

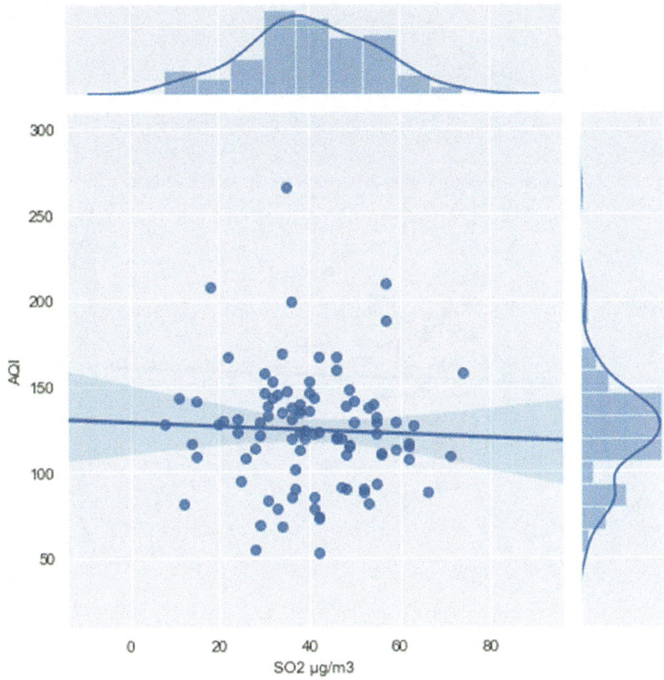

Fig. 2.25 MPCB-NS

2. For Karve Road, most of the points are seen to be concentrated on the regression line, exhibiting a very strong linear relationship between the two variables.
3. For Pimpri-Chinchwad, though the slope is the highest, but the recorded values show high concentration up to 150 for both the variables, where 150 is level of AQI and 150 $\mu g/m^3$ is concentration of PM10.

2.2.5 Pollutant Concentration for the Months of 2018

In this section, the monthly pollutant concentration levels for SO_2, NO_x and PM10 are studied for the year 2018, as not enough recorded values were available for the year 2019. Only 181 records were available for 2019 in comparison to 823 records for 2018. Hence, the study focuses on only the months of the year 2018.

a. **SO_2 (in $\mu g/m^3$)**

The swarm plot is created by plotting the monthly SO_2 concentration for all the months of the year 2018. The code for the same is given below, and the output is presented in Fig. 2.37.

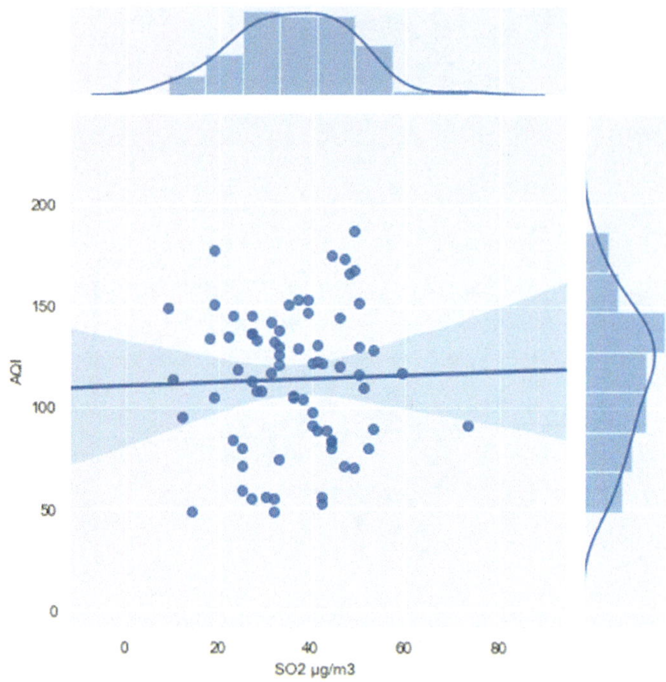

Fig. 2.26 MPCB-SWGT

```
[] sns.swarmplot(x='month', y='SO2 µg/m3', data =
df[(df['year']==2018)],
hue='Location')
```

It is clearly seen that the five locations are colour coded differently as per the legend in the plot. The months are coded as (1-January, 2018, 2-February, 2018 and so on to 12-December, 2018). It is observed that SO_2 levels above 70 (in $\mu g/m^3$) are very less in number. Also, the values recorded for Karve Road show maximum concentration under 30 (in $\mu g/m^3$), and some values recorded for Pimpri-Chinchwad as well as Bhosari show outliers in the range of 70–80 (in $\mu g/m^3$).

b. **NO_x (in $\mu g/m^3$)**

The following code is used to draw the swarm plot to present the monthly NO_x concentration for all the months of the year 2018, and the output is presented in Fig. 2.38.

```
[] sns.swarmplot(x='month',y='Nox µg/m3',data =
df[(df['year']==2018)],
hue='Location')
```

It is observed that NO_x levels are mostly concentrated in the range 25–75 (in $\mu g/m^3$) throughout the year for all the locations.

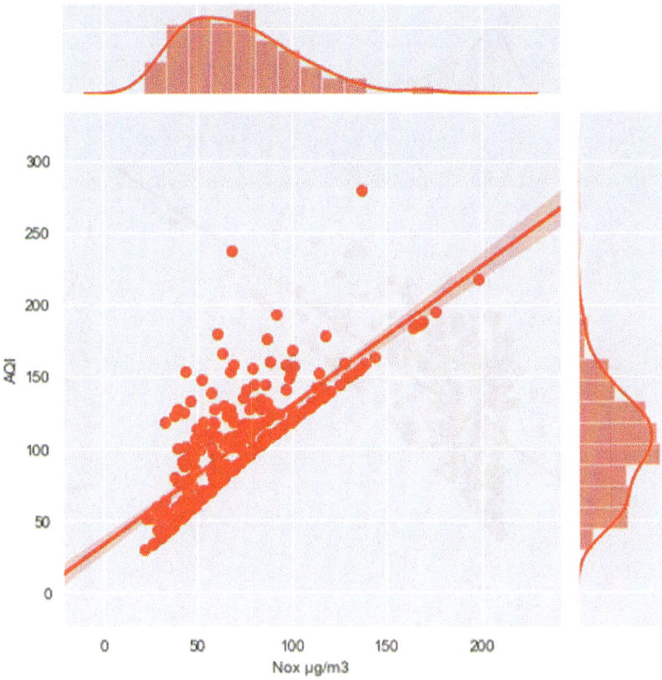

Fig. 2.27 MPCB-PMPR

c. **PM10 (RSPM) (in µg/m³)**

The swarm plot is created by plotting the monthly PM10 concentration for all the months of the year 2018. The code for the same is given below, and the output is presented in Fig. 2.39.

```
[] sns.swarmplot(x='month', y='RSPM µg/m3', data =
df[(df['year']==2018)],
hue='Location')
```

It is observed that the recorded PM10 levels are mostly between 0 and 190 $\mu g/m^3$, whereas the NO_x levels are much higher than the SO_2 levels in general. Most of the NO_x values are between 25 and 125 $\mu g/m^3$ while that of SO_2 are between 10 and 60 $\mu g/m^3$.

2.2.6 AQI Variation for the Months of 2018

Finally, the month-wise analysis of AQI levels at the five locations for the year 2018 is carried out. This is helpful in understanding whether any seasonality or trend exists

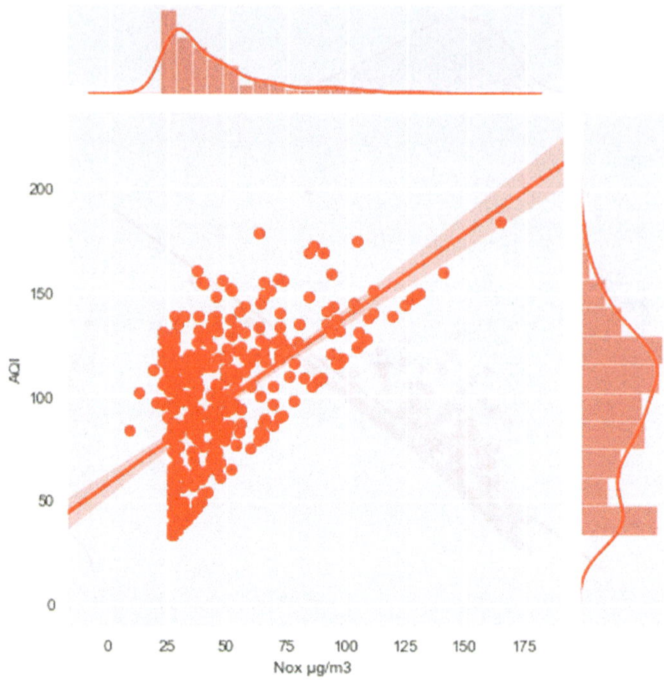

Fig. 2.28 MPCB-KR

and how the AQI changes with the different months and seasons of the year. To achieve this, boxplots are drawn using the following code and the output is presented in Figs. 2.40, 2.41, 2.42, 2.43 and 2.44.

```
[]
sns.boxplot(x='month', y='AQI', data =df[(df['Location']=='MPCB-PMPR') &
(df['year']==2018)])
sns.boxplot(x='month', y='AQI', data =df[(df['Location']=='MPCB-KR') &
(df['year']==2018)])
sns.boxplot(x='month', y='AQI', data =df[(df['Location']=='MPCB-BSRI') &
(df['year']==2018)])
sns.boxplot(x='month', y='AQI', data =df[(df['Location']=='MPCB-NS') &
(df['year']==2018)])
sns.boxplot(x='month', y='AQI', data =df[(df['Location']=='MPCB-SWGT') &
(df['year']==2018)])
```

From Fig. 2.40, it is evident that for Pimpri-Chinchwad, the maximum variation in recorded values was observed in the month of February, while for the other months, it mostly remained between 25 and 150.

From Fig. 2.41, it is evident that for Karve Road, the maximum variability in the recorded values of AQI was observed in the months of November and December.

From Fig. 2.42, it is evident that for Bhosari, the extremes of AQI were recorded in the months of February and December, while September showed the maximum variability in recorded values.

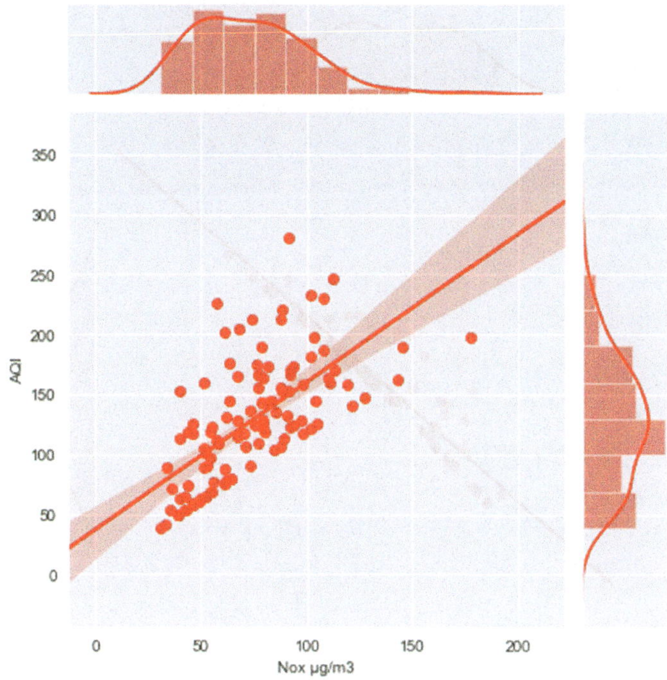

Fig. 2.29 MPCB-BSRI

From Fig. 2.43, it is evident that for Nal Stop, maximum variability in recorded values was observed in the month of September, while the recorded AQI values mostly remained between 60 and 150.

Key findings and interpretations from the above box-plots:

1. The AQI values gradually decreased in the months of March to May, finally forming a sort of plateau in the months of June to September, and then increased in October to December. This type of seasonality was observed for Pimpri-Chinchwad, Karve Road and Swargate.
2. Swargate recorded the most erratic values of AQI throughout the year. June and July were the months where the AQI values recorded were least, while it was maximum for the months of November and December. This might be attributed to the fact that only 65 values were recorded in total the entire year, while 223 values were recorded for Karve Road for the same time span. Not enough data points might be the reason behind this erratic behaviour which could not be truthfully ascertained in this case.
3. The location showing the best or worst AQI over the course of the entire year could not be determined due to incomparable number of records and seasonal variations. Still, Bhosari and Swargate showed the widest variation in recorded

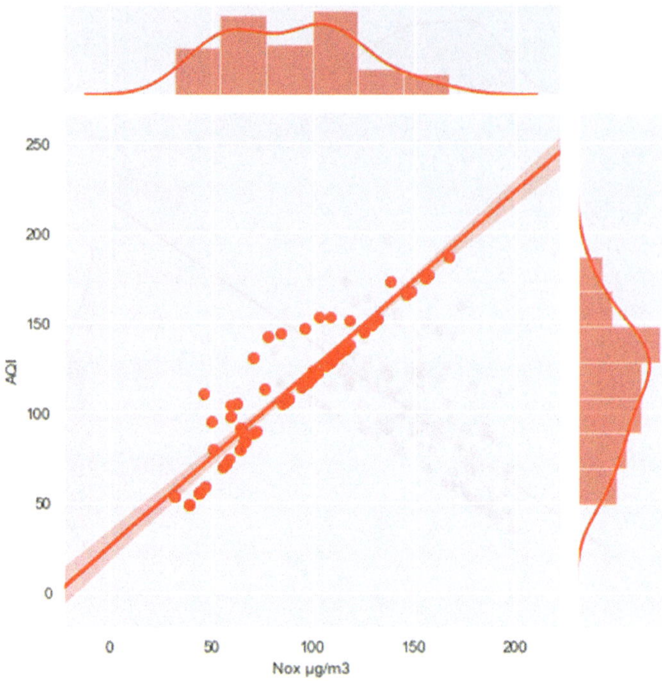

Fig. 2.30 MPCB-SWGT

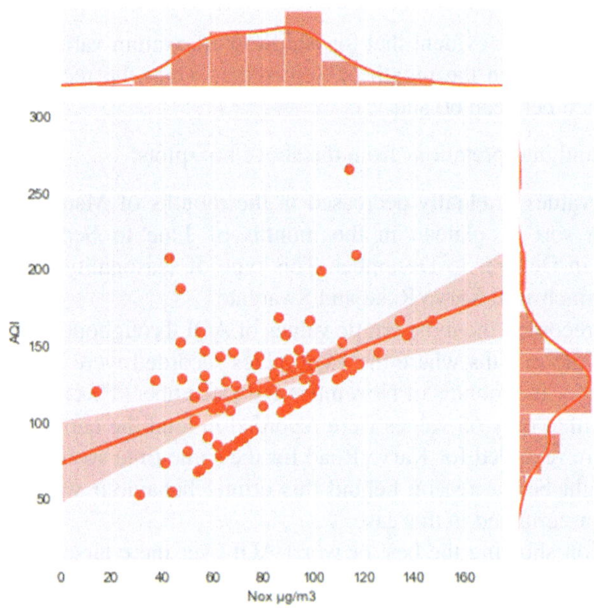

Fig. 2.31 MPCB-NS

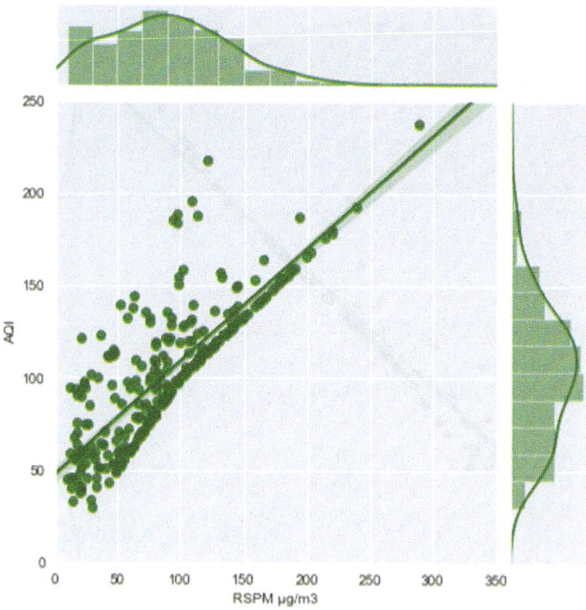

Fig. 2.32 MPCB-PMPR

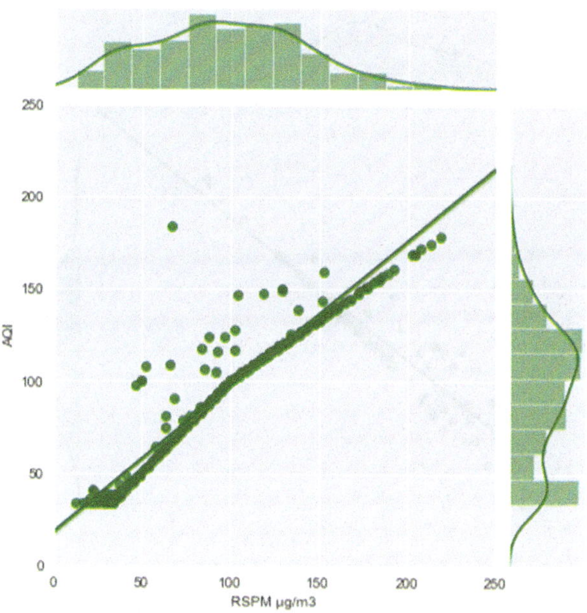

Fig. 2.33 MPCB-KR

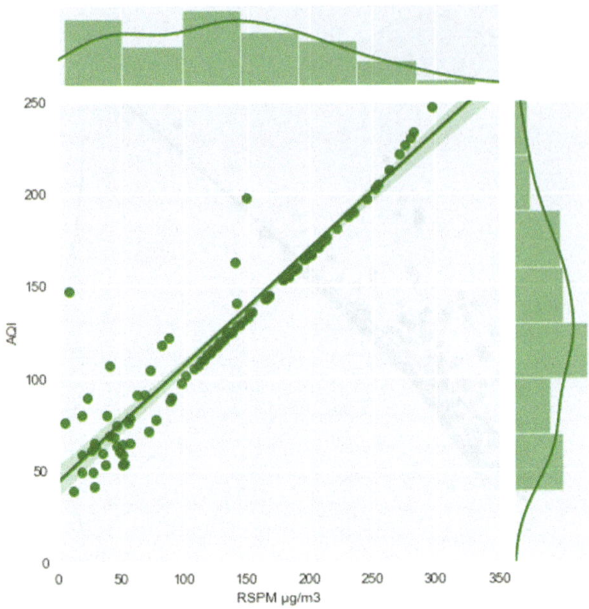

Fig. 2.34 MPCB-BSRI

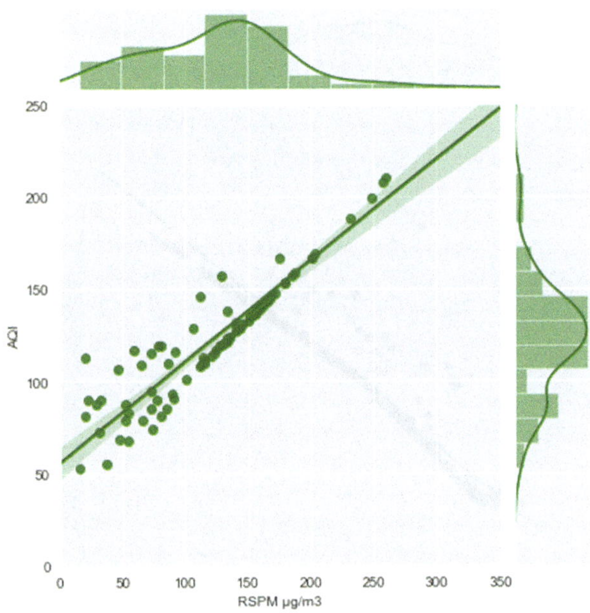

Fig. 2.35 MPCB-NS

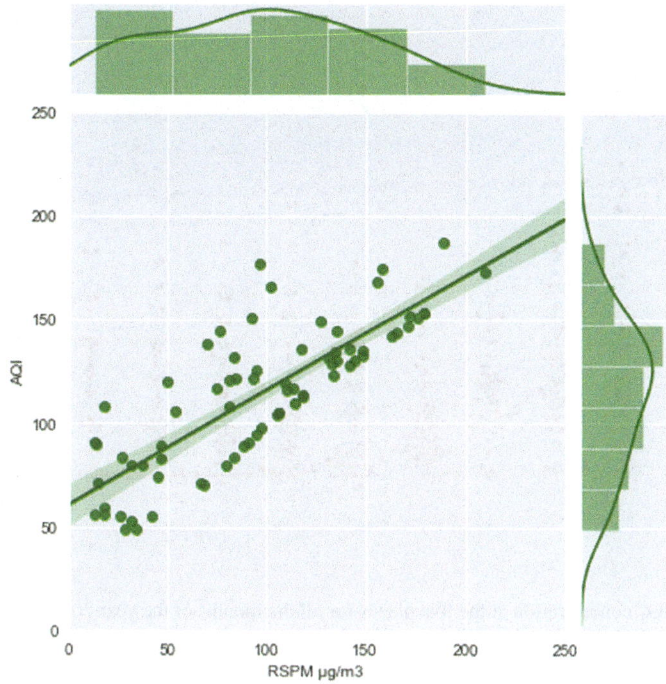

Fig. 2.36 MPCB-SWGT

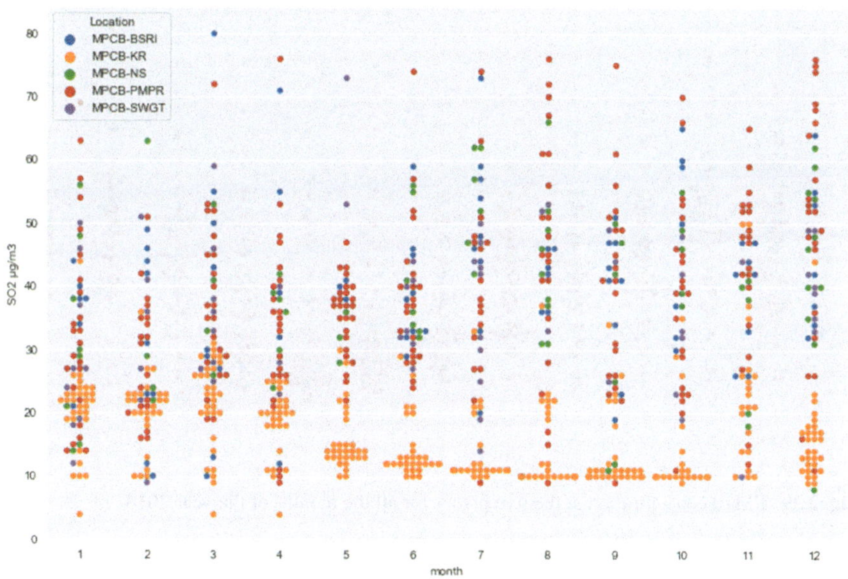

Fig. 2.37 SO_2 concentration at the five places for all the months of the year 2018

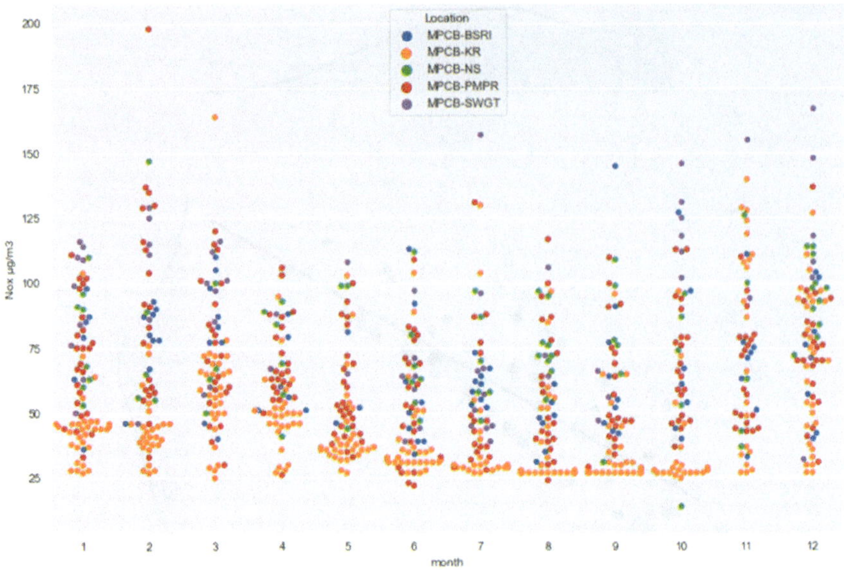

Fig. 2.38 NO_x concentration at the five places for all the months of the year 2018

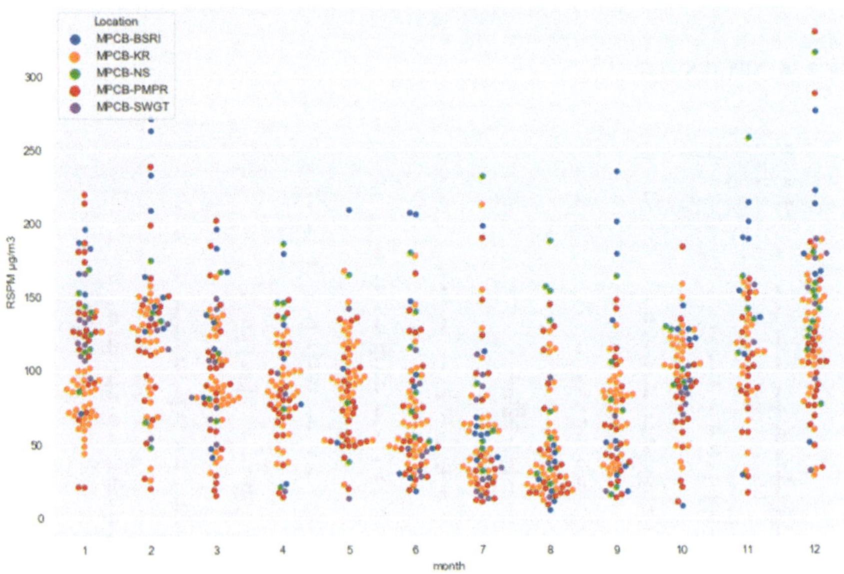

Fig. 2.39 PM10 concentration at the five places for all the months of the year 2018

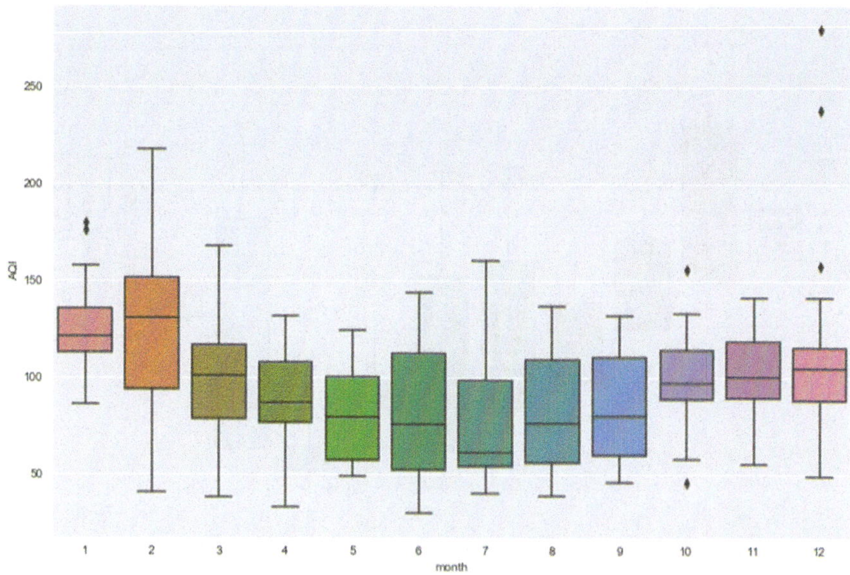

Fig. 2.40 AQI for the months of the year 2018 at MPCB-PMPR

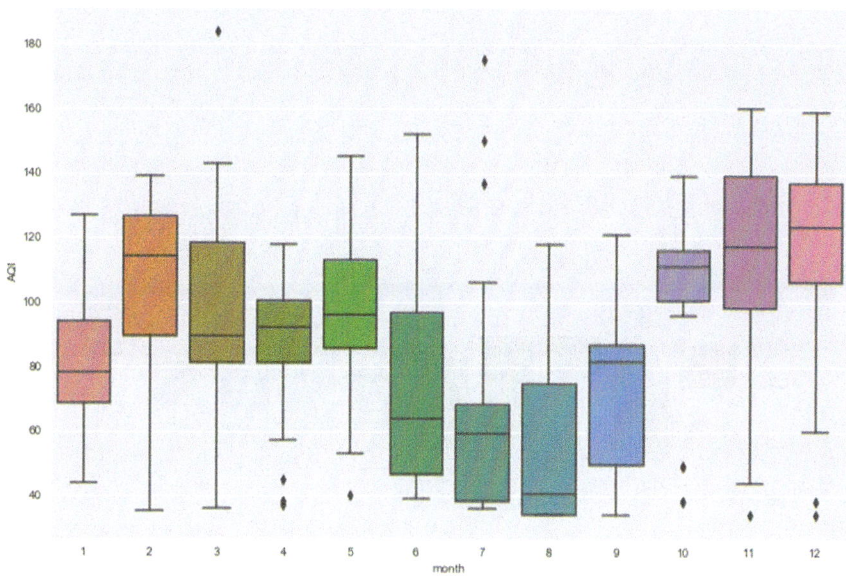

Fig. 2.41 AQI for the months of the year 2018 at MPCB-KR

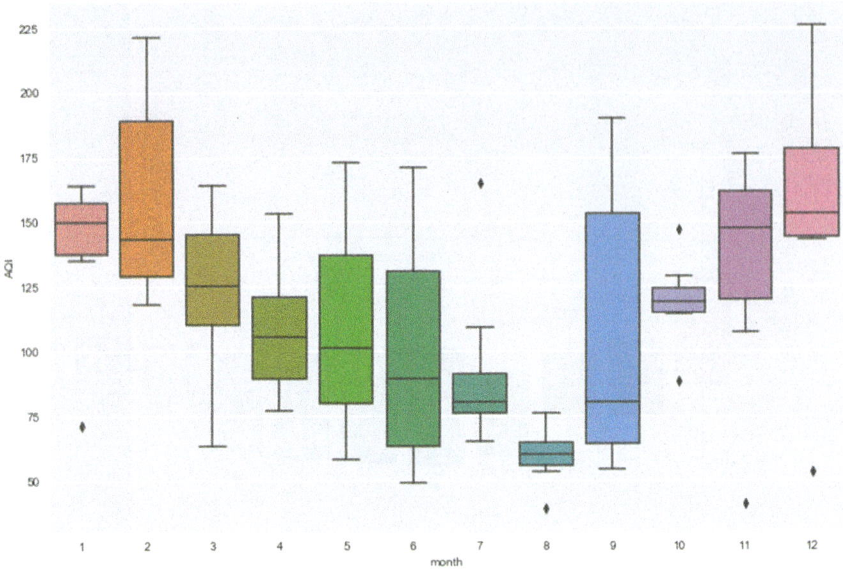

Fig. 2.42 AQI for the months of the year 2018 at MPCB-BSRI

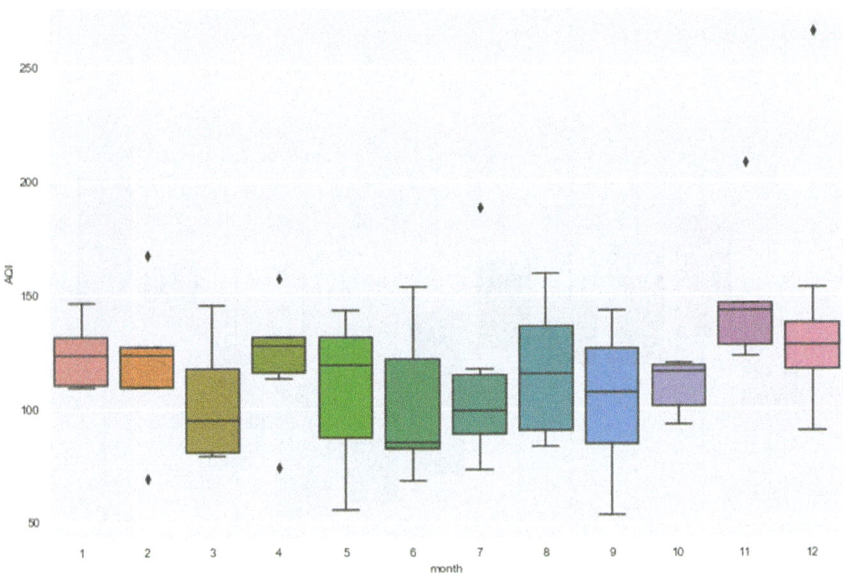

Fig. 2.43 AQI for the months of the year 2018 at MPCB-NS

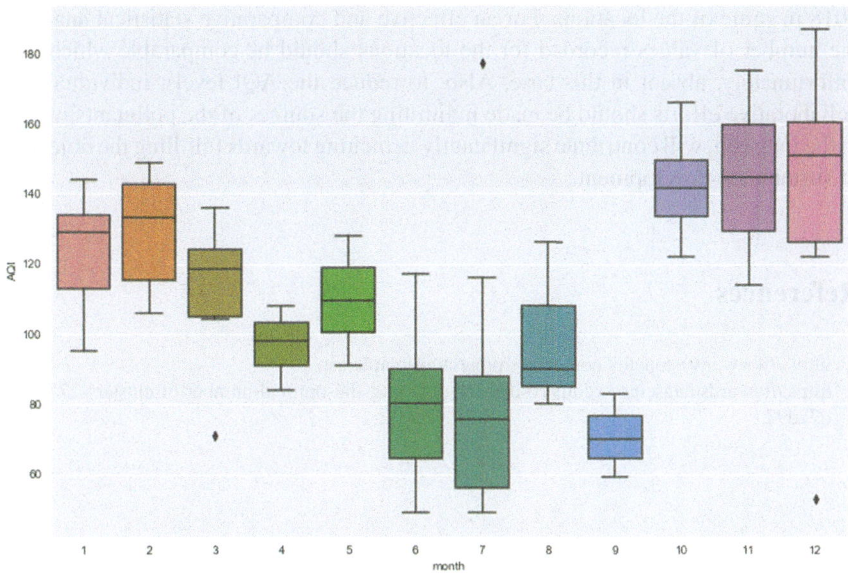

Fig. 2.44 AQI for the months of the year 2018 at MPCB-SWGT

values of AQI throughout the year, while Nal Stop and Pimpri-Chinchwad showed variation within a definite range.

2.3 Conclusion

In this chapter, an attempt is made to comprehend the importance of giving special attention to the city areas that are in dire need of more green cover along with identifying the problems associated with the already growing trees. Researching whether a certain tree species is fit to grow in the environmental conditions offered by the city, regularly checking the balancing of trees especially those growing on high risk grounds, and strategically planting trees near roads, on bridges, along avenues, etc., helps in the proper growth, maintenance and in enhancing the performance of the trees planted for the long run. As the world strives to accomplish the SDGs by 2030, becoming conscious of one's surroundings will help in winning small bits of the bigger fight to tackle the climate change. As the Pareto Principle puts it, '80% of the results come from just 20% of the actions', hence, identifying the areas that need the most attention will help to use the resources efficiently in achieving these goals.

Simultaneously, the air pollutants concentration and the air quality index levels for the five locations were studied in this chapter. The industrial area Bhosari recorded the highest mean values of AQI, SO_2 and PM10 concentration. The AQI levels were found to have a strong linear relationship with respect to the individual pollutants (SO_2, NO_x and PM10) concentration, and it exhibited seasonality patterns for the year

2018 in some of the locations. For an effective and comparative statistical analysis, the number of values recorded for the locations should be comparable which was unfortunately, absent in this case. Also, to reduce the AQI levels, individual and collaborative efforts should be made in limiting the sources of the pollutants, which in the long run, will contribute significantly in inching towards fulfilling the objective of sustainable development.

References

1. https://www.investopedia.com/terms/p/paretoprinciple.asp
2. https://towardsdatascience.com/10-tips-for-choosing-the-optimal-number-of-clusters-277e93d72d92

Chapter 3
Application of Statistical Analysis in Uncovering the Spatio-Temporal Relationships Between the Environmental Datasets

3.1 Introduction

Human activities, such as the burning of greenhouse gases in industries, release harmful pollutant gases and particles into the air. The observed concentrations were significantly higher in crowded urban areas, and countries with huge manufacturing industries—such as China, India and Bangladesh—have the highest recorded levels globally. One of the most studied services of urban forests and trees is their positive effect on air quality, which is expected to improve human health by removing gaseous air pollutants and particulate matter (PM) [1–4]. The forests absorb one-third of global emissions every year at global level. Carbon sequestration, which is the process of capturing and storing atmospheric carbon dioxide, is one way in which trees are reducing the amount of carbon dioxide in the atmosphere [5–8]. Trees take up many different air pollutants, including both ozone and nitrogen oxides, thus reducing their concentrations in the air that one breathe. Particles, odours and pollutant gases such as nitrogen oxides, ammonia and sulphur dioxide settle on the leaves of a tree.

In the atmosphere, nitric oxides are converted to nitric acid, which trees absorb through their pores or stomata. Trees also reduce the greenhouse gas effect by trapping heat, reduce ground-level ozone and release life-giving oxygen. Trees also remove particulate matter from the atmosphere, particularly, small particles which are a major health hazard and can cause chronic health diseases in infants and adults. Trees along urban roadways can reduce the presence of fine particulate matter in the atmosphere within a range of a few hundred yards. Trees with small or hairy leaves are best at removing particles.

Sustainable development is the focus of this century, including the promotion of environment-friendly practices and setting of goals for sustainable development [9–11]. One of the most important areas while studying sustainable development is the analysis of factors affecting the environmental conditions. This study confirms

N. Sharma et al., *Open Data for Sustainable Community*,
Advances in Sustainability Science and Technology,
https://doi.org/10.1007/978-981-33-4312-2_3

the positive changes in the environment, correlated with afforestation, for the city of Pune in the state of Maharashtra, India. In this fact-based study on the open datasets of the Pune Municipal Corporation (PMC), an attempt is made to uncover insights about the condition of trees in the city, air pollution levels recorded by the pollution board and seasonality in the air pollutant levels as recorded at different locations around the city [12, 13]. Along with NGOs and social organizations, the civic body has been involved in conducting various tree plantation drives in the city on a large scale. Due to this, they have helped increase the green cover by 20 lakh trees and documented their progress in tree census surveys.

The city of Pune being one of the first 100 smart cities in India and has been the core subject of study in urban planning and research [14–16]. The enormous biodiversity, fluctuating socio-economic trends and rich history provide an interesting array of research questions for scientists to explore, for better city planning. This chapter is inspired by similar urban planning surveys and studies in other cities around the world that have explored this domain using different techniques and models [17–20].

Data analytics can be integrated into building smart cities by measuring the biodiversity, finding the correlation with major pollutant levels, by observing the urban forest, and suggesting effective sustainable practices [21–26]. In this chapter, an analysis of the data extracted from the Pune Municipal Corporation Open Dataset over 2018–2019 under the category of air pollution and tree census is carried out, and many interesting correlations are presented [12, 13]. These insights may aid the authorities for making positive decisions towards afforestation and environmental conservation. Figure 3.1 depicts the complete process followed for performing data analysis and gaining insights. However, most of it is covered in Chaps. 1 and 2.

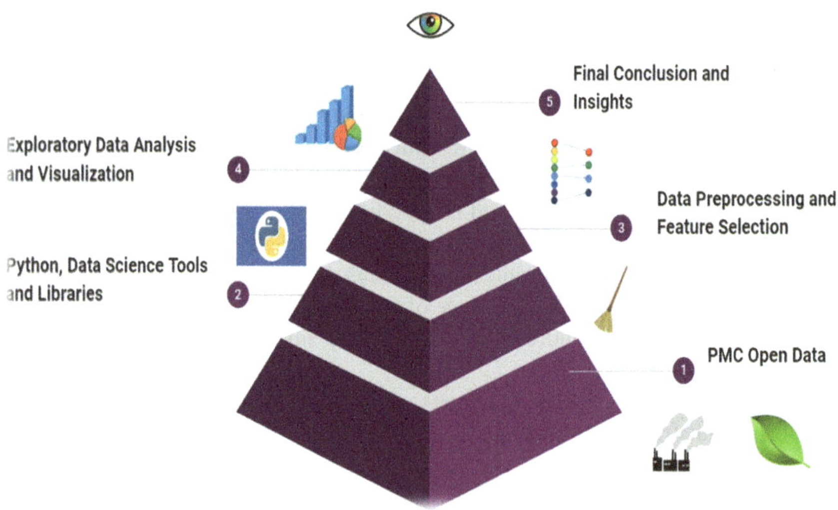

Fig. 3.1 Complete process of data analysis—from dumb data to innovative insights

3.2 Data Sketch for Air Pollution and Tree Census Dataset

The data has been extracted from the Pune Municipal Corporation Open Dataset for the year of 2018–2019 under the category of air pollution and tree census [12, 13]. Air pollution dataset created under the Maharashtra Pollution Control Board, Regional Office (MPCB RO) Pune, captures the data for eight regions. However, for this study, only five regions of interest are considered as remaining three are not in the ambit of PMC. The five regions are MPCB-BSRI, MPCB-KR, MPCB-NS, MPCB-PMPR and MPCB-SWGT, which corresponds to Bhosari, Karve Road, Nal Stop, Pimpri and Swargate which are in Pune city. Air pollution dataset contains the monthly concentrations of air pollutants, found in different regions of Pune city as described in Table 3.1. A snapshot of air pollution dataset is presented in Table 3.2 to provide a glimpse of data values.

The second dataset contains the data collected from the tree census survey conducted by the Pune Municipal Corporation in which they mapped 4,009,623 trees from 70 wards of Pune city in the year 2019. The dataset contains 28 attributes related to the different characteristics of trees (for more details refer Chap. 1), from which seven relevant features are extracted for the correlation analysis. Out of these seven extracted features, the most significant feature observed was the geographical location, which is used to link the air and tree datasets. To pinpoint the exact location of the trees, the Haversine formula is used to measure the distance of the trees from the regions where the pollution levels were measured. The distances thus

Table 3.1 Air pollution data description

Feature	Description
Date	Date of measurement of pollution concentration
SO_2	Sulphur dioxide concentration measured in $\mu g/m^3$
NO_x	Nitrogen oxide concentration measured $\mu g/m^3$
RSPM	Respirable Suspended Particulate Matter concentration measured $\mu g/m^3$
SPM	Suspended Particulate Matter concentration measured $\mu g/m^3$
AQI	Air quality index that lies between 0 and 500
Location	Describes the location at which the air pollution level was recorded

Table 3.2 Air pollution dataset sample

Date	SO_2 ($\mu g/m^3$)	NO_x ($\mu g/m^3$)	RSPM ($\mu g/m^3$)	SPM	AQI	Location
01-02-2018	49	87	263	376	213	MPCB-BSRI
01-05-2018	38	81	209	467	173	MPCB-BSRI
01-09-2018	34	77	201	305	167	MPCB-BSRI
01-12-2018	42	81	165	255	143	MPCB-BSRI
16-01-2018	38	79	196	297	164	MPCB-BSRI

Table 3.3 Tree census dataset description

Feature	Description
girth_cm	Girth of tree in centimetres
height_m	Height of tree in metres
canopy_dia_m	Canopy diameter of tree in metres
condition	Condition of tree in terms of health
common_name	Reported common name of tree
ward	Ward number of the location of tree
is_rare	States whether tree was rare or not
dist_MPCB_KR	Distance of tree from Karve Road
dist_MPCB_NS	Distance of tree from Nal Stop
dist_MPCB_SWGT	Distance of tree from Swargate

obtained are also added to the dataset forming three additional features, namely as dist_MPCB_KR, dist_MPCB_NS and dist_MPCB_SWGT, to represent the distance of a tree from Karve Road, Nal Stop and Swargate, respectively. The feature of tree census dataset that is considered for analysis in this chapter is mentioned in Table 3.3 and the snapshot of dataset with data values is shown in Table 3.4.

The air pollution dataset covers five major zones, and the tree census data covers 70 wards in the Pune city. However, the datasets which are joined on a common feature ‘location’ and ‘ward’, actually have only three regions in common.

Table 3.4 Tree census dataset sample

Feature	Record 1	Record 2	Record 3	Record 4	Record 5	Record 6
girth_cm	10	115	15	13	25	10
height_m	2	10	2	2	7	2
canopy_dia_m	1	4	2	1	1	1
condition	Healthy	Healthy	Healthy	Healthy	Healthy	Healthy
common_name	Yellow bells	Fish tail palm	Subabul	Mango	Subabul	Yellow bells
ward	61	8	29	8	8	61
is_rare	False	False	False	False	False	False
dist_MPCB_KR	6.55428	5.539697	5.07602	5.54111	5.541631	6.55428
dist_MPCB_NS	7.509323	5.708145	3.753423	5.708749	5.709113	7.509323
dist_MPCB_SWGT	4.888684	7.095322	6.40005	7.097067	7.097648	4.888684

3.3 Methodology to Find Correlation

The information about the geographical coordinates of all three stations for which air pollution level is recorded by Municipal Pollution Control Board and that of the individual trees were used as the link between the two datasets. This allowed taking the study further to analyse the correlation and other patterns between the two datasets in greater detail.

All trees within a radius of 1 km of each of the three air pollution control board offices were taken into consideration for finding the correlations between the two datasets. All distances were calculated using the Haversine formula, as given in Eq. 1. In order to take non-overlapping circular areas around the three locations, first, the length of the edges of the triangle (non-Euclidean space) formed by these three locations was found using the same formula. The shortest of these three edges was found to be of 1.41 km (along the great circle of Earth) between Karve and Nal Stop. Next, keeping one of the MPCB office locations (say MPCB-Karve) at the centre, all trees within a virtual circle of radius 1 km around this location were considered as the area under study. The same procedure was followed for Nal Stop and Swargate. Using this procedure and manually choosing the tree girth (in centimetres), tree height (in metres), tree condition, common name of the tree species and ward number, in which the tree was growing, and the boolean variable indicating whether the tree was rare to be found in Pune, among other attributes for the purpose of finding correlations, and the data frame in Table 3.4 was obtained.

3.3.1 *Measuring Correlation*

For understanding the relationships between the two datasets, the first step is to identify the common feature. In this case, 'location' is the common feature between air pollution and tree census dataset that is used to link the two datasets. To measure the relationship between the linked datasets, the statistical measure called correlation is being used for the analysis. Correlation is a bivariate analysis that measures the strength of association between two variables and also the direction of the relationship.

i. **Pearson Correlation**

   This correlation measures the strength of the linear relationship between normally distributed variables. Pearson r correlation is the most widely used correlation statistic to measure the degree of the relationship between linearly related variables.

$$r = \frac{\sum_{i=1}^{N}(x_i - \bar{x})(y_i - \bar{y})}{\sqrt{\sum_{i=1}^{N}(x_i - \bar{x})^2 . \sum_{i=1}^{N}(y_i - \bar{y})^2}} \tag{3.1}$$

where

r is the coefficient of correlation
N is the size of the sample
x_i, y_i are the ith values of the variables X and Y
$\bar{x}, \bar{y}$ are the sample means of the variables X and Y.

ii. **Kendall Correlation**
Kendall rank correlation is a statistic used to measure the ordinal association between two measured quantities. It is a non-parametric test.

$$\tau = \frac{n_\mathrm{c} - n_\mathrm{d}}{n_\mathrm{c} + n_\mathrm{d}} \tag{3.2}$$

where

τ is the Kendall rank correlation
n_c is the number of concordant pairs
n_d is the number of discordant pairs
n is the number of pairs.

iii. **Spearman Correlation**
Spearman rank correlation is a non-parametric test that is used to measure the degree of association between two variables. The Spearman rank correlation test does not carry any assumptions about the distribution of the data and is the appropriate correlation analysis when the variables are measured on a scale that is at least ordinal.

$$\rho = 1 - \frac{6\sum d_i^2}{n\left(n^2 - 1\right)} \tag{3.3}$$

where

ρ is the Spearman rank correlation
d_i is the difference between the ranks of corresponding variables
n is the number of observations.

3.3.2 *Air Quality Index*

The US AQI is the United States Environmental Protection Agency's index for reporting air quality. The AQI lies between 0 and 500. The higher the AQI value, the greater is the level of air pollution. For example, an AQI value of 50 or below represents good air quality, while an AQI value over 300 represents hazardous air quality.

3.3.3 *Time-Series Analysis*

A time-series analysis is a method of understanding a dataset that is in the form of a series of data points indexed in time order. In this research, time-series analysis is used to analyse the trends of the pollution levels in different regions of the city for the year 2018–2019.

3.3.4 *Haversine Formula*

The Haversine formula gives the distance along the great circle of the Earth between any two points on Earth whose latitude and longitude are known.

$$d = 2r \arcsin\left(\sqrt{\sin^2\left(\frac{\varphi_2 - \varphi_1}{2}\right) + \cos(\varphi_1)\cos(\varphi_2)\sin^2\left(\frac{\lambda_2 - \lambda_1}{2}\right)}\right) \tag{3.4}$$

where

φ_1, φ_2 are the latitude of point 1 and latitude of point 2 (in radians), respectively
λ_1, λ_2 are the longitude of point 1 and longitude of point 2 (in radians), respectively
d is the distance between the two points along the great circle of the Earth
r is the radius of the Earth (6378.1 km).

3.4 Analysis of Datasets to Find Correlation

The analysis of air pollution and tree census datasets attempts to uncover the spatio-temporal relationships in the dataset using exploratory data analysis, time-series analysis geographical analysis, statistical analysis and a combined knowledge from all of these individually to gather appropriate insights.

3.4.1 *Exploratory Data Analysis*

The air pollution dataset covers five major zones, and the tree census data covers 70 wards in the Pune city. The aim of the analysis is to study the trends, patterns and to uncover hidden relationships between the datasets. For performing exploratory data analysis on air pollution dataset, the data is grouped by region, and the pollutant levels are plotted to determine the most polluted regions as shown in Fig. 3.2 for SO_2, Fig. 3.3 for NO_x, Fig. 3.4 for RSPM concentration and Fig. 3.5 for AQI for five zones of Pune city.

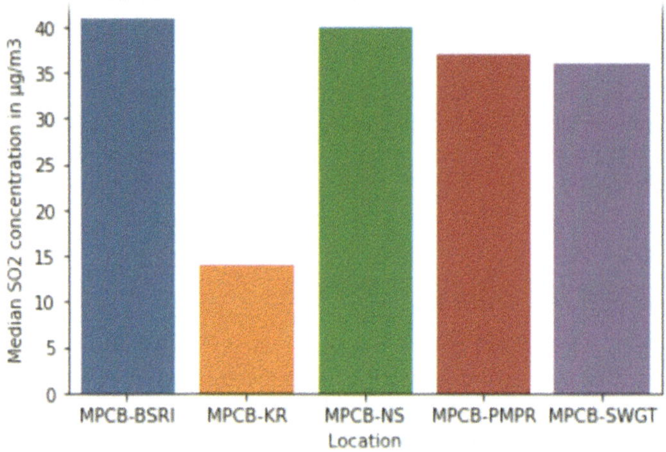

Fig. 3.2 Median SO_2 concentration in five zones

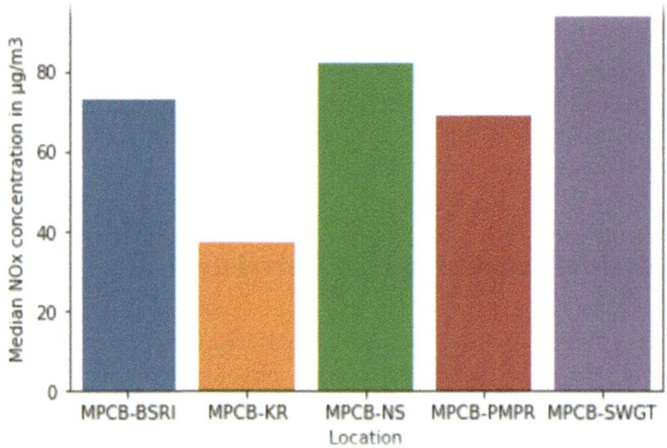

Fig. 3.3 Median NO_x concentration in five zones

From the above bar graphs, it is clearly evident that:

1. Bhosari has the greatest median SO_2 level and Karve road has the minimum.
2. Swargate has the greatest median NO_x level and Karve road has the minimum.
3. Bhosari has the greatest median RSPM level and Pimpri has the minimum.
4. Bhosari has the greatest median AQI level and Karve road has the minimum.

Now, exploratory data analysis on tree census dataset is performed to carry out the comparative analysis of distribution of 'average' and 'dead' tree as well as 'median height' and 'median girth' using joint plot over three regions as shown in Figs. 3.6 and 3.7, respectively.

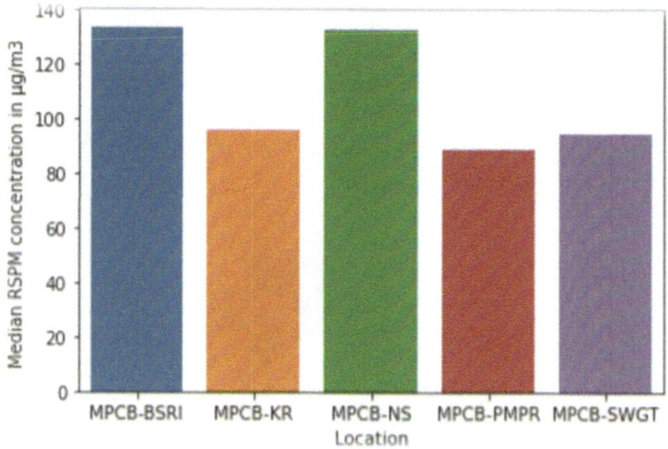

Fig. 3.4 Median RSPM concentration in five zones

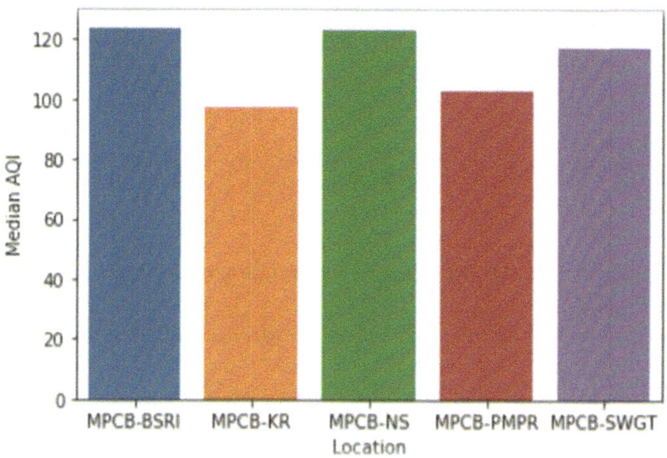

Fig. 3.5 Median AQI in five zones

Further, a heat map is created to find the correlation between air pollution data and tree census data as shown in Fig. 3.8.

Followings are the observations from the joint plots and heat map plotted to find the correlation:

1. Most of the regions have around 800 trees in dead condition and 1000 trees average condition.
2. Most of the regions have around 0.595 cm median girth and 7 m median height.

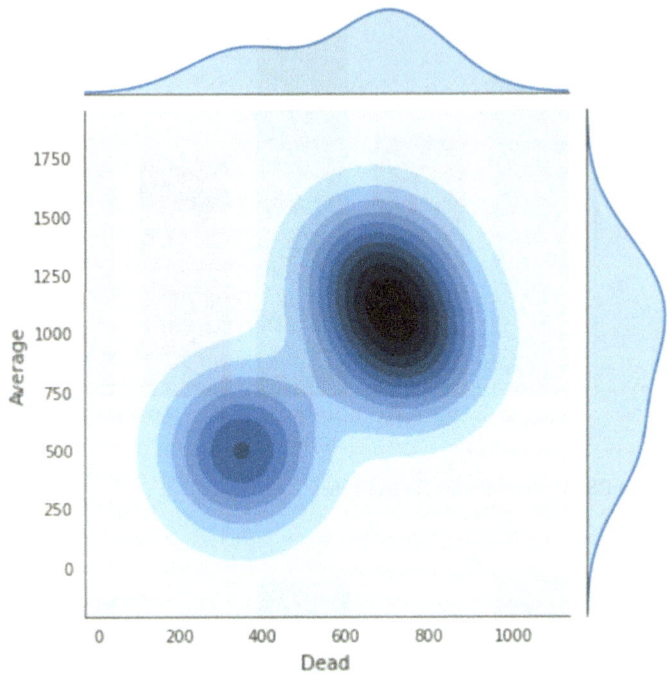

Fig. 3.6 Comparative analysis of average and dead tree distributions in tree dataset using joint plot over three regions

3. The median girth has a strong negative correlation with RSPM levels and positive with tree height.
4. The number of healthy trees has a strong negative correlation with NO_x pollutant levels.

3.4.2 GIS Analysis

A geographical information system (GIS) deals with spatial data, performs proximity analysis or location analysis and presents the visualizations using maps. In this section, a GIS analysis of air pollution and tree census data is presented, and intent is to aid the authorities in understanding the pollution intensities. To visualize the median pollutant levels across Pune, the GIS analysis needs to be done, for which the real-time maps are obtained using Python library called Folium. However, prior to that, for the better comprehension, the median values of all the pollutant concentrations at 03 locations were extracted as presented in Table 3.5. The median values represent the pollution concentrations in the middle of the time period for which the data is collected, i.e., 2018–2019.

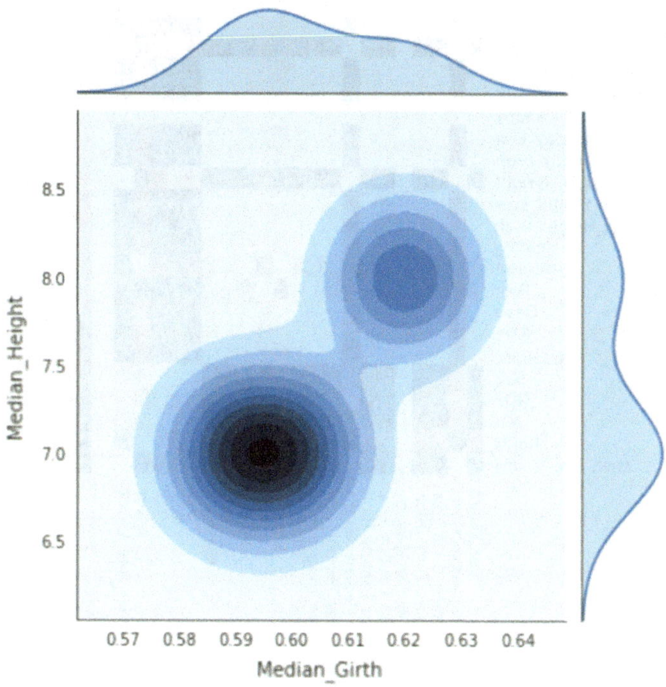

Fig. 3.7 Comparative analysis of median height and median girth distributions in tree dataset using joint plot over three regions

Subsequently, these medians are plotted on the maps. However, for the data visualization through heat maps, there is a need to convert the discrete pollutant concentrations to a range, in order to obtain interpretable heat maps, which bring out the correct magnitude of pollutant concentrations for five regions. Followings are the concentration range for each pollutant:

SO_2 Scale (in μg/m³):
10–20: minimum concentration; 40–50: maximum concentration
NO_x Scale (in μg/m³):
30–40: minimum concentration; 90–100: maximum concentration
RSPM Scale (in μg/m³):
80–90: minimum concentration; 120–130: maximum concentration
AQI Scale:
90–100: minimum concentration; 120–130: maximum concentration.

The heat map for median values of pollutant concentrations for SO_2, NO_x, RSPM and AQI are presented in Figs. 3.10, 3.11, 3.12 and 3.13, respectively. For all the heat maps, the colour coding is as shown in Fig. 3.9.

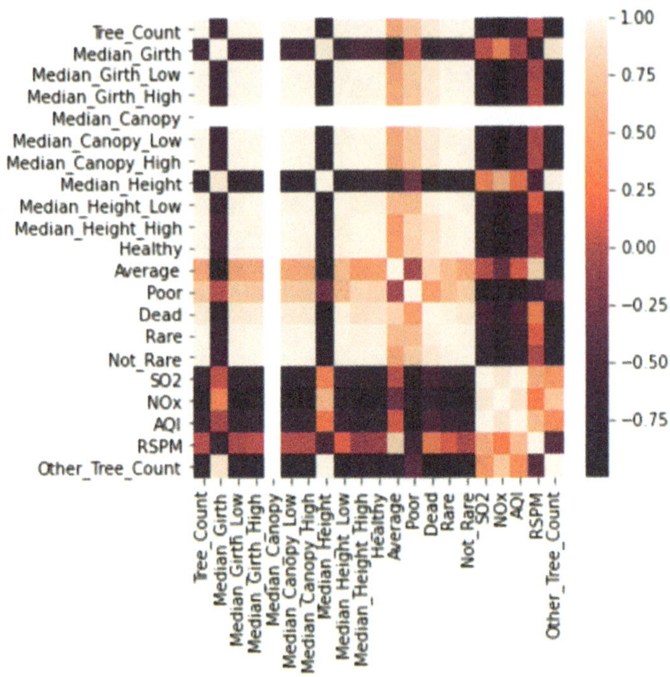

Fig. 3.8 Correlation heat map of tree dataset derived features over three regions

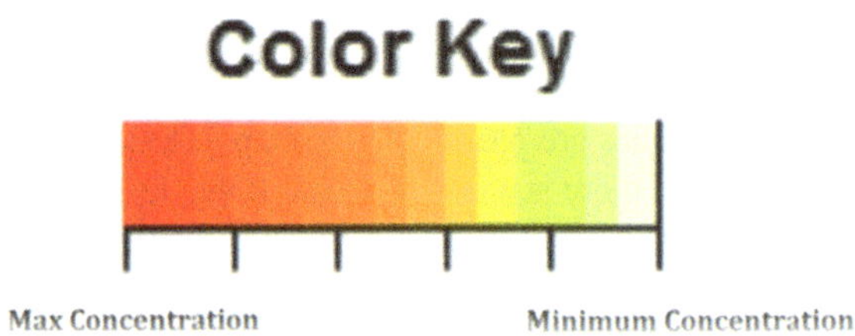

Fig. 3.9 Heat map colour coding

Table 3.5 Median pollutant concentration levels in the five regions

Region	SO_2	NO_x	AQI	RSPM
Karve Road	14	37	97.5	96
Nal Stop	40	82	123	133
Swargate	36	94	117	95
Bhosari	41	73	124	134
Pimpri	37	69	103	89

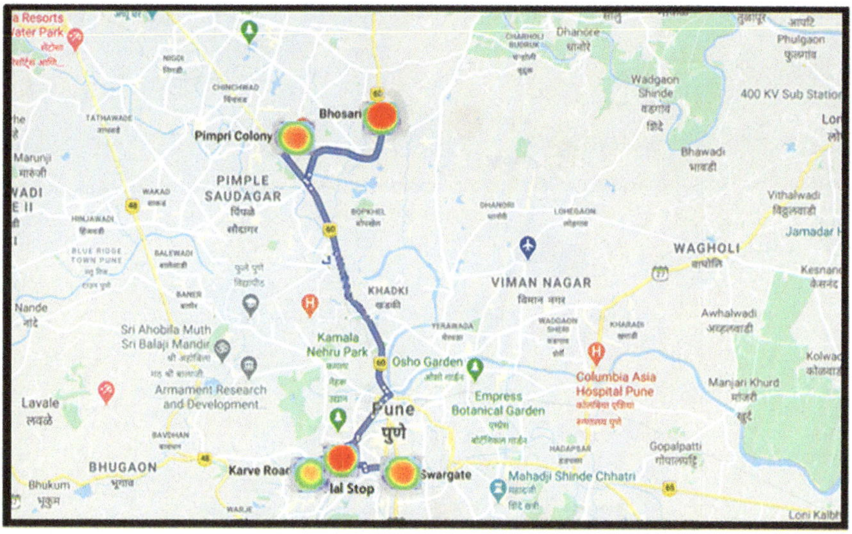

Fig. 3.10 Heat maps of median value of SO_2 levels

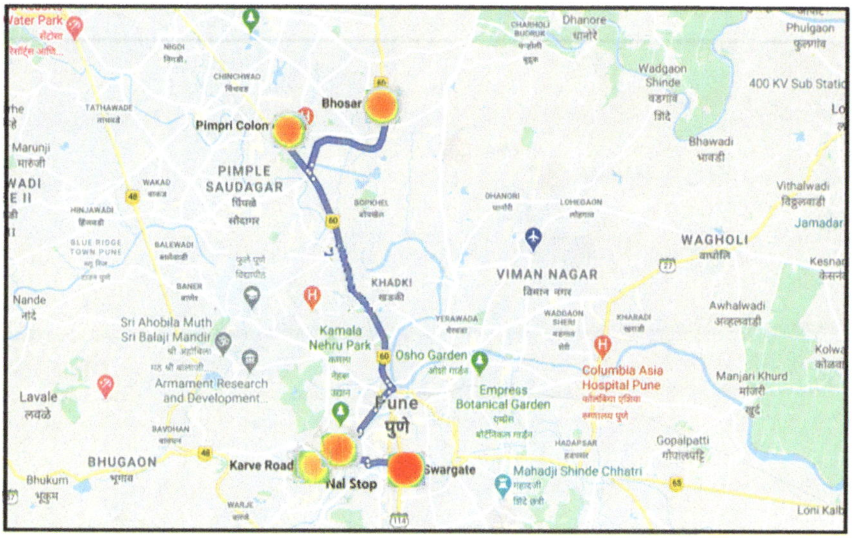

Fig. 3.11 Heat maps of median value of NO_x levels

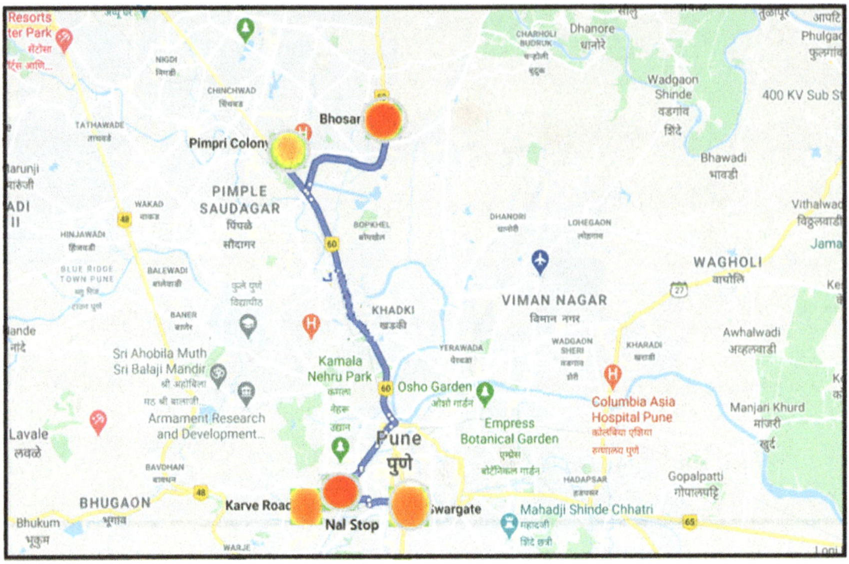

Fig. 3.12 Heat maps of median value of RSPM levels

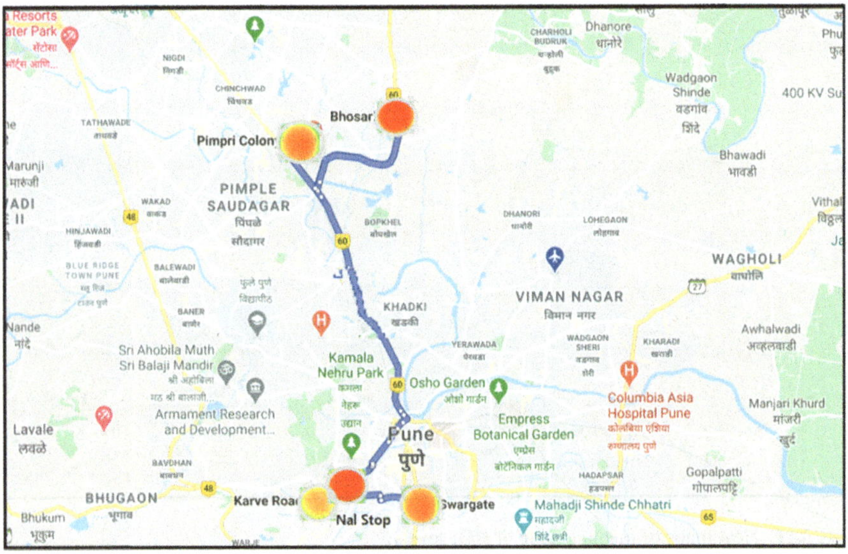

Fig. 3.13 Heat maps of median value of AQI levels

The observations that are drawn from the above GIS analysis are:

1. Nal Stop and Bhosari have high median SO_2 levels.
2. Swargate has high median NO_x level.
3. Nal Stop and Bhosari have high median RSPM levels.
4. Nal Stop and Bhosari have high median AQI levels.

3.4.3 Time-Series Analysis

Time-series analysis deals with the temporal data, performs trend analysis and presents the dynamic changes as line plots over time. To study the pollution levels for the year 2018–2019 from air pollution dataset and to assess dynamic change of the environmental situation over time, the quantities of pollutants found in the air were plotted over time and analysed. Then, the visualizations have been drawn for data pertaining to five regions—Pimpri, Bhosari, Nal Stop, Karve Road and Swargate in Pune city of Maharashtra state in India. Consolidated trends at five locations regarding SO_2 and AQI levels are presented in Figs. 3.14 and 3.18. However, a snapshot of trend is presented for NO_x at Nal Stop in Fig. 3.15, RSPM at Bhosari in Fig. 3.16 and AQI at Pimpri Chinchwad in Fig. 3.17.

The observations that are drawn from the above time-series analysis are:

- The pollution levels seem to have a slight upward trend over time, consistently for all five regions.
- Swargate had a slightly negative SO_2 trend.
- NO_x levels increased in Nal Stop.
- RSPM levels increased in Bhosari.
- AQI levels were almost constant in Pimpri.
- AQI levels for most regions had an upward trend.

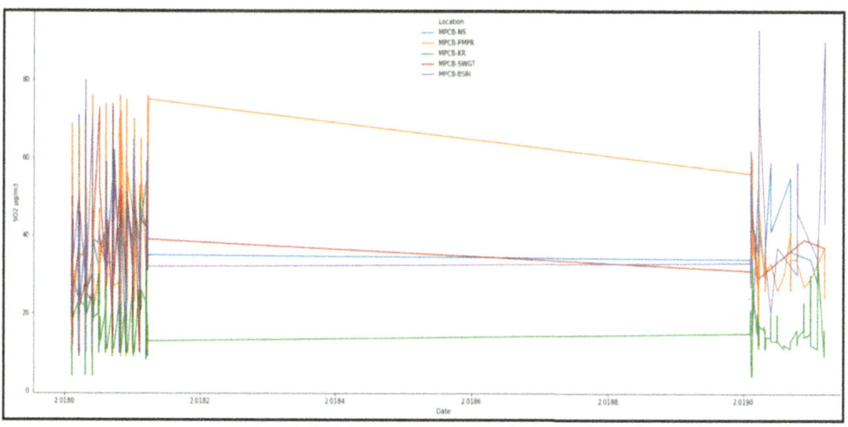

Fig. 3.14 Trends of SO_2 levels for 2018–2019 in five regions

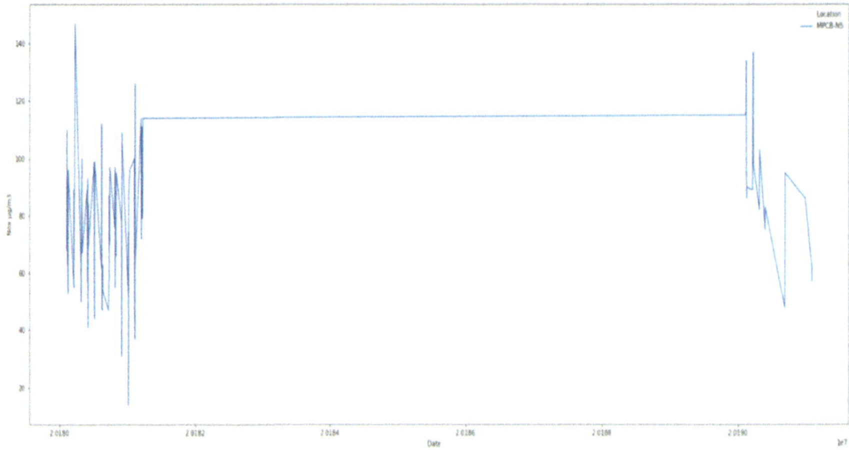

Fig. 3.15 Trends of NO_x levels for 2018–2019 in the Nal Stop Region

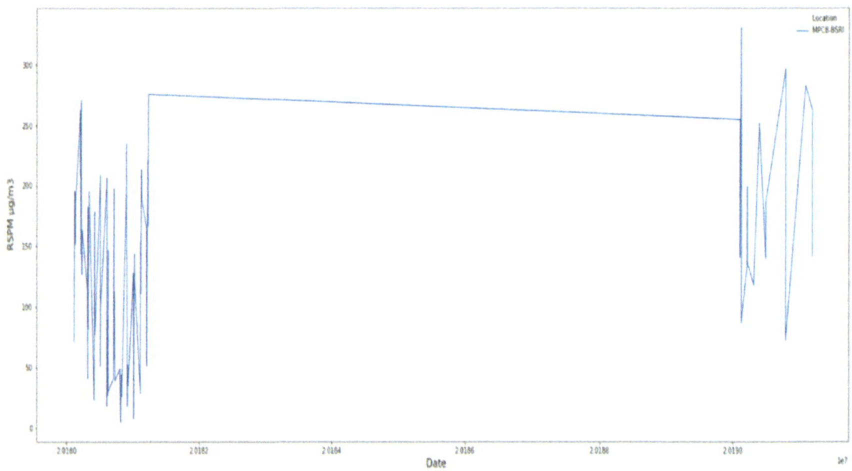

Fig. 3.16 Trends of RSPM levels for 2018–2019 in the Bhosari Region

3.4.4 *Statistical Analysis*

The statistical analysis was divided into three parts, exploratory analysis, correlation analysis and finally biodiversity measurement. The first part consisted of examining the distribution of features in the air pollution dataset using standard measures of central tendency as shared in Table 3.6.

The next part was the correlation analysis that was carried out to determine the nature of relationship between air pollution and tree datasets. For this correlation, a common feature in required to link both datasets. In this case, the spatial data, i.e.,

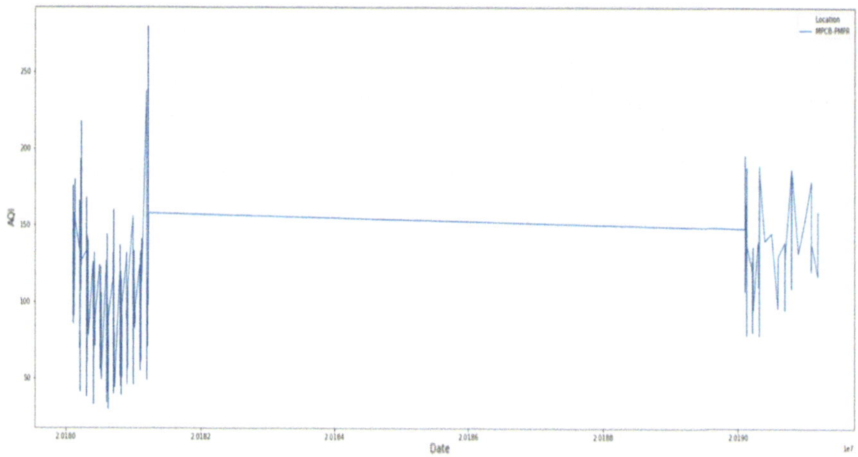

Fig. 3.17 Trends of AQI levels over 2018–2019 in the Pimpri Chinchwad Region

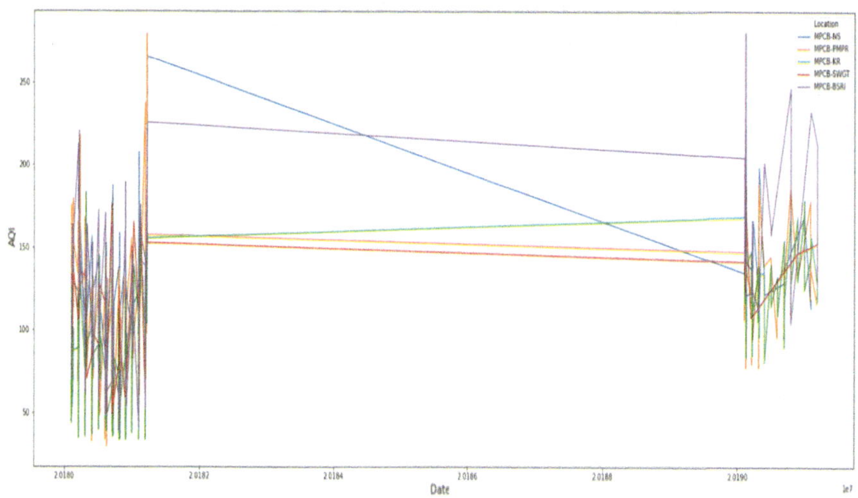

Fig. 3.18 Trends of AQI levels for 2018–2019 in five regions

'location' attribute from air pollution dataset and 'ward' attribute from tree census dataset were convenient for the analysis. The datasets were then merged with inner join using the location as the common column. The 'location' in air pollution dataset is mapped to the distance of a tree from that location, calculated using the Haversine formula and the coordinates of the tree, obtained from the PMC geolocation data available on official site. The joined dataset is presented with the help of Venn diagram in Fig. 3.19.

For the statistical analysis, the first step is to observe the datasets and identify the features that could be useful for the study, by determining their relevance to

Table 3.6 Statistical measures of air pollution dataset

Statistical measure	SO_2 (μg/m^3)	NO_x (μg/m^3)	RSPM (μg/m^3)	SPM	AQI
Count	1004	1004	1004	279	993
Mean	29.51693	63.5757	102.1683	263.7814	105.4209
Std	16.44304	31.46061	54.08161	138.4302	39.92975
Min	4	9	5	11	30
25%	14	38	63	155	78
50%	27	57	98	277	108
75%	41	85	135	352.5	129

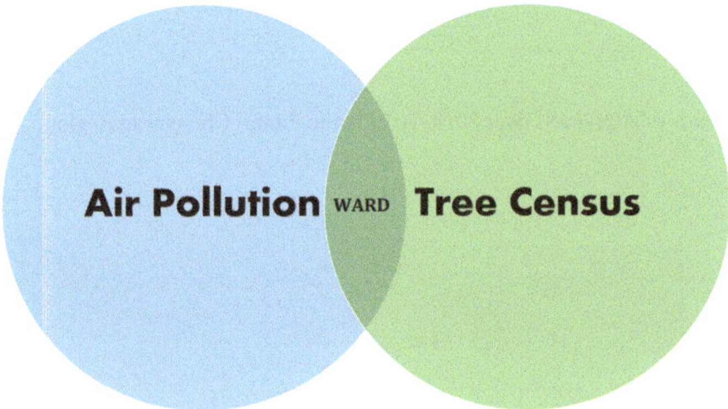

Fig. 3.19 Dataset Venn representation

the research question. A total of 21 features were considered which included the original features as well as some derived features. The original features were the tree count, and count of healthy, poor, average and dead trees along with the air pollutant concentrations. The derived features were obtained by statistical means, like counting the number of trees having height above median height, and those having height below median height. The total number of trees that were rare and not-rare was also counted. From the original and derived features, 21 relevant features were finally obtained, which were grouped as per the location of tree. Upon grouping, the final dataset for the correlation analysis is created and is presented in Table 3.7.

For determining the relationship between the green cover and pollution levels, the final merged dataset is used to calculate correlations. The following correlation coefficients which were explained in Sect. 3.3.1 are used to quantify the correlation:

i. Pearson correlation coefficient is presented in Table 3.8.
ii. Kendall correlation coefficient is presented in Table 3.9.
iii. Spearman correlation coefficient is presented in Table 3.10.

A basic understanding of the overall relationship between the pollutant levels and tree features in Pune over the observed regions can be drawn from above correlation

Table 3.7 Processed and merged features of air pollution and tree census datasets

Feature/Region	Karve road	Nal stop	Swargate
Tree_Count	30,618	25,913	21,536
Median_Girth	0.6	0.59	0.62
Median_Girth_Low	15,255	12,814	10,693
Median_Girth_High	15,363	13,099	10,843
Median_Canopy	4	4	4
Median_Canopy_Low	14,517	11,987	9673
Median_Canopy_High	16,101	13,926	11,863
Median_Height	7	7	8
Median_Height_Low	14,143	12,354	9762
Median_Height_High	16,475	13,559	11,774
Healthy	28,750	23,883	20,564
Average	947	1243	494
Poor	167	118	128
Dead	754	669	350
Rare	494	382	222
Not_Rare	29,707	25,106	21,070
SO2	14	40	36
NOx	37	82	94
AQI	97.5	123	117
PM10	96	133	95
Other_Tree_Count	30,886	110,700	779,540

indices. As expected, the number of healthy trees has a strong negative correlation with the pollutant levels, but the number of average trees does not. Surprisingly, a large number of dead trees do not seem to have a positive correlation with the pollution levels. The overall tree count has a negative correlation with all the pollutants, except for RSPM levels. The number of non-rare trees appears to have a greater correlation than rare, which may be attributed to the sheer majority of the non-rare trees that were able to make a difference.

The most common trees found in each region were also identified to understand the reason of low pollutant levels at a particular area. The trees were sorted according to frequency of each tree type, after being grouped by the location which was calculated from the coordinates using the Haversine formula. The top ten frequently found trees were then separated and analysed and shared in Table 3.11.

It is evident from the above table that the particular commonly found type of tree seems to be consistent among different regions, which was expected due to the geographical conditions like soil condition and weather that are similar all over the city. Thus, the factor of tree type can be eliminated as a probable cause for correlations between features and pollutant levels.

Table 3.8 Pearson correlation between green cover and pollutant levels

Pearson correlation matrix	SO_2	NO_x	AQI	RSPM
Tree_Count	−0.79844	−0.95482	−0.74537	0.002242
Median_Girth	0.046761	0.381246	−0.03683	−0.77084
Median_Girth_Low	−0.8101	−0.96047	−0.75831	−0.01739
Median_Girth_High	−0.78635	−0.94874	−0.732	0.022066
Median_Canopy_Low	−0.80137	−0.95627	−0.74862	−0.00265
Median_Canopy_High	−0.79506	−0.95315	−0.74163	0.007833
Median_Height	0.371154	0.662849	0.292306	−0.51986
Median_Height_Low	−0.71625	−0.90979	−0.65547	0.128167
Median_Height_High	−0.86335	−0.98302	−0.81819	−0.11468
Healthy	−0.84821	−0.97722	−0.80101	−0.08555
Average	0.022911	−0.31595	0.106325	0.81333
Poor	−0.9987	−0.92286	−0.99947	−0.64027
Dead	−0.54897	−0.79892	−0.47725	0.338955
Rare	−0.71897	−0.9114	−0.65841	0.124303
Not_Rare	−0.8085	−0.95971	−0.75653	−0.01466

Table 3.9 Kendall correlation between green cover and pollutant levels

Kendall correlation matrix	SO_2	NO_x	AQI	RSPM
Tree_Count	−0.33333	−1	−0.33333	0.333333
Median_Girth	−0.33333	0.333333	−0.33333	−1
Median_Girth_Low	−0.33333	−1	−0.33333	0.333333
Median_Girth_High	−0.33333	−1	−0.33333	0.333333
Median_Canopy_Low	−0.33333	−1	−0.33333	0.333333
Median_Canopy_High	−0.33333	−1	−0.33333	0.333333
Median_Height	0	0.816497	0	−0.8165
Median_Height_Low	−0.33333	−1	−0.33333	0.333333
Median_Height_High	−0.33333	−1	−0.33333	0.333333
Healthy	−0.33333	−1	−0.33333	0.333333
Average	0.333333	−0.33333	0.333333	1
Poor	−1	−0.33333	−1	−0.33333
Dead	−0.33333	−1	−0.33333	0.333333
Rare	−0.33333	−1	−0.33333	0.333333
Not_Rare	−0.33333	−1	−0.33333	0.333333

Table 3.10 Pearson correlation between green cover and pollutant levels

Spearman correlation matrix	SO_2	NO_x	AQI	RSPM
Tree_Count	−0.5	−1	−0.5	0.5
Median_Girth	−0.5	0.5	−0.5	−1
Median_Girth_Low	−0.5	−1	−0.5	0.5
Median_Girth_High	−0.5	−1	−0.5	0.5
Median_Canopy_Low	−0.5	−1	−0.5	0.5
Median_Canopy_High	−0.5	−1	−0.5	0.5
Median_Height	0	0.866025	0	−0.86603
Median_Height_Low	−0.5	−1	−0.5	0.5
Median_Height_High	−0.5	−1	−0.5	0.5
Healthy	−0.5	−1	−0.5	0.5
Average	0.5	−0.5	0.5	1
Poor	−1	−0.5	−1	−0.5
Dead	−0.5	−1	−0.5	0.5
Rare	−0.5	−1	−0.5	0.5
Not_Rare	−0.5	−1	−0.5	0.5

Table 3.11 Most frequent trees in three regions

Feature/Region	Karve road	Nal stop	Swargate
Tree_1	Ashoka (D)	Ashoka (D)	Ashoka (D)
Tree_2	Coconut	Coconut	Coconut
Tree_3	Mango	Ashoka	Ashoka
Tree_4	Ashoka	Mango	Mango
Tree_5	Subabul	Subabul	Cluster fig tree
Tree_6	Cluster FIG TREE	Neem	Subabul
Tree_7	Sonchapha	Sonchapha	Neem
Tree_8	Neem	Cluster fig tree	Pipal
Tree_9	Jamun	Jamun	Jamun
Tree_10	Indian cork tree	Silver oak	Sonchapha

3.5 Results

This section summarizes the results of analysis carried out in this chapter:

1. The city was observed to have differing pollution levels in different regions in the year 2018–19.
2. Bhosari has consistently high pollution levels, and a dangerous air quality index which must be looked into, as it poses a threat to the residents.

3. Karve Road showed low pollution levels throughout and can be examined in detail by the authorities to find out measures that have helped kerb pollution on Karve Road, in addition to the number of trees. Karve Road has more trees, and these trees also show greater girth, canopy and healthy condition, which show potential for cleaner air, with augmented efforts for taking care of trees.
4. Nal Stop has several pollution hot spots, which can be seen from the deficient tree cover. This conclusion can also be supported by evidence from the city reports that have received excessive complaints about traffic and pollution on Nal Stop.
5. Over time, the pollution levels are steadily increasing, which is a concern as with the worsening air quality, the residents may face health issues like respiratory diseases and even chronic health conditions. In just the span of one year, the pollution levels visibly increased, which presents a growing threat for the future.
6. The overall tree count has a negative correlation with all the pollutants, except for RSPM levels. The authorities must continue to take efforts to boost the tree count and improve the air quality.
7. The biodiversity distribution does not appear to have as significant an impact on pollution as magnitude of the tree count, which has a strong correlation with the air quality.

3.6 Conclusion and Future Work

This study presents a method to democratize the open Pune datasets, for city planning and promoting awareness, as a part of the Smart Cities Mission. It is possible to turn voluminous datasets into precise insights for the city development using statistics, GIS, data science and visualization techniques. The strong correlations found in the analysis, throw light on the spatio-temporal relationships in the datasets and also promises for the formulation of better measures to control pollution.

The current research has a lot of scope in the domain of transformation of cities into smart cities and needs to be explored in future. The effect of policies on all the features and correlations can be examined to develop a better understanding of the city dynamics and response to different measures. In this chapter, only two years data has been considered due to the constraints of the linked datasets, but in a few years, a more detailed analysis can be carried out by collecting extensive data from the next tree census survey. A suggestion to the authorities can be put forward to include the citizens in the policy formation process and to ensure that the tree cover is increased on both private and public lands.

We hope that our research will be applied in future to other cities and villages across the world and will play a substantial role in the open data revolution towards town planning and the development of sustainable cities.

References

1. Samson R et al (2017) Urban trees and their relation to air pollution. In: Pearlmutter D et al (eds) The urban forest. Future city, vol 7. Springer, Cham. https://doi.org/10.1007/978-3-319-50280-9_3
2. Akbari H, Pomerantz M, Taha H (2001) Cool surfaces and shade trees to reduce energy use and improve air quality in urban areas. Sol Energy 70:295–310
3. Escobedo F, Varela S, Zhao M, Wagner JE, Zipperer W (2010) Analyzing the efficacy of subtropical urban forests in offsetting carbon emissions from cities. Environ Sci Policy 13:362–372
4. Zhao M, Kong Z, Escobedo FJ, Gao J (2010) Impacts of urban forests on offsetting carbon emissions from industrial energy use in Hangzhou, China. J Environ Manage 91:807–813
5. Brack CL (2002) Pollution mitigation and carbon sequestration by an urban forest. Environ Pollut 116(Suppl)
6. Moulton RJ, Richards KR (1990) Costs of sequestering carbon through tree planting and forest management in the United States. USDA Forest Service, General Technical Report WO-58, Washington, DC
7. Russo A, Escobedo FJ, Timilsina N, Schmitt AO, Varela S, Zerbe S (2014) Assessing urban tree carbon storage and sequestration in Bolzano, Italy. Int J Biodivers Sci Ecosyst Serv Manage 10(1):54–70
8. Nowak DJ, Crane DE (2002) Carbon storage and sequestration by urban trees in the United States. Environ Pollut 116:381–389
9. https://in.one.un.org/page/sustainable-development-goals/sdg-13/
10. https://in.one.un.org/page/sustainable-development-goals/sdg-15/
11. https://sustainabledevelopment.un.org/sdg15
12. Dataset 1 (Air pollution). http://opendata.punecorporation.org/Citizen/CitizenDatasets/Index?categoryId=17
13. Dataset 2 (Tree census). http://treecensus.punecorporation.org/
14. Butsch C, Kumar S, Wagner PD, Kroll M, Kantakumar LN, Bharucha E, Schneider K, Kraas F (2017) Growing 'Smart'? Urbanization processes in the pune urban agglomeration. Sustainability 9:2335
15. Kantakumar LN, Kumar S, Schneider K (2016) Spatiotemporal urban expansion in Pune metropolis, India using remote sensing. Habitat Int 51:11–22
16. Pune towards Smart City: vision document, Pune Municipal Corporation (2018)
17. UN World Summit on Sustainable Development: Johannesburg plan of implementation. United Nations, New York (2002)
18. Caragliu A, Del Bo C, Nijkamp P (2011) Smart cities in Europe. J Urban Technol 18(2):65–82
19. CBD monitoring and indicators: designing national-level monitoring programmes and indicators. Convention on Biological Diversity, Montreal (2003)
20. Ten Brink B, Tekelenburg T (2002) Biodiversity: how much is left? National Institute for Public Health and the Environment
21. Mora F (2019) The use of ecological integrity indicators within the natural capital index framework: the ecological and economic value of the remnant natural capital of México. J Nat Conserv 47:77–92
22. Czucz B, Horvath F, Molnar Z, Botta-Dukat Z (2008) The natural capital index of Hungary. Acta Bot Hung
23. Scholes RJ, Biggs R (2005) A biodiversity intactness index. Nature 434:45–49
24. Reid WV, McNeely JA, Tunstall DB, Bryant DA, Winograd M (1993) Biodiversity indicators for policy makers. World Resources Institute, Washington DC
25. Nowak DJ, Crane DE, Stevens JC, Ibarra M (2002) Brooklyn's urban forest. Newtown Square (PA): Northeastern Research Station, United States Department of Agriculture, Forest Service, Borough of Brooklyn. General Technical Report NE-290
26. Lawrence AB, Escobedo FJ, Staudhammer CL, Zipperer W (2012) Analyzing growth and mortality in a subtropical urban forest ecosystem. Landscape Urban Plan 104:85–94

Part II
Resilient Agriculture—A War Against Hunger

(SDG2—Zero Hunger)

Chapter 4
Farmer Call Centre Literature Review and Data Preparation

4.1 Introduction

There is a general understanding that mankind began civilization with the discovery of farming, which is arguably the most crucial profession on this planet. The simple yet scientific definition of farming is growing crops for food and nourishment. Farmers are the people who undertake this noble job. At the start of twenty-first century, 1/3rd of the global workforce was employed in the agricultural job [1]. In India, a country with a population of 1.3 billion, 50% of the population was directly or indirectly employed in agriculture as shown in Fig. 4.1 [1]. As per 2011 census, there are 118 million farmers in India. These hardworking men are called Kisan in Hindi, India's most prominently spoken native language. India is a predominantly agrarian economy with agriculture contributing to 18% of the gross domestic product (GDP) [1]. India is also the foremost in net cropped area in the world while also exporting 38 billion$ worth of agricultural products every year [2]. Along with this, as per the 2014 world agriculture statistics by Food and Agriculture Organization (FAO), India is the world's largest producer of many fresh fruits like banana, mango, guava, papaya and vegetables like chickpea, okra, major spices like chilli pepper, ginger, fibrous crops such as jute, staples such as millets and castor oil seed as well as milk [3]. India is the second largest producer of wheat and rice and the world's major food staples [3]. India is also currently the world's second largest producer of several dry fruits, agriculture-based textile raw materials, roots and tuber crops, pulses, farmed fish, coconut, sugarcane and numerous vegetables. India is ranked under the world's five largest producers of over 80% of agricultural produce items, including many cash crops such as coffee and cotton, in 2010 [3]. Figure 4.2 presents production of okra and banana in various countries as a sample and it clearly shows that India is leading producer.

N. Sharma et al., *Open Data for Sustainable Community*,
Advances in Sustainability Science and Technology,
https://doi.org/10.1007/978-981-33-4312-2_4

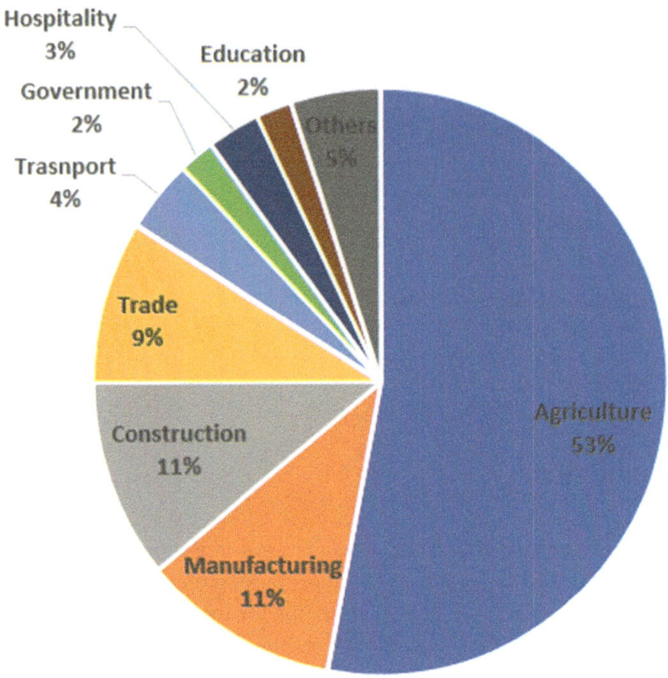

Fig. 4.1 Employment by sector in India

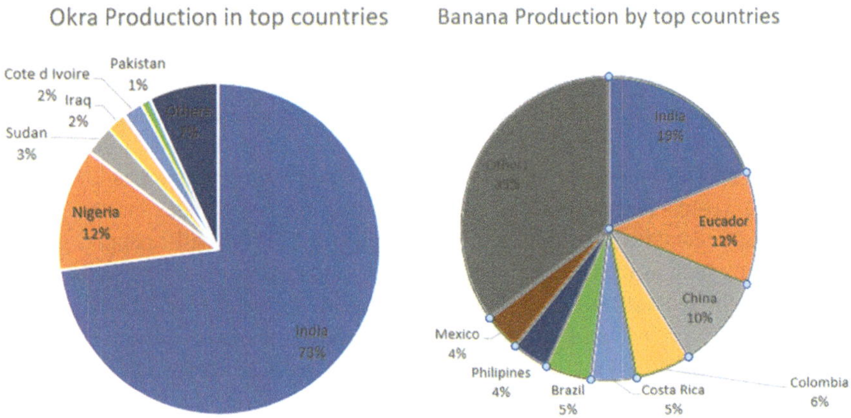

Fig. 4.2 India as a leading producer of okra and banana

Farming, while it appears as a very simple concept has its own various complexities. It is totally dependent on weather and has faced the brunt of bad conditions many times. Moreover, even after suitable weather conditions, market manipulation and middle men have resulted in very low profits on good produce. To combat losses, farmers have had to take loans from banks to buy necessary materials like seeds, fertilizer, etc., for the following year's crop. With more crop failures, farmers are unable to pay bank loans resulting in huge losses for both the farmers and even banks. To combat this, the government introduced many schemes to help agriculture growth and boost farmer morale. Crop insurance was also introduced and supported by the government for healthy relation between farmers and banks. The premiums of the crop insurance are paid jointly by farmers and the government to banks in return for bailouts in case of crop failure. Many private banks from ICICI to TATA AIG have been part of this scheme to uplift our farmers [4, 5]. While banks like ICICI Lombard and Cholamandalam had made some profits in the first year of the scheme as their claim ratio was 79% and 61%, respectively, and they made heavy losses thereafter. TATA AIG made losses in all the three years with over 100 per cent claim ratio [6]. In fact, most of the banks in this scheme suffered huge losses recording over 100% claim ratio. This has led to loss of interest by private banks which have further increased the burden on government associated banks.

The crop insurance has been enacted under Prime Minister Crop Insurance Scheme (Pradhan Mantri Fasal Bima Yojana) and as anticipated the budget allocated has increased drastically every year. Starting at a massive 90 billion INR in 2017, the budget has now skyrocketed to 140 billion INR with no reduction expected in future as well and that is just for one of the many schemes [7]. Including all the schemes taken under the Ministry of Agriculture, the total money spent on farmer's upliftment is a whopping 1125.85 billion INR [7]. The sad part is that even after these massive interventions by both government and private sector, crop failures still rage on, scarcity of food is still very prominent and prices of vegetables still remains a topic of national discussion.

While there are some villains who profit of all this, there is not anyone who takes major blame for this, as government, farmers and even private sectors have put all their resources in helping out and improving the situation. This begs the question, then how do we help and where do we bring out a change? For that, we look towards science. Every science enthusiast can witness and swear upon the miracle of Industry 4.0. The whole world as we see now is the product of the revolution that industrialization has bought forward. Every single field from communication to medicine has reaped its benefits and continues to so. One of the biggest and newest developments in Industry 4.0 has been the field of data science. Data science as the name suggests is the science of data, it is a field that allows improving the machines, using the data that has been collected over time. It is built on one of the simplest principles of all time to learn from our experiences, to analyse the information collected and to form meaningful conclusions that improve our life.

In this unit, we have tried to harness the power of data science and use it to improve the lives of farmers and agriculture sector at large. For data science, the only requisite is ample data, and coincidently, it is available as open data at www.data.gov.in. The Government of India has established a Kisan Call Centre (KCC) in 2004 for addressing farmer queries. An inherent advantage of these centres is that they operate in more than 20 regions with the operators having knowledge of native languages of those regions. They also aim to fulfil the constant need to have consistent information relayed to farmers in the fastest way possible.

In this chapter, an attempt has been made to comprehend the functioning of Kisan Call Centre and the data that is generated by them. The rest of the chapter is organized as follows: Section 4.2 discusses various literatures related to farmer call centre, Sect. 4.3 focuses on understanding operations of KCC and Sect. 4.4 presents data preparation and data pre-processing. Finally, Sect. 4.5 concludes the chapter.

4.2 Literature Review

Farmers are the backbone of the agrarian economy of India. While farming is a cultural and occupational legacy for most, farmers frequently require assistance in making crucial decisions. To resolve their queries and help them in decision making, many call centres have been set up by Government of India. In this section, the intent is to build up our understanding on objectives and operations of farmer call centre. So, the first step is to find out all the relevant literature, for which many research databases such as J-Gate, IEEE explore and Science Direct were searched to find out the research trends from 2010 to June 2020. The advanced search was performed using the terms like 'call centre' and supplementary term like 'farmers', to get the pertinent publications in the above-mentioned research databases. The first search result depicted in Fig. 4.3 clearly shows the growth in publication across databases in the area of agri-call centre.

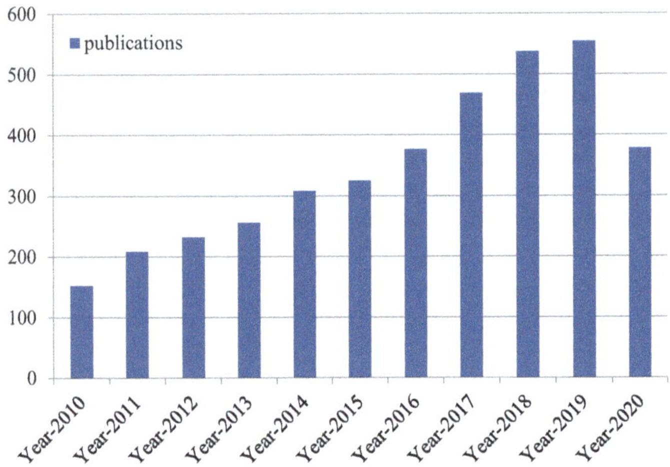

Fig. 4.3 Year-wise growth in number of publications related to farmer call centre

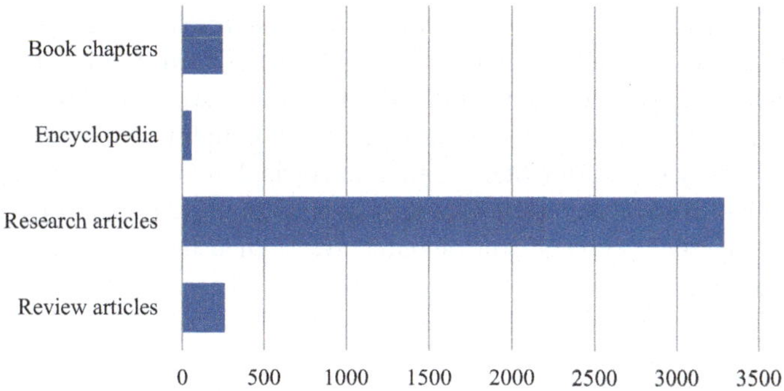

Fig. 4.4 Type of publications related to farmer call centre

Further, the search was performed to know about the most exhaustive type of publication. The search result presented in Fig. 4.4 clearly shows that good amount of research is being carried out to study farmer call centre, which is clearly reflected in the number of research articles.

Finally, a rigorous search was performed explicitly on IEEE explore database using the terms 'call centre' and 'farmers' from 2010 to June 2020. The purpose of this search is to know about various research dimensions related to farmer call centre. The search result is shown in Fig. 4.5 and it highlights the application of artificial intelligence, natural language processing, convolution neural network and data mining on farmer call centre data.

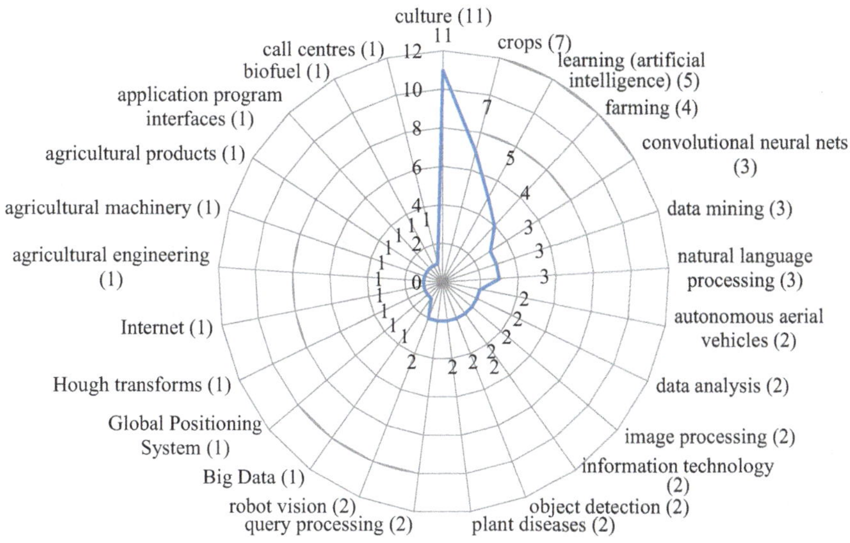

Fig. 4.5 Various research dimensions towards farmer call centre study

Now, it is obvious from the above search of standard research databases that there is ample literature available related to study of queries from farmers registered at call centres. Before proceeding with the actual analysis, we carried out a review of all the available literature on a similar theme. A thorough research study of KCC was carried out by Centre for Management in Agriculture (CMA), Indian Institute of Management Ahmedabad (IIMA) and supported by the Ministry of Agriculture and Farmers Welfare, Government of India. We studied the resulting report titled 'Decision-Oriented Information Systems for Farmers: A Study of Kisan Call Centres (KCC), Kisan Knowledge Management System (KKMS), Farmers Portal and M-Kisan Portal in Gujarat' to get an insight into the general functioning of the centres.

While the centre works all day, the queries received at night are recorded and resolved later [8]. The centre operators, Farm Tele Advisors (FTA), are responsible for resolving the queries of the farmers. This is essentially Level I, while Level II queries are passed on to senior experts, in case it remains unresolved at Level I [8]. Figure 4.6 shows the number of recorded calls as observed from January 1, 2020 to June 1, 2020 [9]. The large numbers of calls placed are indicative of the significance of KCC, and subsequently, their performance for farmers to reap maximum benefits from the available knowledge.

A study was conducted to identify the perception of farmers with respect to KCC in November 2015, in West Bengal [10]. Some major takeaways from the survey are as follows:

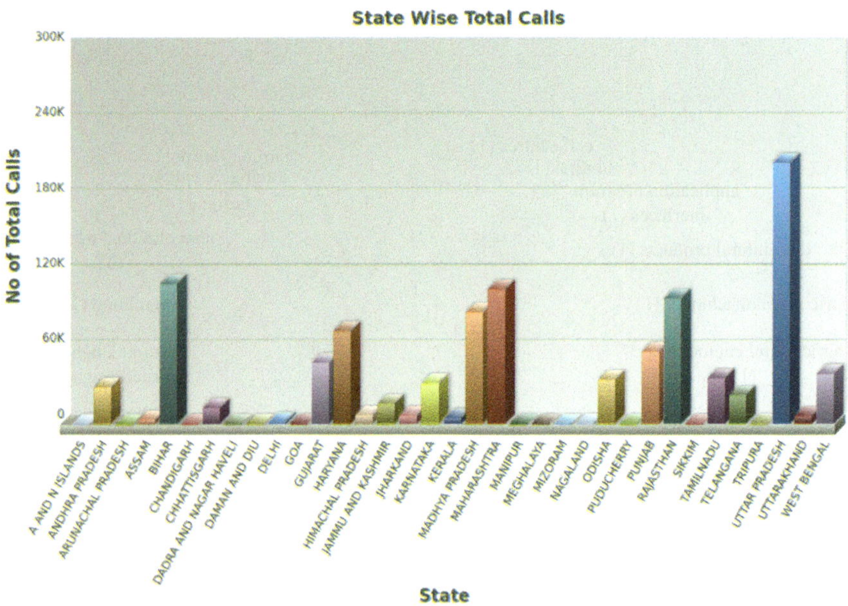

Fig. 4.6 State-wise call count at KCC as given by https://mkisan.gov.in/ [9]

1. Majority of the queries came from farmers, while many were also directed by business owners [10].
2. Most of the queries were from farmers who owned small to medium sizes of land, thus, increasing the importance of KCC resolving their major queries without additional costs [10].
3. Most of the inquirers agreed that the service was readily accessible, easily understandable and beneficial [10].

The above findings are indicative of farmers' positive response to KCC and goes on to show that the improvement of techniques in use at KCC can generate extremely utilitarian results.

Data collected from KCC over the past decade has been studied by several researchers for different purposes. The most common objective of research has been directed towards automatic resolution of the queries in future. This would ensure lesser manual errors, and more accurate information being passed on to the farmers, possibly at faster rates. One such study was conducted with the objective of building a framework which would accept a new query and be able to identify the most appropriate query similar to it in the existing database. After such matching has been done, the system would be able to predict the solution without any manual interruption [11]. The region of interest for this research is Haryana, for the duration of one year. It mainly focuses on the queries and their solutions as outlined in the dataset. Their approach towards the problem includes manual involvement at Step 3, i.e., once the system has recognized the most relevant answer for the query. This research highlights a very important application of this data analysis. The data can be used to identify the most pressing problems region-wise to enable preparedness before time. Gensim model is used for its robustness, and its feature of text similarity. A gensim dictionary of words in the original corpus, prepared after data pre-processing, is created and stored in MongoDB. This dictionary is loaded to create a term frequency-inverse document frequency (TF-IDF) matrix and saved to MongoDB again. Latent Dirichlet allocation (LDA) and latent semantic indexing (LSI) are subsequently used on the TF-IDF model to obtain the required data [11].

One of the major advantages of this model is the rate at which it can generate the list of most relevant solutions. It is able to derive this list at a more efficient pace because the whole corpus is not traversed; instead, a topic-wise distribution is performed to direct the incoming new query towards a specific part of the corpus [11]. However, the major disadvantage is the time taken to create the topic-wise document matrix. The future scope for this study remains to include a voice/audio agent to deliver the most relevant answers to the inquirer, instead of presenting in textual form to the FTA. There have been other research studies along the same lines.

One such research was built on the knowledge of conversational systems, and their role in making human–computer interaction resembles normal conversation as closely as possible. Thus, they propose a conversational agent, called FarmChat for resolution of the farmers' queries [12]. As a preliminary step, the research is focused on identifying four main areas of queries as mentioned below:

1. Plant protection
2. Weather
3. Best practices
4. Unbiased recommendations on products.

The design requirements were identified by observing the challenges in forming suitable and accurate results for the queries posed by the farmers. Some of the challenges are listed below:

1. Specificity of the queries
2. Availability of FTAs
3. Trust in the solutions offered by KCC advisors
4. Prerequisite of local information with the FTAs to provide correct answers.

Thus, keeping in mind the above challenges, the system design of FarmChat was born. Some of the most notable features of FarmChat are:

1. It supports Hindi, a native language spoken by nearly 40% of Indians.
2. It comes with two kinds of interfaces—either an audio-only interface, or an audio + text interface.
3. The knowledge base not only comes from KCC dataset but also from agriculture experts, making it better-rounded.

FarmChat is built using Google's Speech to Text Transcription Service, a Python Flask application and IBM Watson's Conversation Service, along with Google's Text-to-Speech Agent to complete the loop and relay the solution to the query in audio form. The research involved experimental analysis of FarmChat conducted by inviting farmers to use it with a set of tasks structured to identify the usability, ease, responsiveness and comfort in using the application. Nearly 90% of the users strongly agreed that they would use FarmChat in future as well. It was also realized that the users placed a high expectation from FarmChat, leading to a certain factor of dissatisfaction among the users. Another was the unspecificity of the solutions given by FarmChat, thus, failing to achieve localization in some cases [12].

Another study dedicated towards a conversational agent for agricultural applications is highlighted in 'Agribot: A Natural Language Generative Neural Networks Engine for Agricultural Applications' [13]. An added functionality is that the system emphasises on weather forecasting and crop disease detection. AgriBot is an artificial intelligence-based chatbot, which implies that it relies on advanced technologies such as machine learning and natural language processing to improve communication, understanding and interaction. The knowledge base for AgriBot comes from KCC dataset, obtained after performing web scraping. A recurrent neural network-based model, more commonly known as an encoder-decoder model, is used to build the conversational model for AgriBot. Padding is performed in order to handle the variable-sized input. On the other hand, a classification model is used by the disease detection module of AgriBot. Weather forecasting is achieved by using the OpenWeatherMap Application Programming Interface (API), which only requires input of name of city/region and date to get the corresponding response from the API. The conversational model scored an accuracy of nearly 98%. Future scope lies in making the tool available in local languages and to be able to access the user's location without need of specific input.

The KCC dataset for every season and date is available on the official website. It is important to note that the amount of data is large because of the number of queries that KCCs receive on a daily basis in every region that they operate. Thus, it is important for researchers to consider the aspect of the size of data that is to be processed in order to generate any valuable analysis results out of them. Technologies that can deal with big data are recommended to be used in order to evaluate and analyse all the data and produce more accurate results.

An important study in this direction has been outlined in 'Kisan Call Centre Data Analysis Using Hadoop and Hive' [14]. The primary objective of the research was to build a system for prediction using the available Kisan Call Centre (KCC) dataset. The prediction would be done in order to make resolution of future queries relatively easier and more effective. For instance, to identify disease by measuring symptom's similarity and suggest remedies based on historical treatment. The research methodology involved the usage of big data technologies because of the large amount of data entries in KCC dataset. Thus, taking into consideration scalability factors, the Hadoop ecosystem was used. The big data analysis tools used for this analysis, in order of usage, included:

1. Data storage using Hadoop, processing using MapReduce
2. Normalized data stored in Hadoop distributed file system (HDFS)
3. Data analysis using Hive
4. HiveQL used for queries—examples of the queries are—query to study correlation with crop, season and associated problem in a particular state, query to study problem associated with crop and solution told by KCC operator.

The conclusion obtained was that the proposed solution of using Hadoop would be most effective given the large number of queries recorded by KCC every day. The future scope involves making the entire KCC system real time including the prediction analysis [14].

An important application of the KCC dataset is its application in agricultural policy making. A research study solely focuses on data analysis on KCC dataset to identify sequential patterns in the queries, and therefore, gain a good insight into the problems faced by the farmers [15]. The first step involved in this process is the generation of association rules from the dataset. Association rules can be considered to be similar to IF–THEN statements in conditional programming. The researchers attempted to do the same using the Apriori algorithm, but concluded that the volume of rules thereby generated was too large due to the extensive database. Thus, they settled with a technique known as technique for order preference by similarity to ideal solution (TOPSIS) to rank and select only the most relevant rules from the large volume of generated rules. An example of one of the generated rules is—there

Fig. 4.7 Google trend analysis for India towards 'Kisan Call Centre' from 2004 till June 2020

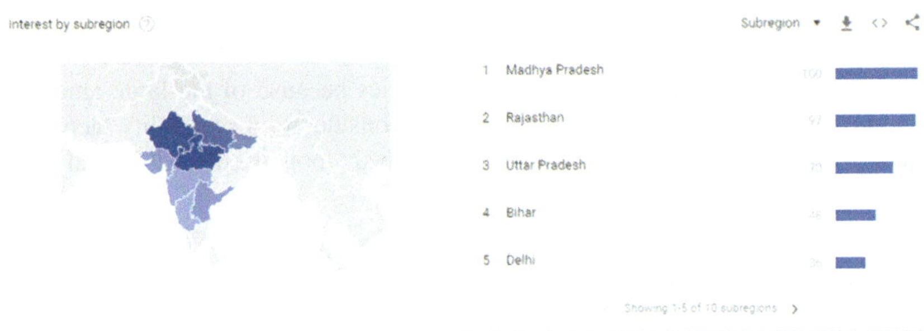

Fig. 4.8 Google trend analysis towards top three states with maximum queries

is found to be a high probability of occurrence of similar kinds of crop problems in adjacent states, which is justified because of common weather and soil conditions.

A Google trend search analysis was also performed to know the interests and curiosity about research on 'Kisan Call Centre'. The result displays the trends from India's perspective for 16 years, i.e., from 2004 to 2020, as shown in Fig. 4.7.

Figure 4.8 depicts the Google trend related to the top three Indian states which registers maximum queries are Madhya Pradesh, Rajasthan and Uttar Pradesh.

4.3 Understanding the Operations of Kisan Call Centre

Kisan Call Centre is an initiative started by the Department of Agriculture and Cooperation, Ministry of Agriculture, Government of India on 21 January 2004 [8]. The primary goal of these centres is to fulfil the real-time resolution of queries directed by the farmers via phone calls to the centre. The call centres are located in 21 locations covering all states and UTs, and the countrywide toll free number 1800-180-1551 is fixed for registering the queries. The farmers' queries are answered in 22 languages and the centres are open from 6 a.m. to 5 p.m. KCC agents who answer these queries are graduates or above in agricultural or related subjects. The major topics handled by KCC experts are as follows:

1. Disease and pest control for different crops grown in the region.
2. Good agricultural practices, livestock management, fishery, etc.
3. Best practices in agriculture in a particular state as well as of other states.
4. Crop-related information in agriculture, horticulture, animal husbandry aromatic plants, spices, plantation crops, cash crops, etc.
5. Vermi-compost, organic farming, including organic plant protection, etc.
6. Information on high yield value (HYV) seeds and nutrient management for different crops.
7. Market-related information for different crops in the state.
8. Farmer support programmes which are being implemented by the government of Madhya Pradesh.
9. Agriculture-related information that impacts farmers, farming practices, etc.

The Kisan Call Centre works on two levels. At the first level, replies/solutions are provided to farmers' queries instantaneously by experts. At the next level, the queries are analysed so that area-specific analysis can be done based on which timely information could be disseminated to farmers through TV, radio, etc., to caution or overcome possible damage to agricultural crops or livestock. It is important to mention here that three Kisan Call Centres have served as an early warning system, for example, during the crisis of drought in 2009 kharif and untimely floods in October 2009. The call centres were transformed into a control room to provide solutions to the emergence and contingent needs of the farmers across the state, segregated on the basis of specific area.

4.4 Data Preparation: Kisan Call Centre Queries

The queries that farmers submit over the years and the data that is generated during the entire process, form the perfect database for all operations of data science. The objective of the project is to perform data analysis on KCC query dataset. There are four phases of this project:

1. Data collection
2. Data pre-processing
3. Data analysis
4. Conclusions of analysis.

Step 1 and 2 together are the part of data preparation process, and it involves many essential steps like exploring the right data source(s), acquiring the data from there and then pre-processing it, so as to make it ready for further analysis.

4.4.1 Data of Kisan Call Centre Queries

This project is about analysing the legitimate queries submitted by the farmers. The data used to create the database for the project was extracted from data.gov.in, which is an official website of the government of India to store open data for citizens to access information for research. The files are categorized on the basis of district, month and year. The data of every single farming district in India over the last few years for the months of April, October and December is present on the website. The three different months also signify the peak of three prominent seasons in India: Summer, Rainy and Winter, respectively.

4.4.1.1 Database Description

For the sake of trial, the files of KCC queries for all 36 districts of Maharashtra (state in western India) in 2019 for all three months, i.e. April, October and December were selected using filters provided by the website. The files were downloaded in Excel format and collated using concat() function from Pandas library in a Python script to form the main database for further processing and later analysing. The database was formed gradually. First of all, the files were collated month wise to check if the analysis had to be done for a specific month, then all the month-wise collated files were collated again to form the main collated database. The complete database consists of 1,10,000 queries and 11 columns to describe those queries, and the same is explained as below:

i. **Season**—This attribute describes the season of the year in which the query was recorded by Kisan Call Centre. Season is one of the most important factors in agriculture. Rains and summers have been very wily as many recent reports clearly indicate that droughts and floods have destroyed the entire crop many times. So, a Query Type analysed according to the season could play a very important role.
ii. **Sector**—This attribute describes the agricultural sectors to which the query belongs. These sectors comprise of 'agriculture', 'horticulture', 'animal husbandry' and 'fisheries'. Defining the sectors is very necessary as they provide an additional level of categorization of queries.
iii. **Category—**This attribute describes the category of the query made by the farmer at Kisan Call Centre. These comprise of 'animal', 'avian', 'bee keeping', 'cereals', 'condiments and spices', 'drug and narcotics', 'fibre crops', 'flowers', 'fodder crops', 'fruits', 'green manure', 'inland', 'marine', 'medicinal and

aromatic plants', 'millets', 'oilseeds', 'others', 'plantation crops', 'pulses', 'sugar and starch crops' and 'vegetables'. Categories are necessary as they encompass multiple Query Types and provide a clear distinction between various Query Types.

iv. **Crop**—This attribute describes the crops for which the query was made by the farmer. There are a total of 222 crop types and few of the crop types are 'cotton kapas', 'onion', 'wheat', 'turmeric', 'others', etc. The crop column is very important as it appears in all Query Types and is more pertinent in the queries related to market information to know more about prices for a specific crop.

v. **Query Type**—The Query Type column is arguably the most important column of the entire database as it is the centrepiece around which all other columns are cross analysed with, so as to define the problem, i.e. query with a specific sector, category, crop, etc. The queries are segregated into a total of 60 different Query Types, most common being 'weather', 'plant protection', 'government schemes', 'market information', 'fertilizer use and availability', etc.

vi. **Query Text**—The Query Text contains the exact query received from the farmer to the Kisan Call Centre. This is completely raw text and entered manually. Some of the Query Text are like—'Farmer asked query on Weather', 'FARMER ASKED ABOUT SPACING OF MAIZE?', 'ASKING ABOUT Attack of White Grub ON GROUNDNUT ? ????? \n', etc. While Query Text does not have major impact on analysis, it is important as it is the primary source of a query directly originating from the farmer and translated by a KCC agent to the best of his/her abilities.

vii. **KCCAns**—The KCCAns records the answers given by Kisan Call Centre on farmers' queries. However, this column is null throughout the database.

viii. **State Name**—This attribute describes the name of the state from where the farmer is registering the query. This is an important attribute as it is the broadest classification for the geographical location regarding the origin of a query.

ix. **District Name**—This attribute describes the name of the district from where the query was recorded by a farmer. It is one of the important properties in dataset as it gives the more detailed geographical information for the analysis. Some of these being 'Ahmadnagar', 'Amravati', 'Beed', 'Nagpur', 'Satara', etc.

x. **Block Name**—The importance of block name is the same as that of district name and state name with the added highlight of getting a pinpoint location to the origin of the query.

xi. **Created On**—This attribute is essential as it signifies the exact date and time of the query. It informs about the season, the weather conditions on the particular day as well as gives a certain proof and accountability to the whole process of query collection and KCC centres.

Fig. 4.9 Text cleaning pipeline

4.4.2 Preparation of Kisan Call Centre Queries

Analysis of any information or data requires a number of prerequisites to be accomplished. In this section, the focus is on KCC data pre-processing, so as to prepare that dataset for analysis.

4.4.2.1 Analysing Presence of Null or NA in Dataset

The presence of null values in a dataset compromises it, since they act as a black hole to average value of the entire dataset. These null values adversely affect the performance and accuracy of any machine learning algorithm. It is generally an error in recording the query by KCC authorities and is invariably classified as a null value or '0'.

4.4.2.2 Text Cleaning Pipeline

Data scientists spend more of their time on text cleaning rather than analysing, because a clean and error-free text can improve any analysis on the said text by leaps and bounds. Uneven text without symmetry will skew any further analysis on it. Unlike integers, characters or texts require more thorough cleaning and many requisites to be adhered to. The entire process of text cleaning adopted in this research is depicted in Fig. 4.9. Since lower-case and upper-case characters have different ascii code and are processed differently in a programme, the first step is to convert all characters in the Query Type to lowercase, using the lower() function. Secondly, all quotes in every Query Type was also removed using re.sub() function. Any and all special characters such as '\t', '\n' were also removed using join() function. Any numbers or extra spaces present were also excluded. Strip() function was used to remove any extra spaces.

4.4.2.3 Handling the Missing Data Values

It is observed that in KCC dataset, a lot of cells in few columns are empty, i.e. filled with null values. The null values skew the entire dataset so it is important to identify them. It is evident from Fig. 4.10 that most of the cells of KCCAns and season columns have null values. Hence, KCCAns and season columns are

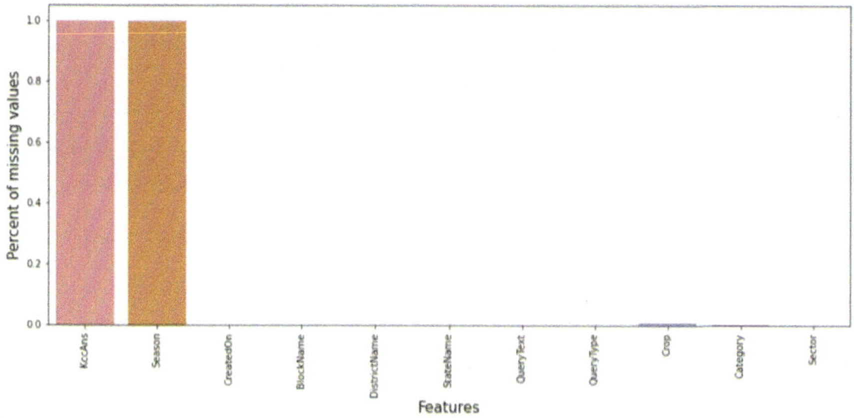

Fig. 4.10 Column-wise per cent of missing values

dropped as recovering these values are highly impossible. There were some columns such as Crop, Query Type and Category with null values less than 5%. For such columns, instead of dropping the entire column, the null values were replaced. This was achieved using isnull() function and matplotlib to determine and visualize the columns with exceeding null values. They were then replaced with the mode value of that particular column. For example, null values in Query Type were replaced with 'weather' values.

It was also observed that the State Name values were missing for some records. In this case, the District Name column values were used to map and identify the correct State Name value and the same was replaced. The same approach is adopted if District Name values are found missing and Block Name values are used to map and identify the correct District Name value. Finally, as Query Type was the core column, the rows with empty Query Types were directly removed.

4.4.2.4 Extracting Data from Created On Column

The column Created On cannot be used directly so string slicing is done in Python to segregate day, month, year and time for better analysis. Separating the month into its own column provided an alternative for the dropped season column, as the month signified the season. For example, the month of April signifies summer and in the same way October signifies rainy and December signifies Winter.

4.4.2.5 Encoding Categorical Data

Any pre-processing for data science would be incomplete without encoding. The two major encoders found in the scikit library are LabelEncoder and OneHotEncode and

```
109467       agriculture
109468       agriculture
109469      horticulture
109470       agriculture
109471      horticulture
Name: Sector, dtype: object
```

```
109467    0
109468    0
109469    3
109470    0
109471    3
Name: Sector, dtype: int64
```

Fig. 4.11 Label encoded data

```
109467       agriculture
109468       agriculture
109469      horticulture
109470       agriculture
109471      horticulture
Name: Sector, dtype: object
```

Sector_agriculture	Sector_animal husbandry	Sector_fisheries	Sector_horticulture
1	0	0	0
1	0	0	0
0	0	0	1
1	0	0	0
0	0	0	1

Fig. 4.12 OneHotEncode data

both of them have been used to convert categorical data or text data into numbers, which serve as an input to any predictive or visualization models like HeatMapping. To convert categorical text data as found in our dataset into model-understandable numerical data, the LabelEncoder class has been used. So, to label encode the first column, import the LabelEncoder class from the sklearn library, fit and transform the first column of the data, and then replace the existing text data with the new encoded data. There might be a situation where after label encoding, the model is confused assuming that a column has data with some kind of order or hierarchy, when it is actually not there. To avoid this, 'OneHotEncode' is applied to that column. OneHotEncode takes a column which has categorical data as well as has been label encoded, and then splits the column into multiple columns. The numbers are replaced by 1 s and 0 s, depending on which column has what value. Figure 4.11 represents the Label Encoding conversion of KCC data and Fig. 4.12 represents the OneHotEncode of 'sector' column of database. Similarly, other column types were also assigned their own separate columns.

4.4.3 Pre-processing of Kisan Call Centre Queries

After completing the initial cleaning process, the following pre-process steps that have been adopted to prepare the KCC dataset:

1. Removing stopwords
2. Stemming
3. Lowercasing.

4.4.3.1 Removing Stopwords

This process involves removing all the words that do not add significant meaning to the queries, like stop words and punctuation marks. In practice, the text is compared against two lists, one with stop words in English and another with punctuation. Every word from the Query Text that appears in the list of stop words should be eliminated. For instance, consider the query mentioned below:

'Farmer asked about virgin coconut oil seller contact number for agriculture use?'.

It contains stop words like—about, for and punctuations like? So these words and punctuations need to be eliminated. The query without stop words and punctuations looks like this

['farmer', 'asked', 'virgin', 'coconut', 'oil', 'saller', 'contact', 'number', 'agriculture', 'use'].

Here, it is important to note that the overall meaning of the sentence could be inferred without any effort. However, do remember that in some contexts punctuations may not be eliminated as they might add important information to a specific data science task like natural language processing (NLP).

At the end of this process, the resulting queries contain all the important information related to its sentiment. To achieve this, the natural language toolkit (NLTK) package, an open-source Python library for natural language processing is used. It has modules for collecting, handling and processing text and the same will be described in the next chapter.

4.4.3.2 Stemming

Now that the query from the above example has only the necessary information, the next step should be to perform stemming for every word. Stemming in NLP is simply transforming any word to its base stem, which could be defined as the set of characters that are used to construct the word and its derivatives. Stemming is the process of converting a word to its most general form or stem. This helps in reducing the size of our vocabulary. Consider the following words:

- **Ask**
- **Ask**ing
- **Ask**ed
- **Ask**s.

All these words are stemmed from its common root Ask. However, in some cases, the stemming process may produce incorrect root words so as to avoid any collision of words. For example, look at the set of words that comprises the different forms of create, and it is obvious that the prefix creat is the incorrect rood word used.

- **Creat**e
- **Creat**ing
- **Creat**ive
- **Creat**ed.

NLTK has different modules for stemming and the PorterStemmer module is used here which uses the Porter Stemming Algorithm, and the code for the same is given below:

```
# Instantiate stemming class
stemmer = PorterStemmer()
```

Now that we have instantiated the stemming class, stemmer.stem() can be called for each word in our Query Text, using the following code.

```
# Stemming each word in the list
stem_word = stemmer.stem(word)
```

Before Stemming

['farmer', 'asked', 'virgin', 'coconut', 'oil', 'saller', 'contact', 'number', 'agriculture', 'use'].

After Stemming

['farmer', 'ask', 'virgin', 'coconut', 'oil', 'saller', 'contact', 'number', 'agricultur', 'use'].

It is observed that words like 'asked' is reduced to its stem 'ask' and words like 'agriculture' is reduced to 'agricultur'.

The vocabulary of final words would be significantly reduced stemming process is performed for every word in the corpus. It reduces the processing time of the model as the vocabulary of words is significantly reduced.

4.4.3.3 Lowercasing

To reduce the vocabulary even further without losing valuable information, the words need to be converted into lowercase. So the word FARMER, Farmer and farmer would be treated as the same exact word.

4.5 Conclusion

In this chapter, the initial two stages of data science, i.e. data collection and data pre-processing, were presented in detail. Now, as the database is prepared and ready to be analysed with maximum efficiency, the aim is to analyse queries submitted by farmers to Kisan Call Centres using data science tools and technique like Python and natural language processing, to form meaningful conclusions that could be used for to help farmers.

References

1. "India economic survey 2018: Farmers gain as agriculture mechanisation speeds up, but more R&D needed". The Financial Express. 29 January 2018. Retrieved 8 Jan 2019.
2. India's Agricultural Exports Climb to Record High
3. "FAOSTAT, 2014 data". Faostat.fao.org. Retrieved 17 Sept 2011
4. https://www.icicilombard.com/crop-insurance
5. https://www.tataaig.com/rural-insurance/crop-insurance/pmfby-maharashtra
6. https://www.downtoearth.org.in/news/agriculture/punctured-cover-india-s-crop-insurance-scheme-loses-sheen-68379
7. https://openbudgetsindia.org/dataset/department-of-agriculture-cooperation-and-farmers-welfare-2019-20
8. Decision-Oriented Information Systems for Farmers: A Study of Kisan Call Centres (KCC), Kisan Knowledge Management System (KKMS), Farmers Portal, and M-Kisan Portal in Gujarat
9. https://mkisan.gov.in/KCC/KCCDashboard.aspx
10. Perception of Kisan call centre (Farmer Call Centre) by the farming community with their socio-economic variable: A study on Coochbehar District
11. Query Answering for Kisan Call Centerwith LDA/LSI
12. FarmChat: A Conversational Agent to Answer Farmer Queries
13. Agribot: A Natural Language Generative Neural Networks Engine for Agricultural Applications
14. Kisan Call Center Data Analysis Using Hadoop and Hive
15. Sequential pattern mining combined multi-criteria decision-making for farmers' queries characterization

Chapter 5
Analysis and Visualization of Farmer Call Center Data

5.1 Introduction

In the last chapter, the focus was on cleaning and pre-processing of the Kisan Call Center (KCC) query data. Just to reiterate that Kisan Call Center is the Indian name for farmer call center. The agenda of this chapter is to perform exploratory data analysis (EDA) or descriptive data analysis. The objective of exploratory analysis is to form meaningful conclusions from the open-source KCC query dataset so as to help the farmers and the government. To begin with, let's define—what is data analysis and how it is being used? Data analysis is a process of inspecting, cleansing, transforming data and preparing a model with the goal of discovering useful information, notifying conclusions and supporting decision-making. While there are various different forms of data analysis, in this chapter the concentration is on descriptive statistics, which is also known as exploratory data analysis and confirmative data analysis. This begs the question as to what is EDA? Putting it in simple words, exploratory data analysis is an approach to understand the data distribution, to discover any specific pattern or trend, to summarize the main characteristics and to explain the results using visual methods. EDA is used to test hypotheses and gather proof to prove the same. Usually, the data analysis is done, when data is too huge to form conclusions humanely, hence programming tools are used to analyse it. However, that is not enough; there is a need for visualization tool to comprehend the conclusions in a better way.

In this chapter, the analysis and visualization are carried out by crossing data columns with each other to test the hypothesis. All the visualizations are done using matplotlib and seaborn libraries and written in a python script. Visualizations helped to easily spot any anomalies or patterns that confirm hypothesis and even allowed to form conclusions from the dataset that could be used to stop and prevent any further errors that are present in the system. It made possible to spot the regularity in specific query types which in a nutshell leads to categorization of the mentioned queries,

N. Sharma et al., *Open Data for Sustainable Community*,
Advances in Sustainability Science and Technology,
https://doi.org/10.1007/978-981-33-4312-2_5

i.e. problems faced by the farmers. This categorization is done by location, season and category, and the three main factors in agriculture. In the following sections, some of these conclusions after performing exploratory data analysis along with the visualization have been illustrated in detail.

5.2 Material Methods Used for Analysis

Before performing the analysis of KCC queries, it would be a great idea to take a look and understand all the libraries and functions used in the analysis. Once the preliminary understanding of these techniques is in place, it would be easier to perform exploratory data analysis on KCC dataset at any point. This part can be thought of as an algorithm behind any case study analysis. Understanding the algorithm in a step-by-step process not only enables present findings with more stature but also allow the readers to analyse the vast KCC dataset to form many more conclusions that can help the farmers and the government. So here is a step by step walkthrough.

5.2.1 Check and Confirm the Pre-processed Data

Once the open data is extracted from the source, it undergoes various steps of data preprocessing like data cleaning and data wrangling. Before starting exploratory data analysis, it is necessary to go through the final preprocessed dataset to ensure that it is ready for analysis that is to be conducted upon it. Few tests may be conducted in python to check if there is still any underlying errors/null values left in dataset. In this case, a pair plot is created on all columns and it did not give out any striking results which lead us to the conclusion that our dataset is stable and ready.

5.2.2 Form an Objective and Acquire Domain Knowledge

The aim as defined in the introduction is 'to find patterns in the open source data', while that is the broader version, it is important to know the current scenario before delving into analysis. For this, a rigorous study of various government documents explaining the missions undertaken by the government to uplift the farmers is carried out. It is noticed that a lot of money is being allocated for these missions but the results have not improved. As mentioned in Chap. 4, the total budget allocated to Prime Minister Crop Insurance Scheme (Pradhan Mantri Fasal Bima Yojana) is 140 billion INR. Likewise, a whooping budget is allocated under Prime Minister Farm Irrigation Scheme (Pradhan Mantri Krishi Sinchayee Yojana—PMKSY) as a part of green revolution at the various sub-missions and national missions undertaken

Table 5.1 Budget allocated by central government (in million)

Sub-mission/years	2017	2018	2019
Rain-fed Area Development and Climate Change	2092.5	2340	2500
National Mission on Oil Seed and Oil Palm	2636.2	4000	No info
National Mission on Horticulture	20,250	21,000	22,000
Sub-Mission on Plant Protection and Plant Quarantine	469.2	410	500
National Project on Agro-Forestry	426.7	400	500
Sub-Mission on Seed and Planting Material	4235.4	3230	3500
National Project on Soil Health and Fertility budget	1940	3000	3250

by the central government [1]. Table 5.1 shows the budget allocated by the central government to be distributed to states for mentioned purposes [1].

5.2.3 *Data Visualization Criteria*

According to the World Economic Forum, the world produces 2.5 quintillion bytes of data every day, and 90% of all data has been created in the last two years [2]. With so much data, its become increasingly difficult to manage and make sense of it all. It would be impossible for any single person to wade through data line-by-line and see distinct patterns and make observations. Data proliferation can be managed as part of the data science process, which includes exploratory data analysis. Data exploration is initial analysis of the dataset to understand characteristics of data, usually a visual exploration.

Data visualization is a graphical representation that expresses the significance of data. It reveals insights and patterns that are not immediately visible in the raw data. It is an art through which information, numbers and measurements can be made more understandable. The main goal of data visualization is to communicate information clearly and effectively through graphical means. It does not mean that data visualization needs to look boring to be functional or extremely sophisticated to look beautiful. To convey ideas effectively, both aesthetic form and functionality need to go hand in hand, providing insights into a rather sparse and complex dataset by communicating its key aspects in a more intuitive way. Data visualization has become an indispensable part of the business world and an ever-increasing part of managing our daily life. Effective data visualization should be informative, efficient, appealing, and in some cases interactive and predictive. Table 5.2 explains basic criteria that data visualization should satisfy to be effective.

Table 5.2 Criteria for effective data visualizations

Criteria	Description
Informative	The visualization should be able to convey the desired information from the data to the reader
Efficient	The visualization should not be ambiguous
Appealing	The visualization should be captivating and visually pleasing
Interactive and predictive	The visualizations can contain variables and filters with which the users may interact to predict results of different scenarios

Table 5.3 Libraries used for visualizations

Library	Description
Matplotlib	As matplotlib is the first Python data visualization library, many other libraries are built on top of it or designed to work in tandem with-it during analysis. While matplotlib is good for getting a sense of the data, it is also useful for creating quality charts quickly and easily [3]
Seaborn	Seaborn harnesses the power of matplotlib to create beautiful charts in a few lines of code. The key difference is Seaborn's default styles and colour palettes, which are designed to be more aesthetically pleasing and modern. Since Seaborn is built on top of matplotlib, we need a basic understanding of matplotlib to tweak Seaborn's defaults [3]
Plotly	Plotly is an online platform for data visualization, but it can also be accessed from a Python notebook. Plotly's forte is making interactive plots as well as it also offers some unique charts like contour plots, dendrograms and 3D charts [3]

5.2.4 *Libraries Used for Visualization*

In computer science, a library is a collection of non-volatile resources used by computer programs, often for software development. Visualization libraries consist of functions that allows user to visualize the dataset for easier and cleaner analysis. Some of the libraries used are depicted in Table 5.3.

5.2.5 *Visualization Charts Used*

(1) **Bar Plot**

A bar chart or bar graph is a chart or graph that presents categorical data with rectangular bars with heights or lengths proportional to the values that they represent. The bars can be plotted vertically or horizontally. A vertical bar chart is sometimes called a column chart. It shows comparisons among discrete categories. One axis of the chart shows the specific categories being compared, and the other axis represents a measured value. Some bar graphs present bars clustered in groups of more than one, showing the values of more than one measured variable [3].

(2) **Stacked Bar Graph**
The stacked bar chart extends the standard bar chart from looking at numeric values across one categorical variable to two. Each bar in a standard bar chart is divided into a number of sub-bars stacked end to end, each one corresponding to a level of the second categorical variable. The main objective of a standard bar chart is to compare numeric values between levels of a categorical variable. One bar is plotted for each level of the categorical variable, and each bar's length indicates a numeric value. A stacked bar chart too achieves this objective, but also targets a second goal. A stacked bar chart is considered in this analysis as it also corresponds with a level of a secondary categorical variable. The total length of each stacked bar is the same as before, but now the contribution of the secondary groups to that total can be seen [3].

(3) **Circle Chart**
A circle chart is a circular statistical graphic, which is divided into slices to illustrate numerical proportion. In a pie chart, the arc length of each slice (and consequently its central angle and area) is proportional to the quantity it represents. While it is named for its resemblance to a pie which has been sliced, there are variations in the way it can be presented [3].

(4) **Radar Chart**
A radar chart is a graphical method of displaying multivariate data in the form of a two-dimensional chart of three or more quantitative variables represented on axes starting from the same point. The relative position and angle of the axes is typically uninformative, but various heuristics, such as algorithms that plot data as the maximal total area, can be applied to sort the variables (axes) into relative positions that reveal distinct correlations, trade-offs, and a multitude of other comparative measures [3].

5.3 Data Exploration and Visualization

In this chapter, the entire process of data cleaning and data pre-processing was described, after which the complete dataset for analysis was ready with 1,10,000 records of queries and 11 columns to describe those queries. The above section has provided enough idea regarding the steps that have been adopted for exploratory data analysis. There is adequate confidence on dataset prepared after pre-processing, and also, there is fair clarity on objectives that need to be accomplished as well as there is proper understanding of domain. Besides, there is sufficient knowledge gained about visualization criterion to be followed, libraries to be used and charts to be plotted. This section explores each and every column of the KCC dataset using the aforementioned libraries and functions to plot various charts and plots, in order to find to find conclusions in the dataset.

5.3.1 Donut Pie Chart Presenting Overview of Query Types

Since the major part of the analysis is against query types, visualization has been created with the help of a donut pie chart to know the percentage of each query types. Following steps are taken to create a chart reflecting the proportion of each query type:

1. Count the number of queries for all the 59 query types.
2. On a closer look, it is revealed that 52 query types made only 10,000 or approximately 10% of all queries. These queries were grouped in to one category called 'other'.
3. The counting was done using value_count() function.

Following code is used to plot the donut pie chart to present the overview of query types and is shown in Fig. 5.1:

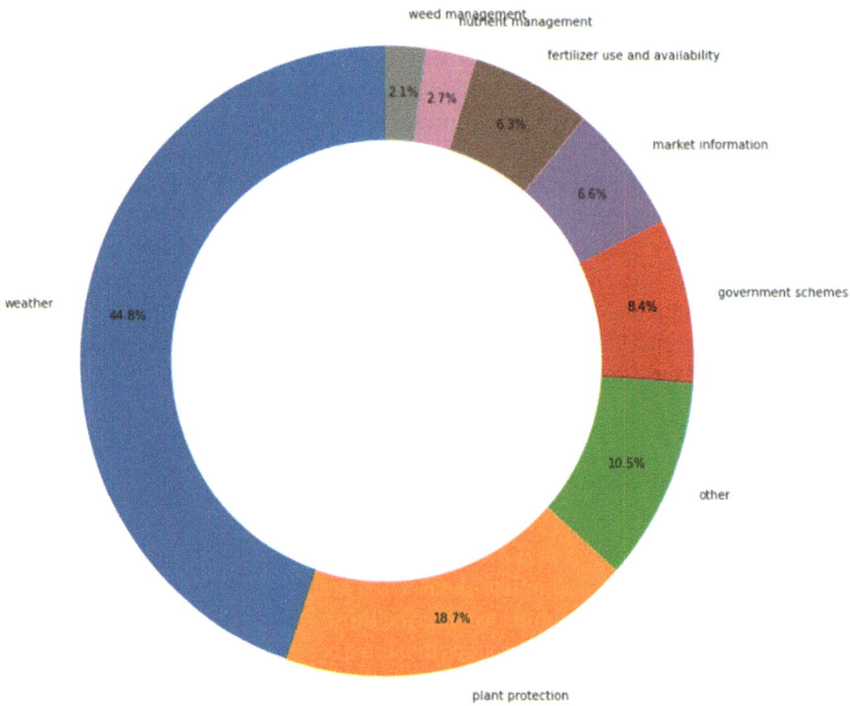

Fig. 5.1 Donut pie chart showing the percentage of all query types

```
import matplotlib.pyplot as plt
# Pie chart
labels = train["Category"].value_counts().index.to_list()    #List
of lables
sizes = train["Category"].value_counts().to_list()    #List of L
able count
#colors
color_series = ['#FAA327','#37B44E','#14ADCF','#1E91CA','#2C6BA0
','#2B55A1','#2D3D8E','#44388E','#6A368B','#D02C2A'] #List of co
lours to be used
explode = (0.05,0.05,0.05,0.05)    #explsion
```

```
plt.figure(figsize=(10,10))    #Figire Size
plt.pie(sizes, labels=labels, autopct='%1.1f%%', startangle=90,
pctdistance=0.85)
#draw circle
centre_circle = plt.Circle((0,0),0.70,fc='white')
fig = plt.gcf()
fig.gca().add_artist(centre_circle)
# Equal aspect ratio ensures that pie is drawn as a circle
plt.tight_layout()
#plt.figure(figsize=(25,25))
plt.show()
```

The donut pie chart is part of the matplotlib library. From the above visualization, it can be concluded that weather is an overwhelmingly important query type, which is understandable since agriculture is dependent on rains and rains are very unpredictable. The second most prominent query type with 18.7% was 'plant protection' which covers pest, animal, locust attacks on plants. The third biggest query type is other, which comprises of all smaller query types.

5.3.2 *Radar Chart and Stacked Bar Graph to Analyse District-Wise Query Type*

One of the most defining parts of agriculture is the location, as it can be used to pinpoint other important factors such as weather and soil type. In a state, location can be classified by districts. Hence the first analysis is to find out district-wise query type, and the radar graph plotted to visualize the result. The radar graph is created using plotly() function which is a part of matplotlib library. The radar graph was chosen as it is the best way to measure performance metrics. From Fig. 5.2, it can be clearly discerned that the districts with most queries were Nanded, Parbhani, Buldhana having more than 8000 queries each. The following code is used to generate the desired chart for visualization of the results:

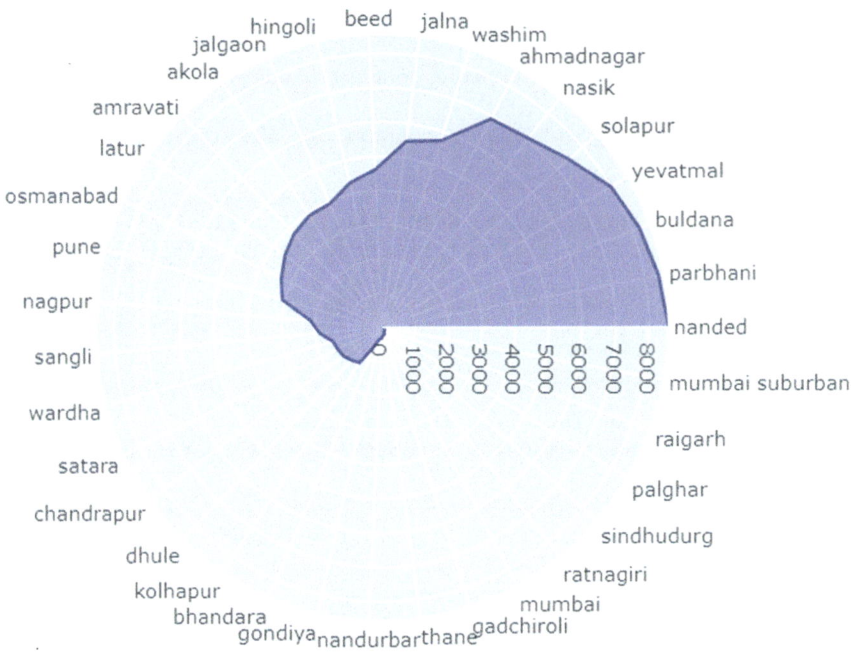

Fig. 5.2 Radar chart for query count per district

```
import pandas as pd
import plotly.graph_objects as go
c = train["DistrictName"].value_counts().index.to_list() #List
containing all district names
d = train["DistrictName "].value_counts().to_list()     #List
ocntainign count of queries
fig = go.Figure()
fig.add_trace(go.Scatterpolar(
        r=d,
    theta=c,
      fill='toself',
      name='plant protection'
))
fig.show()
```

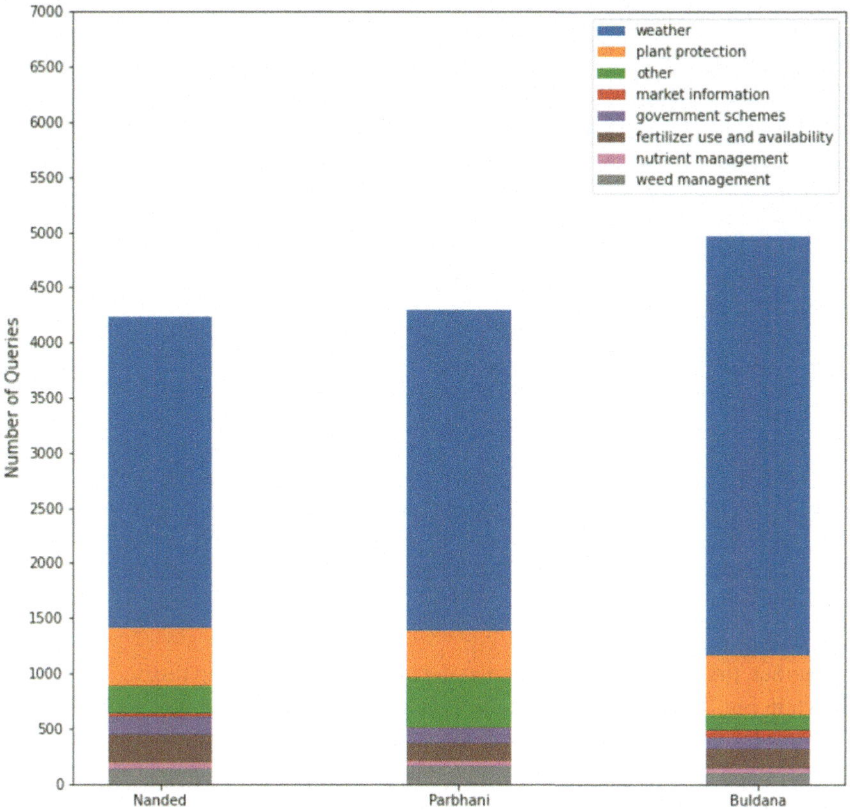

Fig. 5.3 Stacked bar graph for query count of top districts

Figure 5.3 presents a closer look at the 3 major farming districts of Maharashtra State in India: Nanded, Parbhani and Ahmednagar. Over 4000 weather queries and 1500 plant protection queries were registered by Nanded farmers in 2019. Parbhani farmers followed the suit with 4000+ weather and 1000+ plant protection queries. Ahmednagar was not any different with 2000+ weather queries and 1500 + plant protection queries. The code used to plot the preferred graph is mentioned below:

```
ind = np.arange(N)    # the x locations for the groups
width = 0.35          # the width of the bars

p1 = plt.bar(ind, wea, width)
p2 = plt.bar(ind, pp, width)
p3 = plt.bar(ind, oth, width)
p4 = plt.bar(ind, mi, width)
p5 = plt.bar(ind, gs, width)
p6 = plt.bar(ind, fua, width)
p7 = plt.bar(ind, nm, width)
p8 = plt.bar(ind, wm, width)

plt.ylabel('Number of Queries')
plt.xticks(ind, ('Nanded', 'Parbhani', 'Buldana'))
plt.yticks(np.arange(0, 7010, 500))
plt.legend((p1[0], p2[0], p3[0], p4[0], p5[0], p6[0], p7[0], p8[
0]), ('weather', 'plant protection', 'other', 'market informatio
n', 'government schemes', 'fertilizer use and availability', 'nu
trient management', 'weed management'))
plt.show()
```

Further, the analysis is done even at the block level for the top district, Nanded as shown in Fig. 5.4. The code is similar to that of stacked bar graph plotted above.

The analysis of query type per block shows that while weather still dominates the queries, the remaining are a bit more varied with a mixture of queries on market information, plant protection and fertilizer. The analysis provides us specific information that helps us drive appropriate actions accordingly; for example, block 'loha' has lots of queries related to government scheme, whereas block 'himayat nagar' has market information queries trending.

5.3.3 Radar Chart to Present Queries According to Seasons

Agriculture is the main sector driving the economy of most of the above-mentioned districts. All agrarian economies are usually dependent on rainfall and the seasons. This gives rise to many problems centered on seasons which then translates to farmer's queries. Categorizing and analyzing queries according to seasons are the next logical step. The attempt is made to analyse the exact query type according to season. The radar graph shown in Fig. 5.5 depicts that number of overall queries for the month of October and December are same, whereas for the month of April is remarkably low. However, the chart clearly highlights the fact that weather related queries is far more in the month of October and December compared to April. In the summer season, there are just 5000 weather queries compared to winter and rainy season where the number of queries is approximately more than 15,000 and 25,000,

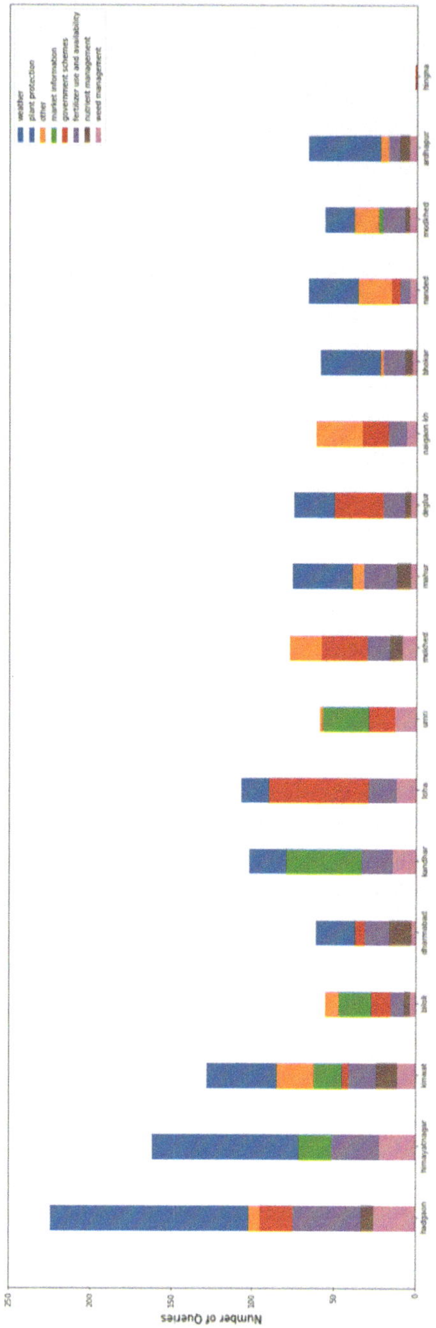

Fig. 5.4 Block-wise queries across Nanded

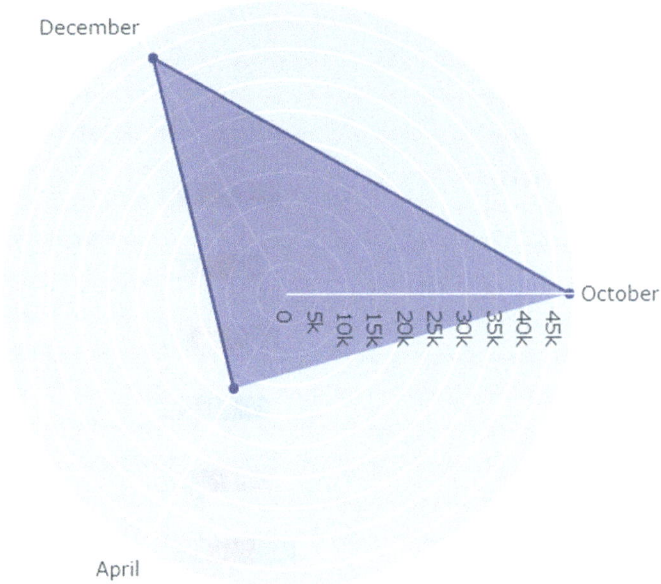

Fig. 5.5 Radar chart for query type per season

respectively. The reason is quite obvious, as rainy season is a major differentiator and can provide the perfect amount of rain for irrigation or destroy the crops with droughts or floods. While not as efficacious as the rainy season, the cold of winter is also not controllable and warrants its 15,000 queries. Since summer is the harvest season, the data reflects that with very few weather related queries. The rise of market information queries in summer is also justified by the harvest season. The following code has been used to plot the radar graph to analyse queries depending on the season in which it has been registered:

```
import plotly.express as px
import pandas as pd
df = pd.DataFrame(dict(
    r = [47204, 44522, 17746],
    theta = ['October', 'December', 'April']))
fig = px.line_polar(df, r='r', theta='theta', line_close=True)
fig.show()
```

The stacked bar graph shown in Fig. 5.6 is also juxtaposition to the previous analysis of radar chart with height of the bar for the month of December falling considerably short in comparison with the bar for the month of October even though both of them have the same number of queries. But that supposed error is quelled by analyzing the graph and realizing the fact that there is a huge reduction in the number of weather queries in winter in comparison to rainy season. This difference is

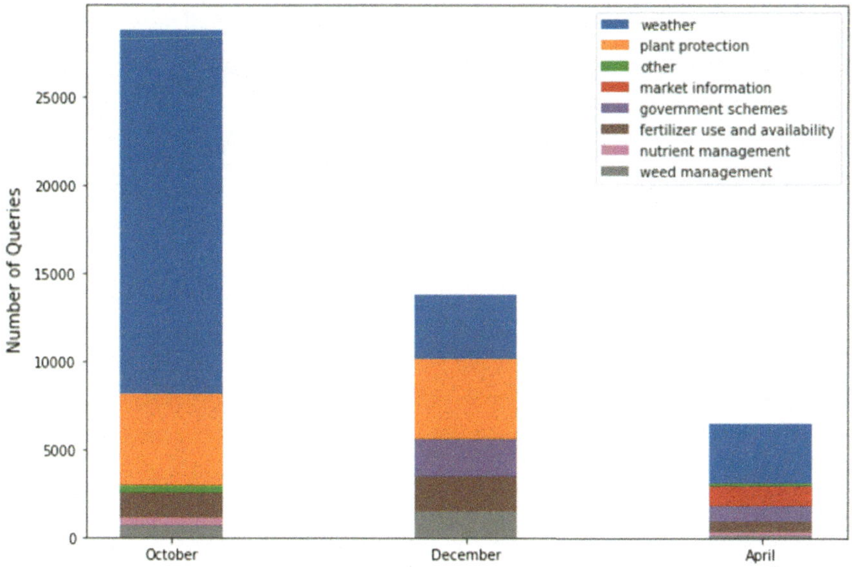

Fig. 5.6 Stacked bar graph reflecting query type variation according to seasons

balanced by the distribution of winter's queries into 10,000+ plant protection queries and 5000 government schemes. The queries regarding fertilizer use and availability also see a spike in the winter, and even weed management queries make a significant appearance in the season.

5.3.4 Radar Chart and Plot Chart to Present Category-Wise Query Type

Analysis of category along with query type acts as the perfect example to show and explain biased data and how it can affect further analysis. Category is the column name for the distribution of crops that the farmer is sowing or the cattle he is rearing when the query is generated by the said farmer. Code for the analysis through visualization in the form of radar chart is mentioned below and depicted in Fig. 5.7.

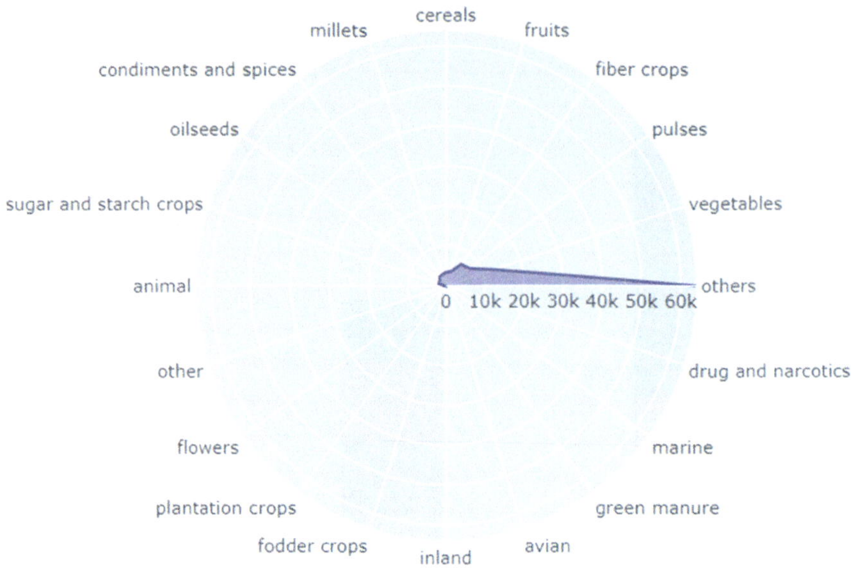

Fig. 5.7 Radar chart depicting category bias

```
import plotly.express as px
import pandas as pd
import plotly.graph_objects as go

c = train["Category"].value_counts().index.to_list()
d = train["Category"].value_counts().to_list()

fig = go.Figure()
fig.add_trace(go.Scatterpolar(
        r=d,
    theta=c,
      fill='toself',
      name='plant protection'
))
fig.show()
```

From Fig. 5.7, it can be analysed that the dataset is skewed towards the category 'others'. The category 'others' comprises of numerous different crops which did not have enough queries to classify as their own category. This is a testament to the diversity of the number of crops in Maharashtra. However, since the analysis does not clarify the query distribution, another radar chart is plotted to shed some more light onto the category-wise query distribution. From Fig. 5.8, it can be concluded that

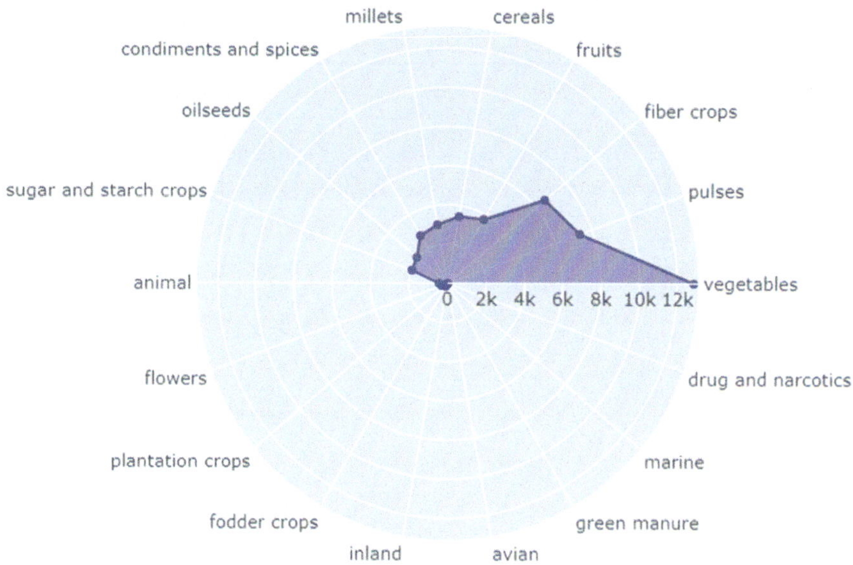

Fig. 5.8 Radar chart on category-wise query type

vegetables with 12,000+ queries have the highest queries related to them, followed by pulses and fibre crops with 10,000 and 8000 queries, respectively.

Now since there is a fair idea on the number of queries per category, a deeper look can be made to understand the distribution of query types per category with the help of plot chart as shown in Fig. 5.9.

While plotting the radar chart shown in Fig. 5.8, the category 'others' has not been considered because of huge number of weather queries, whereas all the other categories have a mixed set of query types. In Fig. 5.9, it is clearly evident that category 'vegetables' has plant protection as a major query type followed by fertilizer use and nutrient management.

5.4 Conclusion and Future Scope

It is very adventurous to work with open data, as dataset structure is not known; it becomes difficult to frame the problem statement at the beginning. Every step from data acquisition to data cleaning and pre-processing is a new revelation in itself. The objectives and goals could only be set after the pre-processing phase is successfully completed. In this chapter, some very firm inferences were drawn from the exploratory data analysis which might encourage readers to use data science to support farmers and improve agriculture with the help of important insights. The KCC data, if construed as the problems that farmers face, suddenly becomes a dataset of all the roadblocks that stand in front of success of the green revolution. The

Fig. 5.9 Analysis of category-wise top query types

government is sparing no resource to help and enable the farmers, and their efforts can be augmented with the use data science. This research is a proof that if data science can be used to visualize and analyse the queries from all perspective, it could lead to smarter and more efficient expenditure of the funds issued by the central government. This chapter presented the understanding of queries and their appropriate categorization with exceeding clarity and highlighted the exact problem faced by farmers. Once the problems are classified aptly, then the resources could be allocated proficiently to ensure their most complete and efficient use.

As mentioned before, the goals of this project were defined upon discovery of problems faced by the system. One such problem we faced while preprocessing and then analysis was the discrepancies that was found within 'Query Text' and 'Query Type'. An attempt has been made to fix that, and the efforts have been documented in the next chapter.

Any research on data science and its applications will be bereft without discussing its future scope since the field is itself constantly evolving. As has been the purpose of the chapter, the reader can follow the algorithm and reference the case studies to form their own algorithms and cases. Our analysis was partially limited due to the absence of exact data from previous years of KCC. This could be substituted with the future years, and an analysis over the years would also be very interesting. While this paper could not exemplify that, it contains sufficient information to create change and to set an example in the otherwise technically orthodox field of agriculture.

References

1. https://openbudgetsindia.org/dataset/department-of-agriculture-cooperation-and-farmers-welfare-2019-20
2. Pittenturf C (2018) What is data visualization and why is it important? https://data-visualization.cioreview.com/cxoinsight/what-is-data-visualization-and-why-is-it-important-nid-11806-cid-163.html
3. https://mschermann.github.io/data_viz_reader/introduction.html#importance-of-data-visualization

Chapter 6
An Approach for Exploring New Frontiers for Optimizing Query Volume for Farmer Call Centre—KCC Query Pattern

6.1 Introduction

Kisan Call Centre (KCC) dataset, as found on data.gov.in, contains a set of attributes, including two of the most important ones—Query Text and Query Type. Since the primary objective of KCC lies in resolution of farmers' queries, the most important part of their communication lies with the operator—in understanding the farmer's query and recording it correctly.

Query Text and Query Type are two attributes in the dataset that are directly related to the query recorded by the KCC operator. As discussed in Chap. 2, Query Text is the raw text as recorded from the phone call, translated directly from the farmer's speech by the operator. Query Type, on the other hand, is the short version of Query Text. It is a direct categorization of the Query Text into a more abstract type of query. Some examples of Query Text and its corresponding Query Type are mentioned in Table 6.1.

According to the assessment and survey by Indian Institute of Management-Ahmedabad (IIM-A) on KCC dataset, the following results were obtained with respect to the knowledge transfer and information availability for farmers and operators [1].

It is obvious from Fig. 6.1, the information provided by the operators is of a satisfactory level for the farmers who raise the queries. On the other hand, a survey conducted amongst the Farm Tele Advisors (FTA) generated the following results, as shown in Fig. 6.2. An important point to note is that the majority opinion on whether 'question details can be easily and quickly recorded' lies in the spectrum of slight agreement to slight disagreement.

There are multiple information sources used by the Farm Tele Advisors while addressing the farmers' queries. Some of them include self-knowledge, Internet search, books and other material. Figure 6.3 depicts the results of the rating provided by the FTAs with respect to their understanding of the quality of these sources. It

N. Sharma et al., *Open Data for Sustainable Community*,
Advances in Sustainability Science and Technology,
https://doi.org/10.1007/978-981-33-4312-2_6

Table 6.1 Snapshot of Query Text and Query Type from the dataset

Query Type	Query Text
Market information	Contact number VFPCK Ernakulam
Plant protection	White fly attack in tomato
Nutrient management	Fertilizer recommendation of banana
Plant protection	Brinjal flower shedding
Plant protection	Fungal attack in banana

	Ratings
Information is available on time	4
Information available is easy to understand	5
Farmers can understand and process it easily	3
Farmers seems to be satisfied with the information provided	3

Opinion Scale: 5-Strongly Agree, 4-Agree, 3-Partially Agree/Disagree, 2-Disagree, 1-Strongly Disagree

Fig. 6.1 Overall assessment of the information and knowledge available [1]

	Strongly Agree	Agree	Partially Agree/ Disagree	Disagree	Strongly Disagree	Average
The software is up-to-date	14.81	66.67	11.11	7.41	0.00	3.9
It is user friendly	18.52	77.78	3.70	0.00	0.00	4.2
The screen interface it shows is good & useful	29.63	66.67	3.70	0.00	0.00	4.3
Calls can be handled easily	37.04	55.56	7.41	0.00		4.3
The voice quality is good & clear	14.81	55.56	22.22	7.41	0.00	3.8
Calls do not get dropped, lost or mishandled by the software	3.70	29.63	22.22	40.74	3.70	2.9
The software can handle heavy call traffic	7.41	37.04	14.81	37.04	3.70	3.1
Software does not frequently crash	3.70	22.22	29.63	29.63	14.81	2.7
Caller details can be easily recorded and registered	22.22	48.15	18.52	11.11	0.00	3.8
Question details can be easily & quickly recorded	22.22	37.04	29.63	11.11		3.7
Repeated Irrelevant calls can be blocked by the software		7.41	7.41	7.41	77.78	1.4
The software meets the requirements	14.81	74.07	11.11	0.00	0.00	4.0

Fig. 6.2 Rating of software [1]

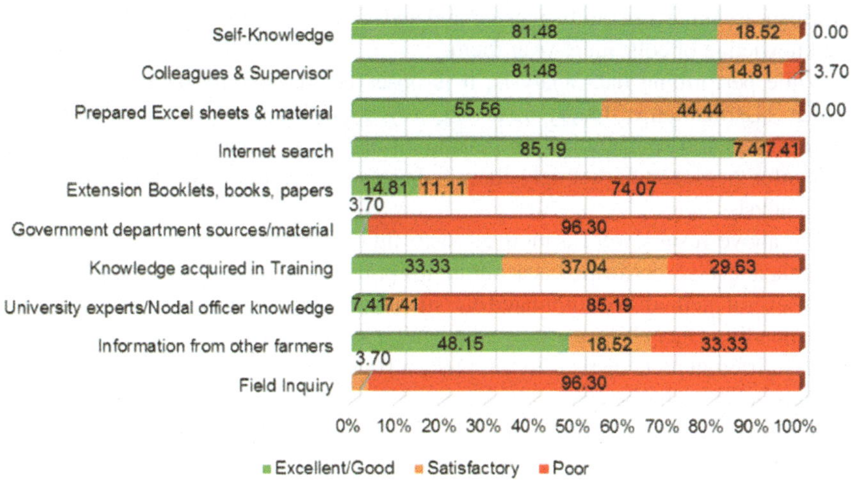

Fig. 6.3 Rating information and knowledge sources used [1]

is important to note that the majority of FTAs rely on self-knowledge and rate the formal sources of information to be poorer.

A reason for the above inference can be found in the next survey question's result. The frequency of updation of the information available from the various sources mentioned above is given below in Fig. 6.4. Here, it is clear that the low rated

	Very Frequently	Frequently	Occasionally	Rarely	Never	Average Rating
Self-Knowledge	22.22	55.56	18.52	3.70	0.00	4.0
Colleagues & Supervisor	3.70	70.37	18.52	7.41		3.7
Prepared Excel sheets & material	0.00	11.11	40.74	37.04	11.11	2.5
Internet search	14.81	44.44	25.93	11.11	3.70	3.6
Extension Booklets, books, papers	0.00	3.70	3.70	14.81	77.78	1.3
Government department sources/material	0.00	0.00	3.70	7.41	88.89	1.2
Knowledge acquired in Training	0.00	3.70	18.52	70.37	7.41	2.2
University experts/Nodal officer knowledge	0.00	0.00	0.00	11.11	88.89	1.1
Information from other farmers	0.00	3.70	22.22	37.04	37.04	1.9
Field Inquiry	0.00	0.00	0.00	0.00	100.00	1.0

Fig. 6.4 Frequency of updating the sources' information [1]

information sources are rarely or never updated, and thus FTAs prefer self-knowledge and knowledge of colleagues over them.

While performing data analysis, it is observed that certain data entries had a wrong classification of Query Text to Query Type. Figures 6.1 and 6.2 and their inferences give a solid indication that this is indeed possible. In accordance with these inferences, certain wrongly classified categories were found while performing data analysis.

The above results were a motivation to find a better alternative to ensure that the 'data analysis' performed is accurate for policy making and future applications. Thus, several methods for automated classification of incoming Query Texts into their corresponding Query Types were studied and researched. Hence, for this project, the entirely different procedures were adopted and that was ranging from classification to text clustering, using *k*-means and also using TensorFlow. In the upcoming sections, the adopted approach is discussed in detail and presents a comparison between all the techniques.

6.2 Different Approaches for Query Text to Query Type Classification

6.2.1 Text Similarity and Clustering

Document or text clustering can be defined as an application of cluster analysis to text documents such that it can be organized into meaningful and topic-specific clusters or groups. It is particularly useful in case of unlabelled data. The entire procedure can be divided into the following steps:

Step 1—Keyword Extraction from the Query Texts

Keyword extraction is the process of identifying the important words, or words that carry the substance of the text, from a sentence or a document. This step is performed to ensure that only relevant words are used to find out the document vectors, and eventually, the similarity between Query Texts.

Step 2—Form TF-IDF vectors of the Query Texts

TF-IDF stands for term frequency-inverse document frequency. It is calculated as follows:

- tf(w) = (number of times w appears in a document)/(total number of words in the document)
  where tf(w) stands for term frequency for a word w
- idf(w) = log_e (total number of documents/number of documents with w in it)
  idf(w) stands for inverse document frequency for a word w
- TF-IDF is the combined product of the above two.

TF-IDF values for all the terms in Query Text can be put together to form a vector representing the particular Query Text. Similarly, such vectors are formed for all the remaining Query Text sentences.

Step 3—Perform *K*-Means Clustering

There are several similarity metrics which can be used to cluster the vectors into meaningful groups. For instance, 'cosine similarity' and 'Euclidean distance' are two frequently used similarity metrics.

Cosine similarity is calculated as follows:

$$\text{similarity} = \sum_{i=1}^{n} A_i B_i \div \left[\sqrt{\sum_{i=1}^{n} A_i^2} \times \sqrt{\sum_{i=1}^{n} B_i^2} \right]$$

Euclidean distance is calculated as follows:

$$d(p, q) = d(q, p) = \sqrt{\sum_{i=1}^{n} (q_i - p_i)^2}$$

K-means clustering makes use of Euclidean distance to find the similarity between vectors (as formed in step 2). In Python, both step 2 and 3 can be accomplished using the library—sklearn [2]. TfIdfVectorizer algorithm to create TF-IDF vector and *k*-means algorithm to create relevant cluster [2].

The results will vary depending on the number of iterations and the number of clusters that is defined while initializing the *k*-means object. Since the number of clusters in the dataset is in the form of the number of distinct Query Types, it can be defined as a part of the algorithm. As the dataset is labelled, other supervised learning algorithms can be used in place of this technique to yield better results.

6.2.2 *Word-Based Encodings*

Word-based encodings take character encodings for each character in a set. A common and simple character encoding is use of American Standard Code for Information Interchange (ASCII) values for encoding. Table 6.2 shows the word 'FARMER' along with its ASCII encoding.

Is it sufficient to encode a word like FARMER using these values? The problem here is that the semantics of the word are not encoded in the letters. This could be

Table 6.2 Understanding word embedding [2]

Character	F	A	R	M	E	R
ASCII code	70	65	82	77	69	82

demonstrated using the word 'ARFMER' which has no verbal meaning, but will be encoded with exact same values. It gives a clear indication that training a neural network with just the letters could be a daunting task. So, to avoid such conflicts, word-based encodings is used. Words can be given a value and those values can be used in training a neural network. Consider these sentences,

Sentence 1 : "**Farmer asked about weather**"
Sentence 2 : "**Farmer asked about pesticide**"

The words '**Farmer**', '**asked**', '**about**' from the sentence 1 can be encoded as **001**, **002** and **003**. These words reappear in sentence 2 as well, where the same word codes can be reused. The words '**weather**' and '**pesticide**' appear only once in sentence 1 and sentence 2, respectively, and can be encoded as **004** and **005**. The final encoded sentences are as follows:

Sentence 1: **Farmer asked about weather** Encoded Sentence 1: **001**, **002**, **003**, **004**
Sentence 2: **Farmer asked about pesticide** Encoded Sentence 2: **001**, **002**, **003**, **005**

The next step is to start training a neural network, based on words. TensorFlow, which is an open-source library and Keras, which is a neural network library, has been used to train the model. Fortunately, both the libraries provide some high-level APIs that makes the job very simple to do. Following are the steps to do the same:

Step 1: Import TensorFlow and Keras libraries. TensorFlow provides the Tokenizer class which handles the heavy lifting by generating the dictionary of word encodings and creating vectors out of the sentences. Following code is used to import the libraries.

```
Import tensorflow as tf
From tensorflow import keras
From tensorflow.keras.preprocessing.text import
Tokenizer
```

Step 2: This step involves instantiating the Tokenizer class. The passive parameter **num_words** signify the maximum number of words to consider based on word frequency. Only the most common **num_words**-1 words are kept.

```
Tokenizer = Tokenizer(num_words = 100)
#Consider the following sentences to experiment with
Sentences = [
"Farmer asked about weather"
"Farmer asked about pesticide"
]
```

Note: **num_words** is a handy shortcut when dealing with lots of data and worth experimenting with while training with data. Sometimes the impact of less number of words can be minimal and lead to training accuracy, but might take huge training time, however, it can be used carefully. It is worth experimenting the trade-offs between accuracy and training time by setting num_words to different values.

Step 3: The fit_on_texts method of the tokenizer then takes in the array of sentences and encodes it. The tokenizer provides a word index property which returns a dictionary containing keyvalue pairs, where the key is the word, and the value is the token for that word. The code for the step 3 is given below:

```
Tokenizer.fit_on_texts(sentences)
Word_index = tokenizer.word_index
Print(Word_index)
```

Word_index = {"farmer": 1, "asked": 2, "about": 3, "weather": 4, "pesticide": 5}

It is important to note that the word 'Farmer' has letter 'F' as capital in the input array of sentences. But the word in dictionary is in lowercase. Tokenizer strips out punctuations and converts all the words to lowercase.

6.2.3 *Text to Sequences*

The next step is to turn the sentences into lists of values based on these tokens. Once the sentences are converted into list, it is preferred to convert every sentence to same length. The reason is that it might be hard to train a neural network with sentences of unequal length.

Fortunately, TensorFlow includes APIs to handle these issues. Following is the code:

```
label_tokenizer = Tokenizer()
label_tokenizer.fit_on_texts(labels)

training_label_seq =
np.array(label_tokenizer.texts_to_sequences
(train_labels)) validation_label_seq =
np.array(label_tokenizer.texts_to_sequences
(validation_labels))
```

Sentence 1 : **"Farmer asked about weather"**
Sentence 2 : **"Farmer asked about pesticide"**
Sentence 3 : **"Farmer asked about the compost"**

Dictionary---

{'farmer': 5, 'asked': 3, 'about': 2, 'weather': 4, 'pesticide': 7, 'the': 9, 'compost': 6}

Sentence 1 is encoded as

Sentence 1 : "**Farmer asked about weather**": [5, 3, 2, 4]

Similarly, sentences 2 and 3 are encoded

Sentence 2 : [5, 3, 2, 7]
Sentence 3 : [5, 3, 2, 9, 6]

6.2.4 *Out of Vocabulary (OOVs)*

The token **OOV** signifies out of vocabulary words that are not in the word index. Any word can be used as an **OOV**, but remember that it should be something unique and distinct that is not confused with a real word. The code to create OOVs is as follows:

```
Tokenizer = Tokenizer(num_words = 100, oov_token =
"<OOV>")
Tokenizer.fit_on_texts(sentences)
Word_index = tokenizer.word_index
```

As the corpus of the word index grows and more words are added and indexed, it provides better coverage to any new sentence or previously unseen sentences.

6.2.5 *Padding*

Building neural networks to handle pictures require uniform sized images. Similarly, handling text also requires the same approach of converting them to uniform size, before training them with neural networks. For images, image generators are used to covert images to uniform sized image. However, for converting text into uniform length text, a popular method of padding is used and the code for the same is as follows:

```
#import
From
tensorflow.keras.preprocessing.sequence import
pad_sequences
```

Once the tokenizer has created the sequences, these sequences can be passed to pad_sequences method in order to have them padded.

```
train_sequences = tokenizer.texts_to_sequences
(train_sentences)

#spliting of data into train and test sets

train_padded = pad_sequences(train_sequences, padding
= padding_type, maxlen = max_length)
```

Now, consider three sentences of unequal lengths to better understand the concept of padding

Sentence 1 : **"Farmer asked about weather"**
Sentence 2 : **"Farmer asked about pesticide"**
Sentence 3 : **"Farmer asked about the compost"**

Length of each sentence:

Sentence 1 : **4**
Sentence 2 : **4**
Sentence 3 : **5**

Dictionary---
{'farmer': 5, 'asked': 3, 'about': 2, 'weather': 4, 'pesticide': 7, 'the': 9, 'compost': 6}

Sequences for the above sentences are:

[[5, 3, 2, 4], [5, 3, 2, 7], [5, 3, 2, 9, 6]]

The list of sentences is now padded out into a matrix so that each row in the matrix has the same length.

Padded Matrix:

$$
\begin{array}{l}
[\\
\quad \begin{bmatrix} \mathbf{0\,5\,3\,2\,4} \end{bmatrix}\\
\quad \begin{bmatrix} \mathbf{0\,5\,3\,2\,7} \end{bmatrix}\\
\quad \begin{bmatrix} \mathbf{5\,3\,2\,9\,6} \end{bmatrix}\\
]
\end{array}
$$

This is achieved by putting the appropriate number of zeros before the sentence. So in the case of the sentence 3: 'Farmer asked about the compost', no change was observed as this is the longest sentence and does not require padding.

```
Model: "sequential_1"
_________________________________________________________________
Layer (type)                 Output Shape              Param #
=================================================================
embedding_1 (Embedding)      (None, 120, 16)           16000
_________________________________________________________________
global_average_pooling1d_1 ( (None, 16)                0
_________________________________________________________________
dense_2 (Dense)              (None, 24)                408
_________________________________________________________________
dense_3 (Dense)              (None, 6)                 150
=================================================================
Total params: 16,558
Trainable params: 16,558
Non-trainable params: 0
```

Fig. 6.5 Output after training the neural network model

Padding is often done after the sentences. This is called as post-padding. The method used above is called pre-padding. Post-padding can be done by adding a parameter **post** in **pad_sequences.** The matrix width is the same as the longest sentence. But this can be overridden with the **maxlen** parameter. For example, if the training set requires only 3 words, the **maxlen** parameter can be set to 3. If the sentences are longer than the **maxlength**, then information is lost from the beginning of the sentence if pre-padding is used. However, with post-padding parameter information is lost from the end.

The final step is to train the neural network using the following code:

```
Model = tf.keras.Sequential([
  tf.keras.layers.Embedding(vocab_size,
  embedding_dim, input_length = max_length),
  tf.keras.layers.GlobalAveragePooling1D(),
  tf.keras.layers.Dense(24, activation = 'relu'),
  tf.keras.layers.Dense(6, activation = 'softmax')])
model.compile(loss = 'sparse_categorical_crossentropy',
optimizer = 'adam',metrics = ['acc'])
model.summary()
```

Different trade-offs can be seen by varying parameters in the following output.

6.2.6 *Visualization*

Graphical representation of high dimensional embedding is possible using the **TensorBoard Embedding Projector**. This is helpful in visualizing, examining and

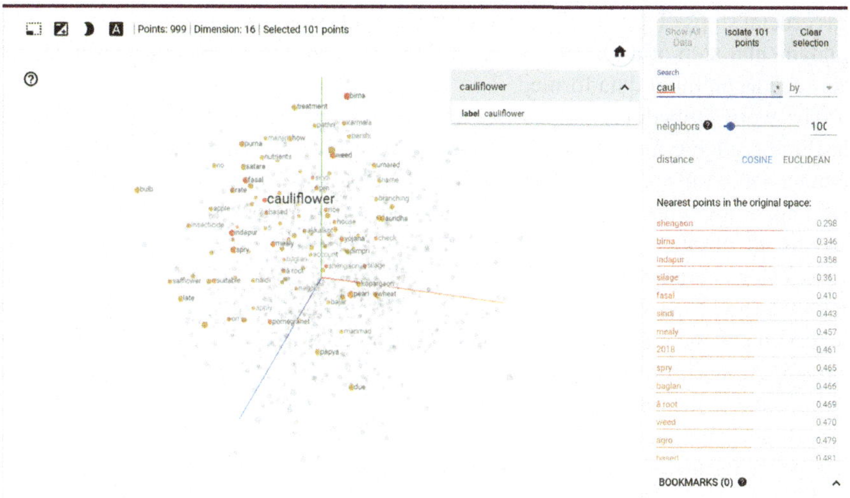

Fig. 6.6 Visualization using TensorFlow projector

understanding the embedding layers. Following code is used for plotting the result on TensorFlow projector and the visualization is presented.

```
out_v = io.open('vecs.tsv', 'w', encoding = 'utf-8')
out_m = io.open('meta.tsv', 'w', encoding = 'utf-8')
for word_num in range(1, vocab_size):
    word = reverse_word_index[word_num]
    embeddings = weights[word_num]
    out_m.write(word + "\n")
    out_v.write('\t'.join([str(x) for x in embeddings]) + "\n")
out_v.close()
out_m.close()
```

6.3 Conclusions

In this chapter, similar Query Types extracted from Query Text are clustered together. Further, improving the Query Types and decreasing it in numbers helped to form discrete clusters to identify the Query Type distinctly.

Several methods were explored before actually coding the above-described approaches. The design methodology for this systematic approach was extremely simple. An attempt was made to understand the input to the system, and the output to be obtained from it. The main hurdle was the 'unstructured data'. The processing, analysing and correctly categorizing it was a complex process as the data input was in the form of text. Data pre-processing, including stemming and lemmatization, was crucial steps of the process, whereas word embedding and tokenization were the fundamental step of the analysis. Several Query Text data entries of uneven length were fed as an input data formed.

The word embedding helped to transform sentences and documents into more machine-readable numeric format, padding helped to fix the issue of the different number of words contained in each document. Tokenization, embeddings and padding, along with text vectorization are mathematical techniques that were made easy by the number of Python libraries that provided these functionalities as simple classes (for example, Keras, TensorFlow and scikit). Objects of these classes were initialized and used to call methods which execute the math functions internally. After the text has been encoded as vectors or word embedding, processing the matrices and/or dictionaries was simple and straightforward. The basic approach remains same irrespective of the method implemented. Words and sentences which resemble more to each other are classified or clustered together, based on whether the approach is supervised or unsupervised. That is the core difference between the approaches.

After having implemented both methods—supervised and unsupervised—a conclusion can be made that the supervised technique is more accurate for this application. The reason is that the most of the data in the original dataset is correctly classified, and the model learns from the already labelled data instead of using unsupervised techniques and algorithms.

References

1. Gandhi V, Johnson N (2008) Decision-oriented information systems for farmers: a study of Kisan Call Centres (KCC), Kisan Knowledge Management System (KKMS), Farmers Portal, and M-Kisan Portal in Gujarat
2. deeplearning.ai

Part III
Demand and Supply Study of Healthcare Human Resource and Infrastructure—Through the Lens of COVID 19

(SDG3—Good Health and Well-Being)

Chapter 7
Sustainable Healthcare in COVID-19 Pandemic—Literature Survey and Data Lifting

7.1 Introduction

COVID-19 is the devastating agent of the current pandemic that has rampaged throughout the planet. It has defined the global health crisis and has become the greatest challenge to the world since World War II. It has been compared to the Spanish Flu pandemic of 1918 that killed millions of people across the globe. Countries have been trying to slow the spread of the virus by increasing testing and curing affected patients, contact tracing, closing schools, colleges and universities, quarantining close relatives of patients and calling off large gatherings. But COVID-19 has not only been a major health crisis, but it has also led to political, social and economic crises that could take much time to heal. The last hundred years, since Spanish Flu, have seen enormous social changes and a full policy analysis to prevent the next pandemic and would now require unprecedented collaboration. The current approach is like the story of five blind people, trying to describe an elephant. In this unit, we try to collate the different aspects of the pandemic, including intervention efforts and its impact on society.

COVID-19 is a part of the family of coronaviruses, which also includes similar, dangerous disease-causing viruses such as the Flu, SARS and MERS. As per Centre for Disease Control (CDC) report, this pandemic can be compared to that caused by H1N1 (*Spanish Flu*) in 1918, which killed 25% of the world's population [1]. This has resulted in a new era of Flu pandemics, and epidemiologically, COVID-19 is the largest of them all. In 1918, antivirals and antibiotics did not exist and medical professionals could only provide supportive health care. The DNA of the Spanish Flu pandemic was published successfully only after 2000, but it only took a few weeks to publish the DNA of COVID-19 [2].

The World Health Organization (WHO) declared this virus outbreak a 'global pandemic' on March 11 [3]. The number of cases has been rising daily in Asia,

N. Sharma et al., *Open Data for Sustainable Community*,
Advances in Sustainability Science and Technology,
https://doi.org/10.1007/978-981-33-4312-2_7

Europe, Americas and Africa. The peak stage of the coronavirus in India has been hindered by the lockdown and it might arrive around mid-November. All public health measures have been reinforced during the lockdown but there might still be a shortage of ICU beds and ventilators. According to a study conducted from an Operations Research Group constituted by Indian Council of Medical Research (ICMR), it is estimated that the lockdown had shifted the peak of the pandemic by an estimate of 34–76 days and had brought down the infection rate by 69–97%. Another model-based analysis of the COVID-19 pandemic in India has shown that by increasing the testing and quarantining of patients, there is a possibility of reducing the number of cases by 70% and the total cases might be reduced by 27% [4]. The country records highest-ever single day spike of 14,516 new cases causing the caseload to close to 4 lakhs as of June 20, while the death toll mounted to 12,948 [5]. The researchers said that increasing the health infrastructure, handling public health measures and the pace of epidemic in different regions can reduce the impact of the virus.

While pharma companies are racing to develop vaccines, non-medical interventions have been compared across countries to see which succeeded and which failed [6]. Heavy emphasis has been placed on acting early, testing extensively, contract tracing and social distancing. One of the major impacts of the current pandemic is the economic crisis. The South Korea model has highlighted how to manage a pandemic without an economic disaster. One of the defining features of the countries that have handled COVID-19 well in Asia, like South Korea, is that they have already faced SARS/MERS-like-pandemics before. Proactive governments that bring in experts for quick decision making make a big difference in handling COVID-19 cases. Since 1980, the number of outbreaks has tripled per year [7], and all one needs to do is to click on any year to see the outbreaks on the WHO website [8] to check the common outbreaks; the increased frequency of outbreaks, the chances of any one of them becoming global like COVID-19. Quotes from a BBC article that are relevant here are, 'For all the advances we've made against infectious disease, our very growth has made us more vulnerable, not less, to microbes that evolve 40 million times faster than humans do'. and 'Climate change is expanding the range of disease-carrying animals and insects like the Aedes aegypti mosquitoes that transmit the Zika virus'.

COVID-19 has been analysed epidemiologically and presented with the help of various forms of visualizations like charts to show the number of infected patients. On the other hand, economic analysis is also being done to estimate the sector-wise impact on the economy. With one of the largest hit areas of the economy being the travel, hospitality and tourism sector, discretionary spend have decreased. And like every pandemic, knowledge workers are the least impacted by the current situation. US unemployment rates have broken the records set during the Great Depression, since this is both a supply and demand side shock, as everyone has been asked to stay home, and maintain physical distance. Microbes evolve 40 million times faster than humans, as mentioned in the BBC article. So the next time this happens, based on recent history, is not far away considering the world has already faced SARS, MERS, H1N1 and Ebola before COVID-19 in the last 20 years. This pandemic is highlighting the digital divide, where conferences are happening online, knowledge workers are working from home. Most of the in-person work has been impacted.

All these problems created an urge to study the spread of the coronavirus in our country. Hence, in this unit, we study and analyse, the distribution of the hospitals, ventilators and ICU beds as well as availability of the doctors, nurses and health workers, across the country during the pandemic. This could give us a brief idea about the amount of human resources and health infrastructure that we would possibly need for the future in order to mitigate the virus. It can also keep us well informed about the condition of all the states of the country and can guide us to take actions accordingly and effectively. This chapter presents the detailed process of data preparation. The rest of the chapter is organized as follows: Sect. 7.2 scans various literatures to understand the demand–supply scenario, Sects. 7.3 and 7.4 present data preparation and exploratory data analysis, respectively. Section 7.5 concludes the chapter.

7.2 Literature Review

COVID-19 has impacted our society on a massive scale. There is no inhabited geographical area which is untouched. As the COVID-19 virus continues to spread globally, the study on the domestic and international upsurges and predicting the future trend has become a latest topic for research. A coherent study on the indeterminate contagion pattern of COVID-19 has been done and efforts made to map the same with the potential features. Since these are early days, an initial step in this direction is to carry out a systematic literature review (SLR) to find the number of publications on various burning issues such as employment, supply chain, models, infrastructure and human resource with respect to the COVID-19 pandemic situation. To do so, an advanced search was performed using above-mentioned terms, in three different research databases, i.e. JGATE, IEEE explore, ScienceDirect, as shown in Fig. 7.1. It is clearly evident that more research is being done in developing various models which will give a better solution to COVID-19 pandemic situation.

Further, the SLR is conducted to find out country-wise publications on search terms like ‘employment’ and ‘model’ in JGATE and IEEE explore research database. The analysis of the same is presented in Fig. 7.2.

Finally, the SLR is carried out to find out the country-wise publications on the search terms like ‘supply chain’, ‘human resource’, ‘infrastructure’ in JGATE and IEEE explore research database. The analysis of the same is given in Fig. 7.3.

The analysis presented in Figs. 7.2 and 7.3 clearly indicates that India is way ahead in research and publication work related to COVID-19. Besides, various related literatures are studied to understand the progression of the virus, and the impact of the virus on infrastructure, human resource, etc. and predictive models developed to predict the spread of the virus. Few literatures give the insight about the district vulnerability index that gives us a lot of information about the parameters of disease and the parts of India susceptible to such parameters. Another paper has been studied that gives us the predictions of an algorithm to find the future probable hotspots that are more severe to the virus. These papers and research works highlight the need

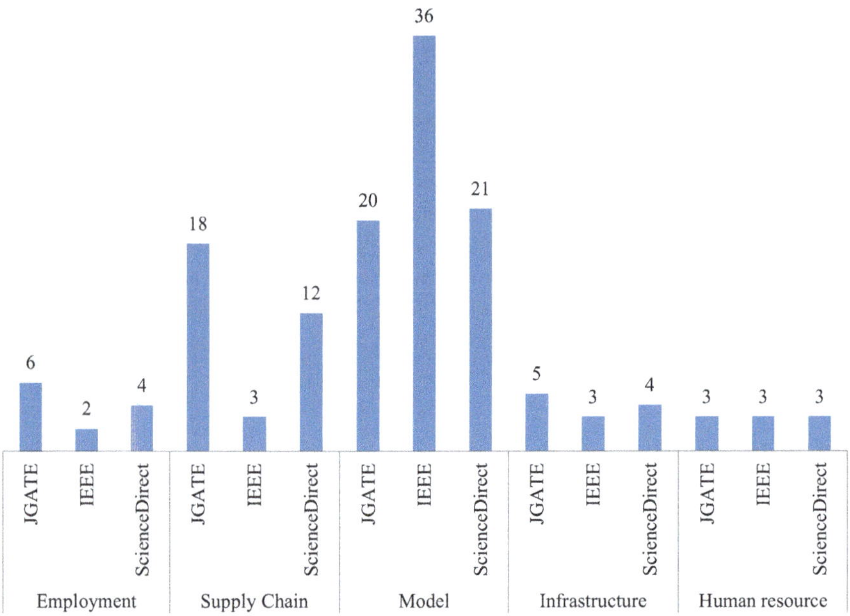

Fig. 7.1 Search terms employment, supply chain, models, infrastructure and human resource in JGATE, IEEE explore, ScienceDirect research databases

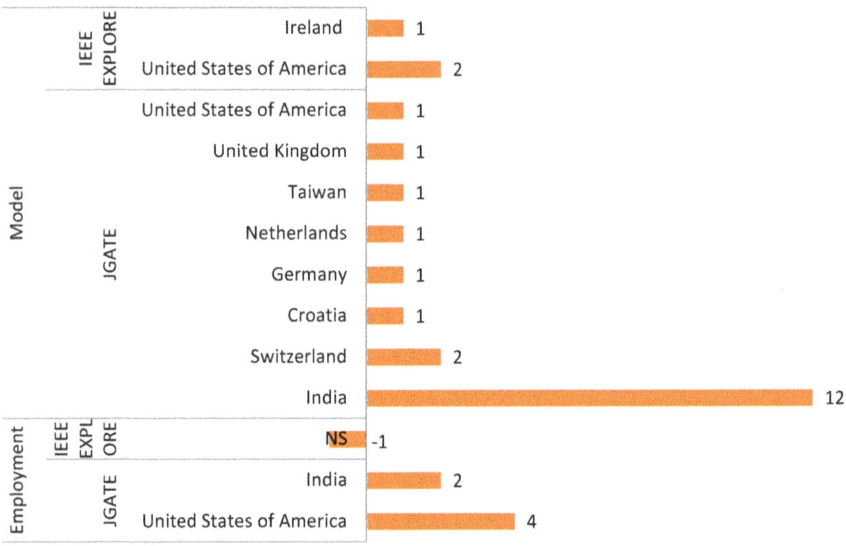

Fig. 7.2 Analysis of country-wise publications for search terms 'employment' and 'models' in JGATE, IEEE explore

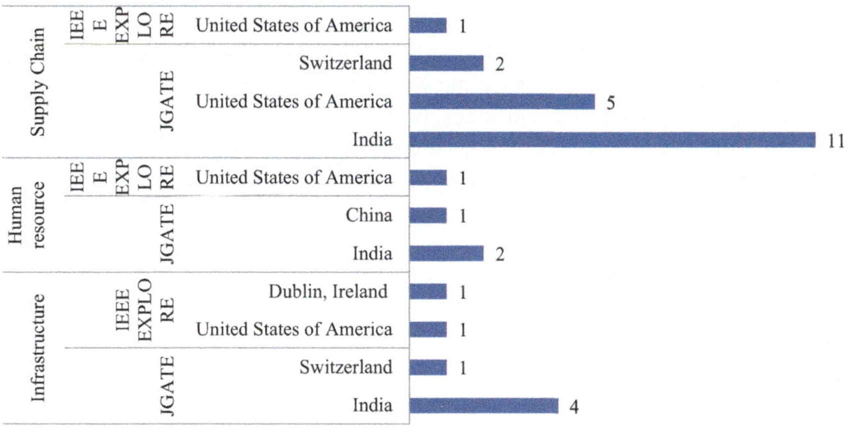

Fig. 7.3 Analysis of country-wise publications for search terms supply chain, human resource, infrastructure in JGATE, IEEE explore research databases

for strengthening and improving the healthcare system that depends critically on the proportion of corona positive cases.

David et al. in his paper 'The SIR Model for Spread of Disease' published in December, 2014, suggested a model to predict the spread of the disease known as Hong Kong flu in the USA during the years of 1968–1969 [9]. No flu vaccine was available at the time, so this study is extremely pertinent to the current situation of the COVID-19 pandemic where vaccine is not available either. The authors built a differential equation model to help predict the spread of the disease in the city of New York. The figures used were the number of excess deaths per week—that is, the number of deaths above the weekly average. The independent variable was time, measured in days. The dependent variables were divided into two related sets. The first set of dependent variables counted people in each of the groups as a function of time—susceptible individuals, infected individuals and recovered individuals. The second set of dependent variables included the same three groups but they were represented as proportions of the total population of New York. Either set of variables gave the same result when it came to the progress of the epidemic. Using these variables, the authors were able to build a model to predict the spread of the disease in New York.

The paper 'impact of COVID-19 on hospital bed days, ICU days, ventilator days and deaths by US State' was written by Christopher J. L. Murray in March, 2020. The study showed us the first set of estimates of predicted health service utilization and deaths due to COVID-19 by day for four months for each state of the USA [10]. The aim of the study was to estimate the span and timing of deaths and the huge demand of hospital services due to coronavirus in the USA. The data for the study was on confirmed COVID-19 deaths day wise from WHO websites and local and national governments accompanied with hospital capacity and utilization data for US states; also observed COVID-19 utilization data taken from selected localities in order to develop a statistical model that predicts the death and hospital utilization against capacity by state for the USA over 4 months—April, May, June and July. The paper predicted that the peak of the pandemic would arrive in the second week of April and varies up till May for different states. It predicted a total of almost 82,000 deaths and estimated that the deaths per day would fall below 10 between May 31 and June 6. The main conclusion was that the load placed on the US healthcare system by the pandemic would be far beyond its current capacity.

The study 'India fights Corona—understanding the coronavirus epidemic data' was carried out by Rubiscape in India [11]. The study took into account the variables time (recorded in days), total number of confirmed cases, number of recovered patients and the number of deaths. Rubiscape used this data in various forecasting and time-series analyses to predict the cases in different states. All the data recorded was up to 18 April 2020. Rubiscape recovered many insights from this data. On a national level, the most affected age group was people below the age of 40. Percentage of positive cases is highest in Tamil Nadu, Delhi and Maharashtra. Hospitals and beds in each state were also analysed. A Weibull distribution was fitted to the data to predict the number of cases. A COVID-19 model estimator was also built using the Rubiscape data application which estimated the symptom to disease confirmation period in days.

Melinda Moore et al., in their paper 'Discovering Future Disease Hot Spots' published by the Rand Corporation, have proposed four ways in which a robust approach towards vulnerability assessment is represented: an additionally comprehended evidence base, a more robust set of factors potentially contributing to outbreak vulnerability and associated proxy measures, the usage of adjustable weights for the parameters, and an analysis of all countries worldwide [12]. According to the report, the assessment algorithm described in it is inherently applicable to all outbreak-prone infectious diseases. This paper designs a tool known as the infectious disease vulnerability index—a number between 0 and 1 which highlights how vulnerable a particular nation is to the outbreak of an infectious disease where 0 is most vulnerable and 1 is least vulnerable. It also helps identify and assess potential hotspots where the disease outbreak can be most severe. The algorithm was designed to help US federal agencies, national and international health planners to help identify and raise alertness of those countries that might be most vulnerable to infectious disease outbreaks. Based on literature review, the authors arrived at a set of dependent variables across seven domains—demographic, health care, public health, disease dynamics, political-domestic, political-international and economic. The results found that out of the top 25 most vulnerable states, 22 were located in sub-Saharan Africa. The other three countries were Haiti, Afghanistan and Yemen. India ranks 71st in the list of most vulnerable countries with a normed score of 0.4938.

Milan Batista coded in MATLAB using the data from https://ourworldindata.org/coronavirus-source-data and applying data-driven method for the prediction model causing the prediction to change with the change in data [13]. The model is dependent on the daily cases of COVID-19. The evolution of COVID-19 is not completely random. Like other pandemics, it follows a life cycle pattern from the outbreak to the acceleration phase, inflection point, deaccelerating phase and eventual ending.

The life cycles of the pandemic vary from country to country, and different countries might be in different phases of the life cycles at a same point of time. Such decisions and planning can be rationalized by well knowing where our own country is. The result for the world stated that the impact of the virus would end 97% around May 30, 100% around December 2. As per the research, for India, it would end by 97% around May 27 and 100% around August 9.

The objective of the paper 'COVID-19 district vulnerability index' was to find parameters which have higher risk due to COVID-19. India did 2 lockdowns in that first one was 21 days and second one of 18 days as everyone knows the impact of lockdowns as it responds strongly [14]. The methodology includes using tableau for every parameter created graph to compare the situation of all states and the data is mix of 2011, 2015–2016 and it might exclude some area where census was not conducted. The results are as follows:

1. ***Age risk***: This visualizes percentage of population in district which is greater than 60 years of age. Here top 5 districts according to graph which are at the highest risk-Pattanamtitta (Kerala), Sindhudurg (Maharashtra), Kottayam (Kerala), Alappuzha (Kerala) and Ratnagiri (Maharashtra). Bottom 5 districts which are at the lowest age risk and have a lot of young population are: Papum Pare (Arunachal Pradesh), Kurung Kumey (Arunachal Pradesh), Upper Subansiri (Arunachal Pradesh), East Kameng (Arunachal Pradesh) and Lower Subansiri (Arunachal Pradesh).
2. ***Hypertension risk***: This visualizes the percentage of adults which have very high blood pressure. Top 5 districts which are at the hypertension risk are: West Siang (Arunachal Pradesh), Tawang (Arunachal Pradesh), East Siang (Arunachal Pradesh), Anjaw (Arunachal Pradesh) and Mokokchung (Nagaland). The bottom 5 districts which are at the lowest hypertension risk are: Mirzapur (Uttar Pradesh), Tehri Garhwal (Uttarakhand), Kottayam (Kerala), Lalitpur (Uttar Pradesh) and Bhind (Madhya Pradesh).
3. ***High blood sugar (diabetes) risk***: This visualizes the percentage of adults with high blood sugar level, i.e. which have hyperglycemia and diabetes. Top 5 districts which are at the highest high blood sugar risk are: Wayanad (Kerala), Kolkata (West Bengal), Guntur (Andhra Pradesh), Cuddapah (Andhra Pradesh) and Puri (Odisha). The bottom 5 districts which are at the lowest high blood sugar risk are: Kargil (Jammu and Kashmir), Ramban (Jammu and Kashmir). Auraiya (Uttar Pradesh), Lahaul and Spiti (Himachal Pradesh) and Nandurbar (Maharashtra).
4. ***High BMI (obesity) risk***: This visualizes the percentage of adults with high BMI, i.e. which are overweight and obese. Top 5 districts which are at the highest obesity risk are: Krishna (Andhra Pradesh), Guntur (Andhra Pradesh), Kolkata (West Bengal), Hyderabad (Telangana), Mahe (Puducherry). The bottom 5 districts which are at the lowest obesity risk are: Simdega (Jharkhand), Dantewada (Chattisgarh), Narayanpur (Chhattisgarh), The Dangs (Gujarat) and Dindori (Madhya Pradesh).
5. ***Acute respiratory infections (ARI) Risk***: This visualizes the percentage of children under 5 years with prevalent symptoms of acute respiratory infections in

the last 2 weeks preceding the NFHS survey. Evidence suggests children below 5 years are less vulnerable and this is an indication of high prevalent ARI infections. This might evolve and might be reconsidered in the risk index. The top 5 districts which are at the highest ARI risk are: Ramban (Jammu and Kashmir), Kishtwar (Jammu and Kashmir), South Garo Hills (Meghalaya), West Garo Hills (Meghalaya) and Kannauj (Uttar Pradesh). The bottom 5 districts which are at the lowest ARI risk are: Sheopur (Madhya Pradesh), Longleng (Nagaland), Hailakandi (Assam), Karimganj (Assam) and Bolangir (Odisha).

6. ***Current status factor***: Along with considering the above risk factors, considering there are still a lot of factors which affect the spread and vulnerability, consider the current status factor which gives us what is the situation today—accounting a lot of random events and other possible factors.

The above paper has also introduced the terms 'district vulnerability index' and 'population density' [14]. The index which is attempted here factors in the above parameters, weightages are given appropriately (more weightage for comorbidity and age) and composite index number is generated. This visualizes the district vulnerability index. This of course is not an accurate projection or prediction, but a measure and interpretation of risk according to our public data. It does not take into account tons of other parameters like travel, religious events, policy changes, etc. and it may not reflect the reality on the ground today. Population density also plays a big role in the spread of COVID-19. Along with population density, members per household also play a crucial role since following social distancing is very difficult. Identifying the most vulnerable through targeted surveys and district level data collection: The government has multiple online citizen channels it can leverage like Aarogya Setu and MyGov plus and it has the district level administration which can effectively undertake a quick data collection exercise to identify people who are/have: A. greater than 60–65 years of age and B. existing conditions like hypertension, diabetes, HIV, cancer, COPD, etc. Of course, these suggestions need to be very carefully evaluated and considered but one should at least know where the vulnerable population is and start getting ready to protect them first.

Finally, an attempt is made to find the search trend of the coronavirus spread in order to know people's curiosity regarding the disease. The search was made in Google trends for three terms, i.e. Covid 19, covid-19 and Coronavirus, and is presented in Fig. 7.4. It is evident that in the March month the search regarding 'COVID-19' was at peak while it is comparatively negligible by 31 May.

We have discussed various aspects of the current pandemic and historical references. The 1918 pandemic led to global transformation in improved healthcare systems and pandemics of the twenty-first century has destroyed many myths about the resilient modern society. We must design mechanisms, for not only acting early, testing and contract tracing extensively but also economic transactions that include physical distancing by design. We must also improve the application of artificial intelligence in the areas it is being used in the fight against COVID-19, including drug repurposing and drug discovery. Major clinical trials are not only running with new vaccines, but existing drugs identified through literature analysis, for reducing

Fig. 7.4 Search term Covid 19, covid-19 and Coronavirus in Google trends

the impact of COVID-19. However, in the twenty-first century, artificial intelligence is helping with the war with COVID-19 in different aspects of accelerating testing, drug discovery and social monitoring [15]. Technologies like drone [16] and 3D printing [17] are contributing to the fight against the pandemic.

7.3 Data Preparation: COVID-19, Infrastructure, Human Resource, State Population Data

The battle against corona has raged on for a few months, and its unwieldy spread in some regions may affect the entire generation. The entire society is set to follow new protocols for daily human life and various business processes by adopting non-technical or technical solutions to reform themselves. COVID-19 has made a drastic and dramatic change to the way the world operates. This change has restructured the political-economic, social, technological, legal and environmental (PESTEL) framework across boundaries. This restructuring was a basic need to be stringently followed by public, private and socials sectors. With the colossal entry of the coronavirus in each country, strategies to deal with it are being decided at medical, operational, emotional and technical levels. Government officials are taking measures to fight the virus depending on the sensitivity, spread, scope and speed of transmission in the respective country. The government officials across countries, medical organizations and the World Health Organization have planned various PESTEL strategies to overcome the problem. Every nation will undergo five stages of this 5R pandemic journey, going from being an infected country to a 'normal one'. These five stages are: resolve, resilience, return, reimagination and reform.

At this stage, it is critical to understand how the coronavirus is progressing and spreading in various countries and what healthcare facilities are available to

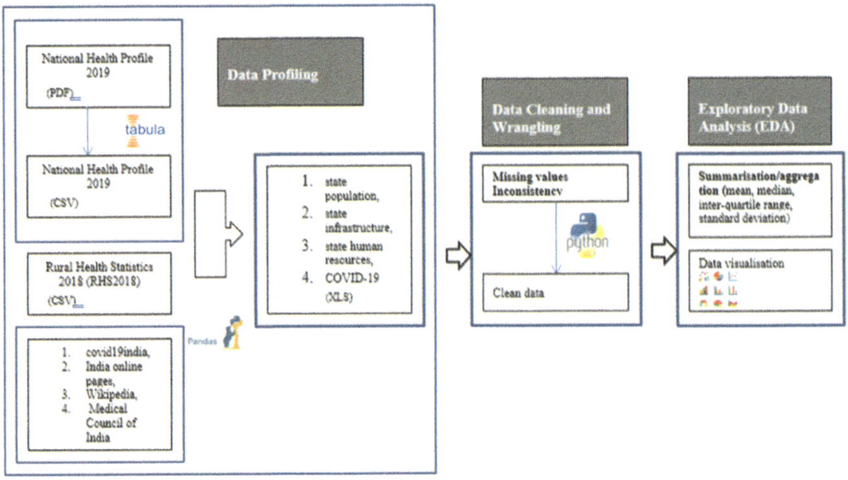

Fig. 7.5 Systematic presentation of the of data collection, pre-processing and analysis

manage this crisis—it is essentially a question of demand and supply. Both sides of this economic equation must be thoroughly analysed so that people understand the damage being caused by the virus rampaging through the world, and what can be done to limit it. Figure 7.5 depicts a step-wise representation of the entire journey of data preparation, from data collection, pre-processing, to analysis. This explains the methodology in a manner that is easy to interpret and understand, before a deeper dive into each step is taken.

7.3.1 Data Source Identification and Data Acquisition

Beginning with the demand side of the equation, it is important to understand the effects of the coronavirus and how it has led to significant increase in the demand for healthcare facilities. The patients who test positive for coronavirus can be either symptomatic or asymptomatic. It is extremely difficult to track the progression of the virus when the patients are asymptomatic—most do not even know that they have contracted the virus. They can only be traced as contacts of symptomatic patients—this is what makes the virus so dangerous. A person carrying the virus may not show any symptoms but can spread it effectively nevertheless. The incubation period for the virus is also quite long—approximately 5–6 days. This means that an individual who has contracted the virus will, on average, take 5–6 days to display any symptoms. Therefore, in the initial 5–6 days prior to showing any symptoms, this individual is perfectly capable of infecting individuals around him.

It is also worth noting that the test to detect coronavirus in an individual takes 1–2 days to show results. Therefore, a person who gets tested on a particular day might catch an infection after taking the test and spread the virus before his test result returns as negative. All these possibilities make this extremely contagious virus very hard to contain.

Symptoms of the virus include coughing, chest pain, loss of smell and taste, etc. and can even lead to severe lung damage and death. This virus primarily spreads by entering the body through the respiratory tract (nose or mouth) and attacks the respiratory system. Since this virus was only discovered in December 2019, its effects are yet to be fully understood. But, through the increasing number of infections, it has been understood that many cases are asymptomatic, while patients who are advanced in age and patients with comorbidities (such as heart disease, diabetes, obesity, etc.) are more susceptible to the virus. This is the demand side of the equation. To quantify this, data has been collected on the daily infections added each day, the total number of infections, the number of people cured and the number of deaths occurring each day (for India). This data has also been segregated for each state. This process of collecting relevant data for our study is known as 'data acquisition'. Similarly, data must also be collected for the supply side.

Now, not only is the virus hard to contain, but it is also dangerous to those that catch it. Asymptomatic and mildly symptomatic cases are quarantined (at home, or at a government facility) as soon as they are detected. But for more serious cases, there is a larger demand for hospital beds, and for ventilators. Isolation wards have been set up in hospitals to prevent the spread of the infection within the hospital. With increasing infections, more doctors are required to treat more patients, more health workers to assist these doctors. Dispensaries would need to be well stocked and functioning to provide medicines to those quarantined at home. Blood banks would also have to be ready to supply patients that require blood. Figure 7.6 presents the hierarchy of the healthcare system in India.

In the case that there is a shortage of doctors and nurses, medical students may be required to assist with the growing number of cases. Thus, it is important to know how many medical students are there in India—or more specifically, students enrolled in medical post-graduate courses since they are most likely to be called up for assistance. Similarly, if there is a shortage of hospitals and hospital beds, medical colleges can be used to treat the patients. Therefore, data for the healthcare infrastructure—hospitals, hospital beds, blood banks, dispensaries—and data for the health human resources—MCI registered doctors, health workers and assistants, post-graduate students employed in various courses, pharmacists, etc. were collected. This data was also collected at the state level and represents the supply side of the equation. Data on the estimates for populations of each state of India in 2020 was also collected, in order to analyse the proportions of the population that have been affected in different ways by the COVID-19. Since the census was last conducted 9 years ago in 2011, this data might be outdated and provide biased figures. Thus, population estimates for 2020 were used.

The data used in this study is secondary data. The data has been collected from various sources and collated together. This data was readily available online. The various sources that were explored to collect the data are given below:

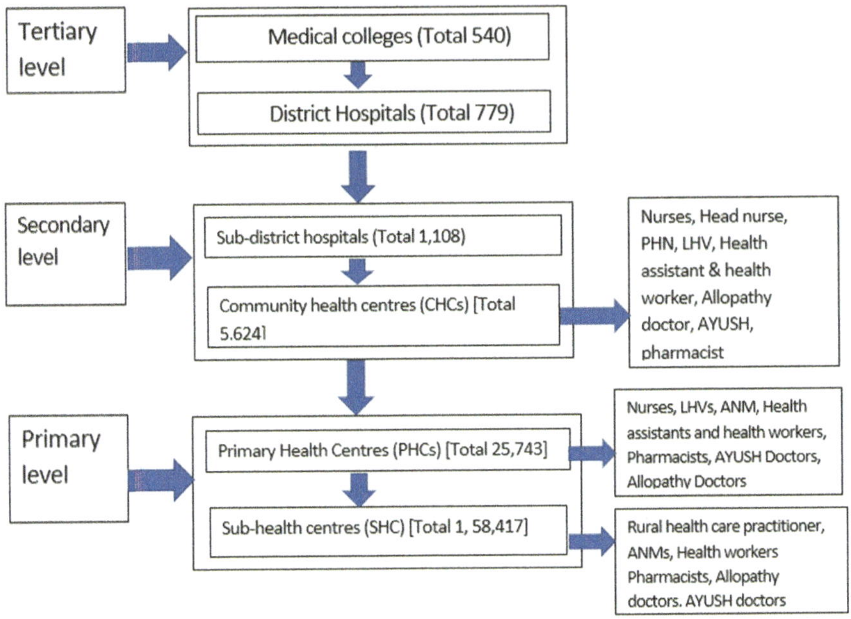

Fig. 7.6 Systematic presentation of the Indian healthcare system

1. COVID-19 data
   (a) www.covid19india.org
   (b) www.wikipedia.org
   (c) www.kaggle.com/sudalairajkumar/covid19-in-india
   (d) www.mohfw.gov.in
   (e) www.worldometer.com.
2. Infrastructure data
   (a) National Health Profile 2019
   (b) Rural Health Statistics 2018
   (c) www.data.gov.in.
3. Human Resources data
   (a) National Health Profile 2019
   (b) Rural Health Statistics 2018
   (c) www.data.gov.in
   (d) Medical Council of India—www.mciindia.org.

4. State population data
   (a) Census 2011
   (b) For 2020 estimates—http://www.indiaonlinepages.com/population/india-current-population.html.

The data used is at most 3 years old, with the oldest data coming from 2017. Relevant data for each state was collected—population estimates for 2020, MCI registered doctors, nurses and medical students in different post-graduate courses, primary healthcare centres, community healthcare centres, government hospitals and beds in rural and urban areas, etc. It is an extremely comprehensive dataset containing maximum information about healthcare systems of each state—both in terms of infrastructure and human resources.

7.3.2 Data Profiling: COVID-19, Infrastructure, Human Resource, State Population Data

Data profiling is the process of collating the data collected from the various sources in the previous step. Since the data was collected from so many different sources, it had to be collated together in one place for ease of use and analysis purposes. The followings steps were followed to prepare the data in desired format:

1. The data from the National Health Profile 2019 (NHP2019) was in the form of tables within a PDF file. In order to extract this data, the software Tabula was used. Tabula reads tables from a PDF file and converts them to CSV text files which can be used in Excel for analysis.
2. Data from Rural Health Statistics 2018 (RHS2018) was available from data.gov.in in the form of CSV files.
3. Data from various websites such as covid19india, India online pages, Wikipedia, Medical Council of India, etc. were copied from the respective websites into Excel sheets directly.

7.3.2.1 COVID-19 Data Profiling

Using the powerful Pandas library in Python, these data are combined from the different sources. For the COVID-19 dataset, the main functions used were joining and concatenating functions—which are used to combine the different data frames appropriately. For example, the data for cured persons from a state, say Maharashtra, must only be combined with data for infected persons from Maharashtra. Pandas enabled us to do this effortlessly without having to manually join each row and column of the data frame. This resulted in a final dataset where each row was a date, and contained information about the confirmed, cured and death cases recorded in a state on that day. Aggregate measures could now easily be used to compare data across states and could be filtered for dates. This was now a time-series data. Time-series data refers to information gathered on several factors and organized by dates. This enables to not only analyse change in variables over time, but also use various time-series techniques to build prediction models to forecast change in demand side in future.

7.3.2.2 Healthcare Infrastructure and Human Resource Data Profiling

Similarly, Pandas was also used to prepare the healthcare system. The data from the different sources was available for each state. Therefore, using repeated left joins and merges, the data was joined. A left join is used to combine two datasets, and it keeps all the data from the first dataset while adding data from the second dataset wherever there are matches in the specified column (in this case, the State/UT). In order to keep the datasets more uniform, the healthcare infrastructure and healthcare human resources were built as two separate datasets. In the healthcare infrastructure dataset, each row referred to a state, and contained information on the hospitals, hospital beds, medical colleges, blood banks, dispensaries and so on. Similarly, the human resources dataset was built with each row referring to a state and containing information on MCI registered doctors, health workers and assistants, medical college students, pharmacists, medical college intake capacities, etc.

7.3.2.3 State Population Data Profiling

The fourth and final dataset consisted of the different state population estimates for 2020. This dataset was good to go from the start and did not require any changes.

This process resulted in four Excel sheets—one each for state population, health infrastructure, health human resources and COVID-19 data. The variables in the final datasets are explained in Table 7.1 for the convenience of understanding.

7.3.3 *Data Cleaning and Wrangling for Analysis*

The data that was collected and collated was not appropriate and ready for analysis. This data consisted of many missing values and inconsistent entries. Missing values refer to a cell in the dataset for which there is no data available and which could happen due to several reasons. There are three types of missing data that can normally be classified into—missing completely at random (MCAR), missing at random (MAR) and missing not at random (MNAR). These can be explained as follows:

i. MCAR refers to data that is missing completely at random and there is no logical reasoning as to why the data is missing. Therefore, this kind of data can

Table 7.1 Describing the variables of the datasets

S. No.	Variable	Domain
1.	State/UT	All 28 states and 8 union territories of India
2.	Population (Census 2011)	Population of each state according to the Census 2011
3.	Population (2020 Estimates)	Estimates of the population of each state for the year 2020
4.	MCI registered doctors	Number of doctors registered with the Medical Council of India
5.	Number of government allopathic doctors	Number of allopathic doctors appointed by government
6.	Rural doctors	Two columns with number of doctors at primary healthcare centres and number of doctors at community healthcare centres
7.	Rural health workers	Four columns with numbers of health workers—male, health workers—female, health assistants—male, health assistants—female
8.	AYUSH doctors	Number of AYUSH doctors working in each state in different fields—Unani, Ayurveda, Siddha, Naturopathy and Homoeopathy
9.	Total number of registered nurses	Number of registered nurses in each state including auxiliary nursing midwifery (ANM), registered nurse (RN), registered midwife (RM) and lady health visitor (LHV).
10.	Pharmacists	Number of pharmacists in each state
11.	Human resource in Employees State Insurance (ESI) corporation	Columns with number of medical officers, specialists, super specialists, dental surgeon, total number of nurses (RN and RM) and pharmacists in the ESI corporation
12.	DM/MCH courses	Several columns containing the number of medical students in different super specializations courses like Doctorate of Medicine (DM) and Master of Chirurgiae (MCH) courses Note: 1 column for each specialization
13.	Diploma courses	Several columns containing the number of medical students in different specializations of post-graduate diploma courses Note: 1 column for each specialization

(continued)

Table 7.1 (continued)

S. No.	Variable	Domain
14.	MD/MS courses	Several columns containing the number of medical students in different specializations of post-graduate Doctor of Medicine (MD)/Master of Surgery (MS) courses Note: 1 column for each specialization
15.	AYUSH college admission capacity	Several columns containing the admission capacities of medical colleges in different specializations of post-graduate AYUSH courses Note: 1 column for each specialization
16.	Government hospitals and beds	Six columns detailing the number of rural government hospitals, rural hospital beds, urban government hospitals, urban hospital beds, total government hospitals and total hospital beds in each state
17.	Sub-centres, PHCs, CHCs	Five columns detailing the number of sub-centres, primary healthcare centres (PHC), community healthcare centres (CHC), sub-divisional hospitals and district hospitals in each state
18.	Infrastructure at ESI corporation	Three columns containing the no. of dispensaries, hospitals and beds in the Employees State Insurance (ESI) corporation in each state
19.	AYUSH hospitals and dispensaries	Several columns detailing the number of Ayurveda, Unani, Siddha, Yoga, Naturopathy, Homoeopathy, Sowa-rigpa and total hospitals and dispensaries
20.	Medical colleges	Two columns detailing the number of medical colleges in each state and the corresponding number of beds in the attached hospitals of the medical colleges
21.	AYUSH colleges	Six columns detailing the number of Ayurveda, Unani, Siddha, Yoga, Naturopathy, Homoeopathy and Total AYUSH colleges in each state
22.	AYUSH post-graduation colleges	Several columns detailing the number of Ayurveda, Unani, Siddha, Yoga, Naturopathy, Homoeopathy and total AYUSH post-graduation colleges and their corresponding admission capacities in each state
23.	Licensed blood bank	Three columns detailing the number of private, public and total blood banks in each state

(continued)

Table 7.1 (continued)

S. No.	Variable	Domain
24.	Confirmed	Number of total confirmed COVID-19 cases in each state
25.	Active	Number of active COVID-19 cases
26.	Recovered	Number of cured/discharged/migrated cases in each state
27.	Deaths	Number of deaths due to COVID-19 in each state
28.	Date[a]	The date, starting from differing dates for each state up till the current date
29.	Confirmed[a]	Cumulative number of confirmed COVID-19 cases corresponding to the date in the date column
30.	Deaths[a]	Cumulative number of deaths due to COVID-19 cases corresponding to the date in the date column
31.	Recovered[a]	Cumulative number of recovered COVID-19 cases corresponding to the date in the date column
32.	Total samples tested[a]	Cumulative number of samples tested for COVID-19 corresponding to the date in the date column
33.	Positive[a]	Cumulative number of samples that have tested positive for COVID-19 corresponding to the date in the date column
34.	Negative[a]	*Cumulative number of samples that have tested negative for* COVID-19 corresponding to the date in the date column
35.	State/UT[a]	The state/union territory for which the above data has been collected corresponding to the date in the date column

[a]Available only in the COVID-19 dataset which is updated daily

be imputed—the process of replacing missing values with estimates obtained from various statistical techniques.

ii. MAR refers to data where observations on a variable are missing, and this absence is related to another variable. For example, suppose there is a data that is dependent on temperature and pressure, and it is realized that there are missing values for pressure. On further examination, it is observed that the pressure measurements are missing when the temperature is very low. Therefore, this missingness in pressure has occurred because the instrument used to record the pressure values malfunctions at lower temperatures. Data imputation techniques can be used here in some cases.

iii. MNAR refers to data that is missing due to some logical reason. For example, if there are attributes to record temperature and pressure data, and the temperature data is missing because the thermometer used to record the temperature malfunctioning, then this data is not missing at random. Data imputation techniques cannot be used in this case.

When it came to missing data in the COVID-19 dataset, the only missing values arose due to the different dates on which the virus was detected in each state. For example, the first coronavirus case was detected in Delhi on 2 March 2020. But the first case in Gujarat was only detected on 20 March 2020. Therefore, for all preceding dates for Gujarat, the data was technically missing. This is a case of data missing at random—data for Gujarat is missing because of data recorded for Delhi. However, since there were no cases detected before 20 March, these missing values could be substituted with 0.

For infrastructure and human resource datasets, there were many missing data. But the data was missing because it was simply not recorded by the investigators—making this a case of data missing not at random. Therefore, no imputation techniques could be used here. Cases with missing data were simply excluded from the final analysis from these two datasets.

Regarding the state population data, there were no missing values. However, in the COVID-19 dataset, Daman & Diu, and Dadra & Nagar Haveli, two very small union territories of India were combined into a single row. Therefore, the population estimates for the two union territories were also combined for analytical purposes.

Inconsistent entries refer to data values that do not make sense or violate some logical conditions. For example, a state cannot have more positive COVID-19 tests on a given date than the number of samples tested on that day. This needed to be corrected before an analysis of the different datasets could be performed and a predictive model for COVID-19 cases could be built. Using Python, the data was cleaned and made ready for analysis. For cases where there were such inconsistencies in the infrastructure and human resources data, the data was simply considered to be missing instead.

7.4 Exploratory Data Analysis (EDA)

Once the data is cleaned and ready for analysis, it is explored using Python. Exploratory data analysis is conducted to get a sense of the data one is working with. In this step, different aspects of the data are summarized using descriptive statistics such as calculating the mean, median, inter-quartile range and standard deviation for the quantitative variables. Data is also visualized to see how the different variables are distributed individually, as well as to see how different variables are related with each other. These techniques give us better insights into the dataset, resulting in a more efficient analysis and a better model.

The initial exploratory data analysis is conducted as follows, using Jupyter notebook which is powered by Python:

1. First step is to import the required libraries, using the following code:

   ```
   import numpy as np
   import pandas as pd
   import matplotlib.pyplot as plt
   import seaborn as sns.
   ```

   Numpy and Pandas are used for data importing and manipulation, while Matplotlib.pyplot and Seaborn are used for data visualization. Data visualization is the process of using graphs and images of variables to better understand our data and it is much easier to interpret than plain numbers.
2. The different datasets are imported with some initial manipulation to make it easier to write the Python code.

   ```
   infra = pd.read_csv('State Infrastructure.csv').set_index('State/UT)
   hr = pd.read_csv('State Human Resources.csv').set_index('State/UT)
   covid = pd.read_csv('COVID-19 Cases—17th April.csv').set_index('State/UT)
   pop = pd.read_csv('State Population.csv').set_index('State/UT).sort_index()
   ```

   Here, the Excel sheets (in the form of text files) were imported into Python in the form of Pandas data frames. The snapshot of datasets is mentioned below:

   (a) Dataset of healthcare infrastructure is depicted in Fig. 7.7.
   (b) Dataset of healthcare human resource is presented in Fig. 7.8.
   (c) Dataset of COVID-19 data (aggregated for each state) is shown in Fig. 7.9.
   (d) Dataset of state population data is demonstrated in Fig. 7.10.

```
#A quick look at the datasets
display(infra.head(), hr.head(), covid.head(),pop.head())
```

State/UT	Rural Hospitals (Govt.) no.	Rural hospitals (Govt.) beds	Urban Hospitals (Govt.) no.	Urban Hospitals (Govt.) beds	Total Hospitals (Govt.) no.	Total Hospitals (Govt.) beds	Sub Centres	PHCs	CHCs	Ayurveda Hospitals	...	Naturopathy Hospitals	Naturo Dispens
Andaman & Nicobar Islands	27	575	3	500	30	1075	123	22	4	6	...	0	
Andhra Pradesh	193	6480	65	16658	258	23138	7458	1147	193	3	...	0	
Arunachal Pradesh	208	2136	10	268	218	2404	312	143	63	10	...	0	
Assam	1176	10944	50	6198	1226	17142	4644	946	172	1	...	0	
Bihar	1032	5510	115	6154	1147	11664	9949	1899	150	5	...	0	

5 rows × 27 columns

Fig. 7.7 Snapshot of healthcare infrastructure dataset

State/UT	Anaesthesiology	Aviation Medicine/ Aerospace Medicine	Bio-Chemistry	Dermatology, Venereology & Leprosy	Forensic Medicine/ Forensic Medicine & Toxicology	General Medicine	Hospital Administration & Community Health Administration	Immuno Hematology & Blood Transfusion	Microbiology	Pa
Andaman & Nicobar Islands	57.0	NaN	3.0	13.0	4.0	65.0	NaN	5.0	14.0	
Andhra Pradesh	148.0	NaN	42.0	53.0	22.0	247.0	6.0	1.0	62.0	
Arunachal Pradesh	NaN	NaN	NaN	NaN	NaN	NaN	NaN	NaN	NaN	
Assam	39.0	NaN	16.0	11.0	10.0	75.0	NaN	NaN	18.0	
Bihar	45.0	NaN	19.0	14.0	8.0	79.0	NaN	NaN	32.0	

5 rows × 70 columns

Fig. 7.8 Snapshot of healthcare human resource dataset

State/UT	Total Confirmed cases (Including 76 foreign Nationals)	Cured/Discharged/Migrated	Death
Andaman & Nicobar Islands	11	10	0
Andhra Pradesh	572	36	14
Arunachal Pradesh	1	0	0
Assam	35	5	1
Bihar	83	37	1

Fig. 7.9 Snapshot of COVID-19 dataset

State/UT	Population (2020 estimates)
Andaman & Nicobar Islands	379944
Andhra Pradesh	53206421
Arunachal Pradesh	1382611
Assam	31169272
Bihar	126750326

Fig. 7.10 Snapshot of state population dataset

3. Some quick transformations were done, like adding the populations to each of the individual datasets to make it easier to write codes for visualization.

```
infra = pd.merge(infra, pop, on = 'State/UT', how = 'left')
hr = pd.merge(hr, pop, on = 'State/UT', how = 'left')
covid = pd.merge(covid, pop, on = 'State/UT', how = 'left')
```

4. Missing values were checked for in each dataset and the same is presented in Fig. 7.11.
5. Application of basic statistics to compute quick summary for each dataset.
   (a) Healthcare Infrastructure:
   It is evident from Fig. 7.12 that the mean and median are quite far apart for all the columns and this suggests that the data contains outliers. Hence, it is advisable to use median in this case.
   (b) Healthcare Human Resource:
   Figure 7.13 presents the statistics of human resource data that includes the number of PG students in different courses and specialities, doctors, health assistants and health workers.
   (c) COVID-19 data:
   The mean and median are far apart as shown in Fig. 7.14, which suggests there are outliers. This is indeed true—as Maharashtra state has almost 3000 cases, whereas some states are yet to report cases. Therefore, it is better to use the median in this case.

```
Rural Hospitals (Govt.) no.          0
Rural hospitals (Govt.) beds         0
Urban Hospitals (Govt.) no.          0
Urban Hospitals (Govt.) beds         0
Total Hospitals (Govt.) no.          0
Total Hospitals (Govt.) beds         0
Sub Centres                          0
PHCs                                 0
CHCs                                 0
Ayurveda Hospitals                   0
Ayurveda Dispensaries                0
Unani Hospitals                      0
Unani Dispensaries                   0
Siddha Hospitals                     0
Siddha Dispensaries                  0
Yoga Hospitals                       0
Yoga Dispensaries                    0
Naturopathy Hospitals                0
Naturopathy Dispensaries             0
Homeopathy Hospitals                 0
Homeopathy Dispensaries              0
Sowa-Rigpa. Hospitals                0
Sowa-Rigpa. Dispensaries             0
Total Hospitals                      0
Total Dispensaries                   0
No. of medical colleges              0
No. of beds in attached hospital     8
Population (2020 estimates)          0
dtype: int64
```

```
Anaesthesiology                                          8
Aviation Medicine/ Aerospace Medicine                   35
Bio-Chemistry                                            9
Dermatology, Venereology & Leprosy                      11
Forensic Medicine/ Forensic Medicine & Toxicology       10
                                                        ..
District Hospital - Sanctioned                           5
District Hospital - In Position                          1
Sub District/ Sub Divisional Hospital - Sanctioned       7
Sub District/ Sub Divisional Hospital - In Position      4
Population (2020 estimates)                              0
Length: 71, dtype: int64
```

```
Total Confirmed cases (Including 76 foreign Nationals)    0
Cured/Discharged/Migrated                                 0
Death                                                     0
Population (2020 estimates)                               0
infection rate                                            0
dtype: int64
```

Fig. 7.11 Snapshot of state datasets

	Rural Hospitals (Govt.) no.	Rural hospitals (Govt.) beds	Urban Hospitals (Govt.) no.	Urban Hospitals (Govt.) beds	Total Hospitals (Govt.) no.	Total Hospitals (Govt.) beds	Sub Centres	PHCs	CHCs	Ay Ho
count	36.000000	36.000000	36.000000	36.000000	36.000000	36.000000	36.000000	36.000000	36.000000	36.
mean	594.527778	7368.750000	121.527778	12464.194444	716.055556	19832.944444	4400.472222	715.083333	156.222222	88.
std	900.548991	9775.545864	170.074351	15429.825672	998.043341	23754.460661	4997.461707	842.327876	186.929339	353.
min	0.000000	0.000000	0.000000	0.000000	5.000000	240.000000	12.000000	0.000000	0.000000	0.
25%	23.750000	623.500000	13.250000	1559.500000	41.250000	2860.000000	287.500000	49.000000	7.750000	1.
50%	240.500000	5290.000000	54.500000	5649.000000	348.000000	11012.000000	2769.500000	400.000000	91.000000	3.
75%	729.250000	10161.000000	139.000000	17778.250000	934.000000	23449.250000	7771.500000	1153.000000	211.250000	16.
max	4442.000000	40179.000000	760.000000	58882.000000	4635.000000	78566.000000	20521.000000	3621.000000	822.000000	2104.

8 rows × 28 columns

Fig. 7.12 Simple statistics on healthcare infrastructure dataset

	Anaesthesiology	Aviation Medicine/ Aerospace Medicine	Bio-Chemistry	Dermatology, Venereology & Leprosy	Forensic Medicine/ Forensic Medicine & Toxicology	General Medicine	Hospital Administration & Community Health Administration	Immuno Hematology & Blood Transfusion	Microbiology	Pae
count	28.000000	1.0	27.000000	25.000000	26.000000	29.000000	6.000000	14.000000	28.000000	28.
mean	94.428571	17.0	24.111111	30.640000	16.730769	116.413793	8.166667	4.785714	33.321429	65.
std	88.324917	NaN	22.616082	27.517994	15.240886	114.710666	2.714160	3.238674	32.983782	63.
min	4.000000	17.0	2.000000	3.000000	2.000000	2.000000	4.000000	1.000000	2.000000	1.
25%	35.500000	17.0	5.500000	11.000000	5.250000	30.000000	6.500000	2.000000	7.750000	17.
50%	58.500000	17.0	19.000000	21.000000	13.500000	75.000000	9.000000	4.000000	23.000000	43.
75%	155.250000	17.0	34.000000	40.000000	20.750000	181.000000	10.000000	5.750000	47.250000	99.
max	317.000000	17.0	100.000000	111.000000	72.000000	405.000000	11.000000	12.000000	124.000000	231.

8 rows × 62 columns

The data represents the number of PG students in different courses and specialities, doctors, health assistants and health workers.

Fig. 7.13 Simple statistics on healthcare human resource dataset

	Total Confirmed cases (Including 76 foreign Nationals)	Cured/Discharged/Migrated	Death	Population (2020 estimates)	infection rate
count	32.000000	32.000000	32.000000	3.200000e+01	32.000000
mean	431.781250	54.781250	14.125000	4.307182e+07	0.000010
std	690.172477	76.863404	35.438042	5.192679e+07	0.000015
min	0.000000	0.000000	0.000000	3.799440e+05	0.000000
25%	10.500000	4.000000	0.000000	2.903444e+06	0.000001
50%	71.500000	25.000000	1.500000	3.048865e+07	0.000005
75%	614.750000	67.250000	13.250000	7.072726e+07	0.000014
max	3205.000000	300.000000	194.000000	2.370950e+08	0.000081

Fig. 7.14 Simple statistics on COVID-19 dataset

6. Finally, some visualizations were created to help understand the data better. From the above plot in Fig. 7.15, it is quite evident that the smaller states and union territories are doing well but the larger and more densely populated states such as Maharashtra and Uttar Pradesh are in trouble.

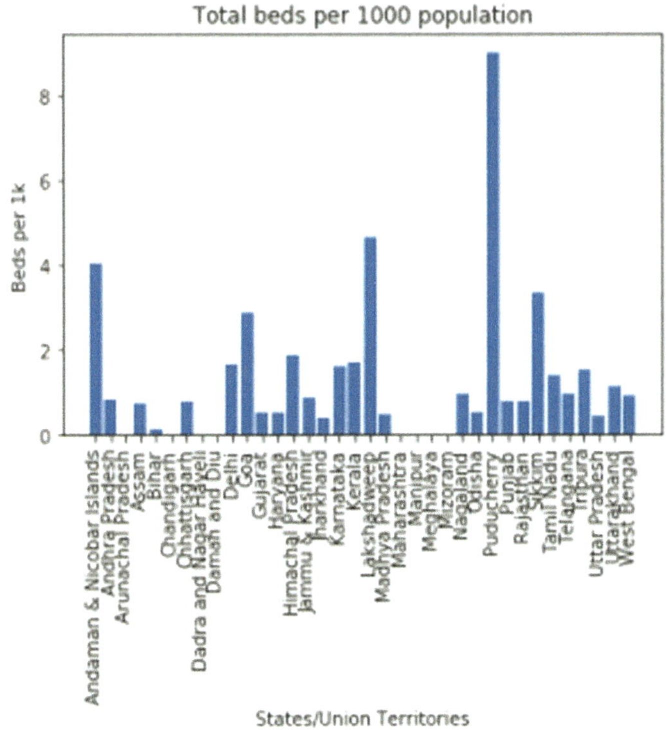

Fig. 7.15 Bar chart of total beds per 1000 population

Infection rates for every state are still very low (less than 0.001% for each state) as depicted in Fig. 7.16. However, Delhi has the highest infection rate.

From Fig. 7.17, it is evident that in some states and union territories, the infected per 1000 population and the doctors per 1000 population are already the same, and this could mean trouble for these states if the infections were to keep rising.

Thus, the exploration of the data enables to have a better understanding of the information that was collected on the supply and demand side of this delicately balanced pandemic equation. This EDA was conducted in July 2020, and therefore, figures might appear out of date at the time of reading.

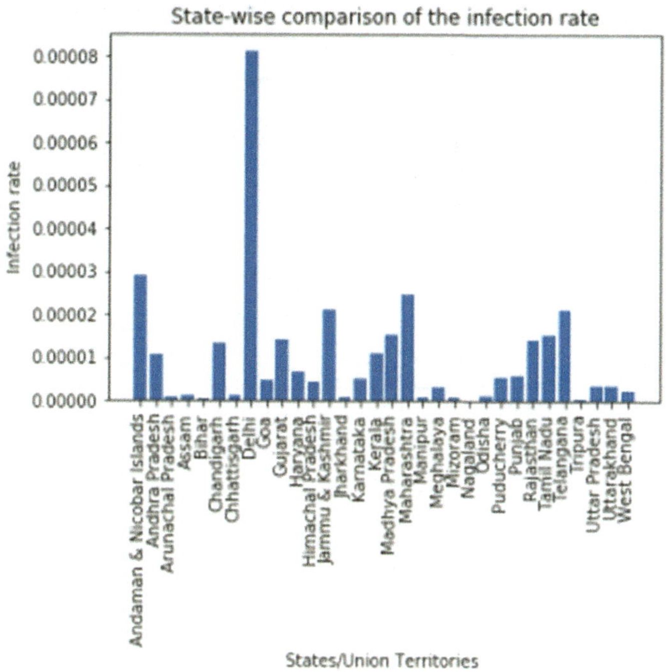

Fig. 7.16 Bar chart of COVID-19 infection rate

7.5 Conclusion

The coronavirus has clearly emerged to be an international community health problem. Because of the large scale community spread, countries around the globe need to increase their attention towards their health operating systems and scale up hospital infrastructures and testing centres. All of the political leaders have closed international flights and sealed their borders and unlike the past scientists have created a global association. The researchers all across the world have halted their own researches and have united on one single topic simultaneously. The whole world seeks for a solution that can eradicate the virus and save people and the best way to achieve it is to collaborate. All have to stand by each other, support the system and maintain all norms essential for the survival like wearing a mask, washing hands and social distancing. The world can become virus-free one day if we all fight against it together, support doctors, nurses and other people in the frontline and be smart enough to follow the rules and regulations issued by WHO and other health organizations.

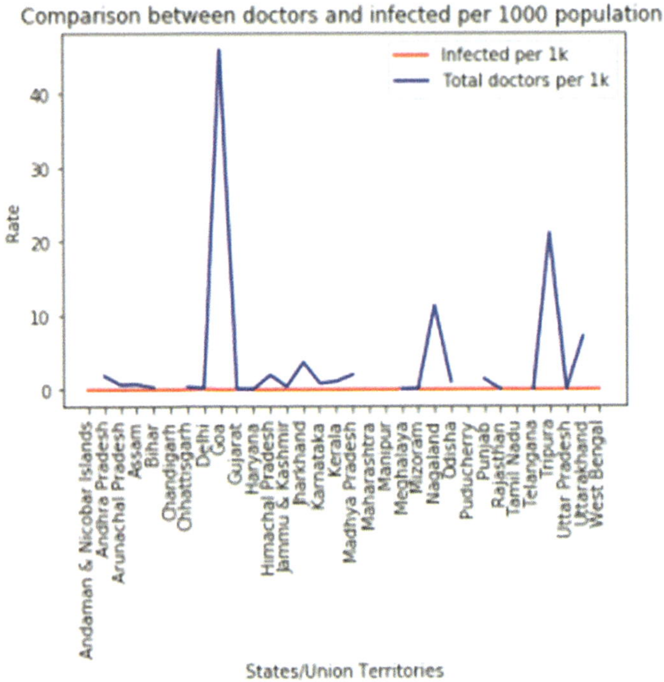

Fig. 7.17 Line charts of infection rate and total doctors per 1000 population

References

1. https://www.cdc.gov/flu/pandemic-resources/reconstruction-1918-virus.html
2. https://www.cdc.gov/flu/pandemic-resources/basics/index.html
3. https://www.euro.who.int/en/health-topics/health-emergencies/coronavirus-covid-19/news/news/2020/3/who-announces-covid-19-outbreak-a-pandemic
4. https://www.msn.com/en-in/news/other/covid-19-peak-in-india-may-arrive-mid-nov-paucity-of-icu-beds-ventilators-likely-study/ar-BB15ssZE?ocid=XMMO
5. https://www-indiatoday-in.cdn.ampproject.org/v/s/www.indiatoday.in/amp/india/story/india-records-highest-spike-of-14516-new-coronavirus-cases-in-24-hours-caseload-nears-4-lakh-1690875-2020-06-20
6. https://www.abc.net.au/news/2020-03-26/coronavirus-covid19-global-spread-data-explained/12089028
7. https://www.bbc.com/future/article/20200325-covid-19-the-history-of-pandemics
8. https://www.who.int/csr/don/archive/year/en/
9. Smith D, Moore L (2014) The SIR model for spread of disease -the differential equation model. https://www.maa.org/press/periodicals/loci/joma/the-sir-model-for-spread-of-disease-the-differential-equation-model
10. Murray CJL (2020) Forecasting COVID-19 impact on hospital bed-days, ICU-days, ventilator days and deaths by US state in the next 4 months. http://www.healthdata.org/research-article/forecasting-covid-19-impact-hospital-bed-days-icu-days-ventilator-days-and-deaths
11. “India fights Corona” by rubiscape.2020.[PDF file]

12. Moore M, Gelfeld B, Okunogbe A, Paul C (2017) Identifying future disease hot spots: infectious disease vulnerability index. https://www.semanticscholar.org/paper/Identifying-Future-Disease-Hot-Spots%3A-Infectious-Moore-Gelfeld/d4c3a71ca19a28c9a5fc39f57cac46a445c73604
13. https://ddi.sutd.edu.sg/
14. https://www.linkedin.com/pulse/covid-19-district-vulnerability-index-mitul-jhaveri/
15. https://arxiv.org/abs/2003.11336
16. https://enterprise.dji.com/news/detail/fight-covid-19-with-drones
17. https://formlabs.com/covid-19-response/

Chapter 8
COVID-19 and Indian Healthcare System—A Race Against Time

8.1 Introduction

The novel coronavirus (or COVID-19) has impacted the entire world. This disease is highly contagious and caused by the severe acute respiratory syndrome coronavirus 2 (SARS-CoV-2). It was first identified in December 2019, in Wuhan, China. However, the first confirmed case was recorded much earlier on 17th November, but it was believed to be a case of pneumonia at the time. Its common symptoms are dry cough, shortness of breath, fatigue, high fever and loss of smell and taste. Most of the cases result in mild symptoms, but some may progress to more severe situations. The incubation period for this disease is normally 5 days, but it can range from anywhere between 2 and 14 days. A major way that this virus is spread is through close contact via respiratory droplets—thus, social distancing is used as a preventative measure. At the time of writing, there are no vaccines available to cure or prevent the disease.

Most nations have imposed lockdowns in their respective jurisdictions, restricting the movement of citizens with the hope that this will curb the spread of the disease and advising social distancing at every step of the way. This has brought the global economy to a stand-still. Confirmed cases of COVID-19 are rising steadily across the globe and in India as well. India currently has the third highest number of cases in the world.

COVID-19 cases are growing steadily in India, and the number of cases added is rising every day. Scientists, researchers and epidemiologists are hard at work as they look for a cure or vaccine for the disease. But in the meantime, several preventative measures have been suggested—these include social distancing, regular hand-washing for at least twenty seconds at a time, having an alcohol-based sanitizer on hand at all times, staying at home as far as possible and always wearing a mask when outdoors. However, new infections are still on the rise which has placed a large stress on the Indian healthcare system. States have set up several isolation centres

N. Sharma et al., *Open Data for Sustainable Community*,
Advances in Sustainability Science and Technology,
https://doi.org/10.1007/978-981-33-4312-2_8

across the country for patients who have tested positive and their high-risk contacts. For cases with more severe symptoms, hospitalization is required and sometimes, the use of a ventilator is necessary for survival. Private health care can prove to be expensive for poor households in India and using government healthcare services is the only option available to them. At the time of writing, all positive cases are referred to government hospitals, but testing for the virus is being carried out in both government and private laboratories. Therefore, analysing the current healthcare infrastructure of India is not a futile exercise and will help point out areas that need significant improvements.

Doctors and health professionals are at the forefront in the fight against COVID-19. Their efforts and spirits are being wildly acclaimed at all levels. The skills and spirit of service among these professionals' places them in a unique position to save people from this disease. The capacities of human resources including the medical manpower who help patients in hospital, as well as non-medical personnel, front line workers who may be involved in non-medical duties such as logistics and surveillance are built up to manage the COVID-19 pandemic in the best possible way. They are also provided with possible role assignments and proper training requirements. In India, human resources can be drawn from AYUSH doctors to man the care-centres. The Ministry of AYUSH has been carrying out training sessions for this purpose. An allopathic doctor may be used as a guide in the process. Therefore, another aspect of analysing the current state of India during this pandemic is to have a look at the human resources available to fight this pandemic.

This chapter aims to analyse the current situation and the progression of COVID-19 in India, the situation in the most affected states of the country, and compare the situation in India to that of the most affected countries in the world such as the USA, Brazil, Russia, Spain and the UK. A deeper look is then taken at the state of the healthcare system in India at present. In this paper, an analysis is conducted of the healthcare infrastructure—such as hospitals, available beds, medical colleges and dispensaries—and human resources—such as doctors, nurses and medical students.

The primary objectives of this paper are:

1. To assess how well India's healthcare system is equipped to handle a global pandemic like COVID-19
2. To analyse the current situation and the progression of COVID-19 in India
3. To compare the situation in India to that of the most affected countries in the world.

8.2 Material Methods Used for Analysis

The data collected for COVID-19 was mostly secondary and readily available online. The data was quantitative in nature, including number of confirmed cases in India, in different states and in different nations. It also included the number of deaths, recoveries, active cases, dates, tests sampled, etc. Much of the data for different nations had been compiled from various national government websites on Kaggle, although the most recent data was not available for some of the nations, such as Spain.

For the medical infrastructure, after many layers of net searching, information about the private and government hospitals along with the number of beds attached to each hospital of various states across India was gathered. The databases were collated and matched with the respective states using Python and cross-checked several times. Most of the information was collected from the National Health Profile 2019 [1]. The database contains details about government beds and hospitals, rural hospitals, Employees State Insurance (ESI) Corporation, medical colleges and Ayurveda, Yoga and Naturopathy, Unani, Siddha and Homeopathy (AYUSH) hospitals and colleges, along with details of dispensaries and blood banks. Additionally, population of all the states in the year 2019 [2] and the state-wise distribution of red, orange and green zones [3] are considered. Total hospitals in each state were calculated by adding all the government, rural, ESI, AYUSH and medical hospitals and the total number of beds in each state by adding all the government, ESI and medical beds.

The human resources dataset consists of the state-wise details of Medical Council of India (MCI) registered [4], allopathic, rural and AYUSH doctors and ESI medical officers. It also consists of the details of nurses, health workers and pharmacists. The admission capacity of some AYUSH colleges is stated. Apart from these, the dataset also contains the state-wise admission capacities of Doctorate of Medicine (DM)/Master of Chirurgiae (MCH), Doctor of medicine (MD)/Master of Surgery (MS) and DIPLOMA courses of colleges. The total number of doctors is calculated by adding the number of MCI registered doctors, allopathic doctors, rural doctors, ESI medical officers and specialists and AYUSH doctors. Similarly, the total number of health workers is calculated by adding the male and female health workers and male and female health assistants. The total number of nurses are computed by adding the ESI nurses and the registered nurses as of 31/12/2017 which includes auxiliary nurse midwives (ANM), lady health visitors (LHV), registered nurses (RN) and registered midwives (RM). The total number of pharmacists is calculated by adding the number of ESI pharmacists and the pharmacists. There are several types of courses under admission capacities of DM/MCH, MD/MS and DIPLOMA courses. In order to simplify the study, various fields of respective courses have been added up and analysed.

For the sake of simplicity, the analysis has been divided into three parts—firstly, the analysis of the COVID-19 progression in India and comparison to similarly affected nations in the world; secondly, an analysis of the healthcare infrastructure in India and finally an analysis of the healthcare human resources.

8.3 Data Analysis and Visualization

Data analysis is a methodical process of extraction patterns and trends in the collected data, whereas data visualization is a process of translating the findings of analysis into various types of graphs, charts and visual representations.

Table 8.1 COVID-19: confirmed, active, recovered and deceased cases in India as on 18 July 2020

Confirmed cases	Active cases	Recovered	Deceased
10,79,618	3,73,379	6,77,423	26,816

8.3.1 Progression of COVID-19 in India

8.3.1.1 Current Situation in India

As of 18 July 2020, India has just ended the phase known as Unlock 1 and moved into Unlock 2. In this stage, lockdown restrictions are being relaxed significantly in all areas except containment zones. Non-essential services are slowly resuming, and public places such as places of worship, parks and restaurants are slowly being reopened to boost the struggling economy and recover from the impact of the virus. The current situation in India looks as shown in Table 8.1.

The above table shows that there are currently 10.8 lakh confirmed cases in India, of which 3.7 lakh cases are active cases and 6.7 lakh patients have recovered from the disease. 26,816 people have died of COVID-19. 0.07% of the total population of India is already been infected with COVID-19 as on 18 July 2020. Of those infected, 62.93% have recovered from the disease and 2.53% have died [5].

8.3.1.2 Plot/Chart to Present the Total Cases in India Over Time

Figure 8.1 is a line plot which is normally used to observe the behaviour of a quantitative variable with respect to time [6]. Here, the graph details the trend in the total

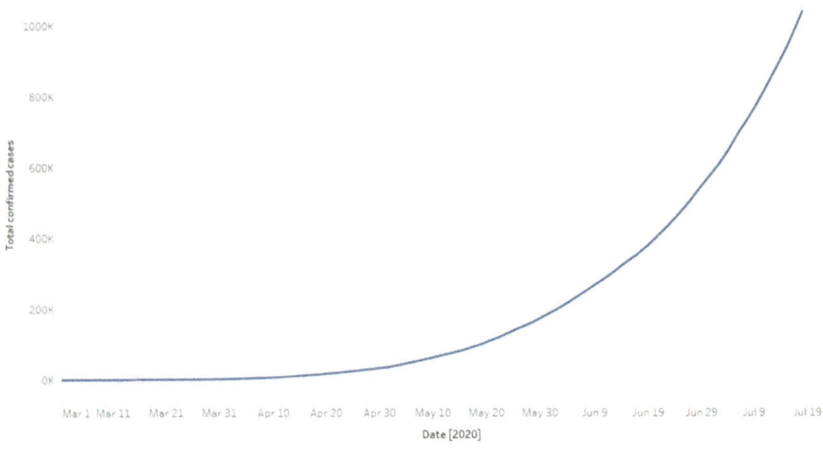

Fig. 8.1 Trend of total confirmed cases in India

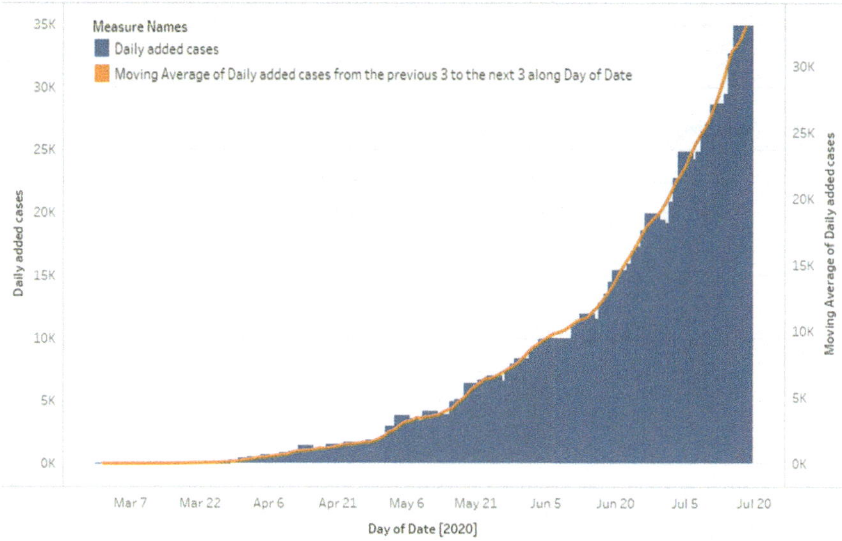

Fig. 8.2 Trends of daily added cases and moving average of daily added cases from the previous 03 to next 03 along with day of date

confirmed cases of COVID-19 in India. The total numbers of new cases every day have steadily been rising since March, despite the implementation of a nation-wide lockdown.

Figure 8.2 is a bar chart where each bar represents a date and the height of the bar represents the number of additional COVID-19 cases in India confirmed on that date. A line plot representing the 7-day moving average has been added to the graph, which helps analyse the trend in daily new cases over time. The 7-day moving average (or weekly moving average) is the arithmetic mean of 7 days of daily new cases. The daily new cases have also been increasing very consistently, as shown by the 7-day moving average.

8.3.1.3 Analysis of the COVID-19 Situation in Indian States and Union Territories

Figure 8.3 consists of two horizontal bar charts, where each bar represents a state or union territory in India and the length of the bar represents the total confirmed cases and the active cases in that state [5].

Layered on top of one bar chart is another bar chart, where each bar represents the same state and the height of the bar represents the currently active cases of COVID-19 in that state. From Fig. 8.3, it is evident that Maharashtra, Tamil Nadu and Delhi are leading the states in terms of the both the total confirmed cases and the active cases, followed closely by Karnataka, Andhra Pradesh and Uttar Pradesh. Maharashtra is

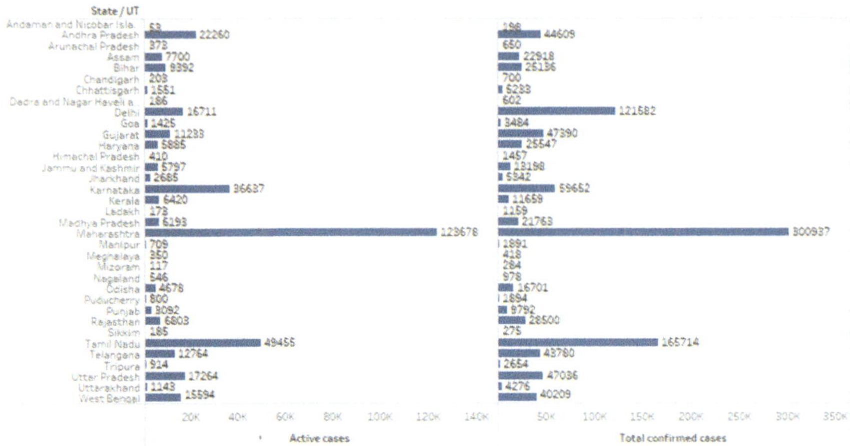

Fig. 8.3 Active cases and total confirmed cases for each State/UT

a comfortable leader in these categories and stands out in the figure, making up for almost a third of India's total confirmed cases, with more reported cases than China.

8.3.1.4 Analysis of Infection Rate and Death Rate

Figure 8.4 presents an analysis of infection rate and death rate with the help of a scatter plot, which is normally used to observe if there is any relationship between two quantitative variables [5, 7]. Each point on the scatter plot represents a state, and its position is found by using the infection rate as the X coordinate and the death

Fig. 8.4 Infection rate versus death rate

rate as the Y coordinate. The infection rate and death rate have been computed as follows:

$$\text{Infection rate} = \frac{\text{Total confirmed cases in the state}}{\text{Population of the state}} \times 100$$

$$\text{Death rate} = \frac{\text{Total deaths in the state due to COVID-19}}{\text{Total confirmed cases in the state}} \times 100$$

To make the graph more readable, a logarithmic scale has been applied, which helps to better highlight the outliers in the data. The outliers have been highlighted and named so that attention can be focused on these special case states.

It is obvious from the above figure that the infection rate is the highest in Ladakh, Delhi, Uttar Pradesh and Maharashtra. From among those infected, the states with the highest death rates are Gujarat, West Bengal, Madhya Pradesh, Delhi and Maharashtra. Uttar Pradesh and Delhi have simultaneously high infection rates and death rates.

8.3.1.5 Segmentation of Districts into Hot Spots

In order to understand and then subsequently control the spread of COVID-19 in each state of India, the districts of the states are categorized into three zones—red, orange and green—based on the number of positive cases in each district. This helped in controlling the spread of the virus in districts that are more severely affected (hot spots) while providing some relief in the restrictions of less affected zones to help boost the economy. The three zones are defined as follows:

1. **Red**: The districts with substantial numbers of positive cases would fall under the red zone. No activities can be conducted in these zones. Also, areas that show a high rate of infection and number of cases double in less than four days will also come under this category.
2. **Orange**: The areas with limited number of cases in the past and with no surge in positive cases recently would be included under the orange zone. Only restricted activities such as limited public transport and the harvesting of agricultural produce are allowed in coronavirus orange zone.
3. **Green**: The districts with no coronavirus positive cases would fall under the green zone.

Figure 8.5 is a horizontal coloured bar chart where bars represent states, the colour represents the colour of the zone and the length of the bar represents the number of districts in that category. The graph is arranged in decreasing order of the number of red zones of the respective states and demonstrate state-wise breakdown into red, orange and green zone that can help to have better control over the critical situation.

The graph clearly shows that Uttar Pradesh has the highest number of red zones followed by Maharashtra and Tamil Nadu. Delhi and Chandigarh have no orange and green zones. North-eastern states are performing far better than others. There are 14

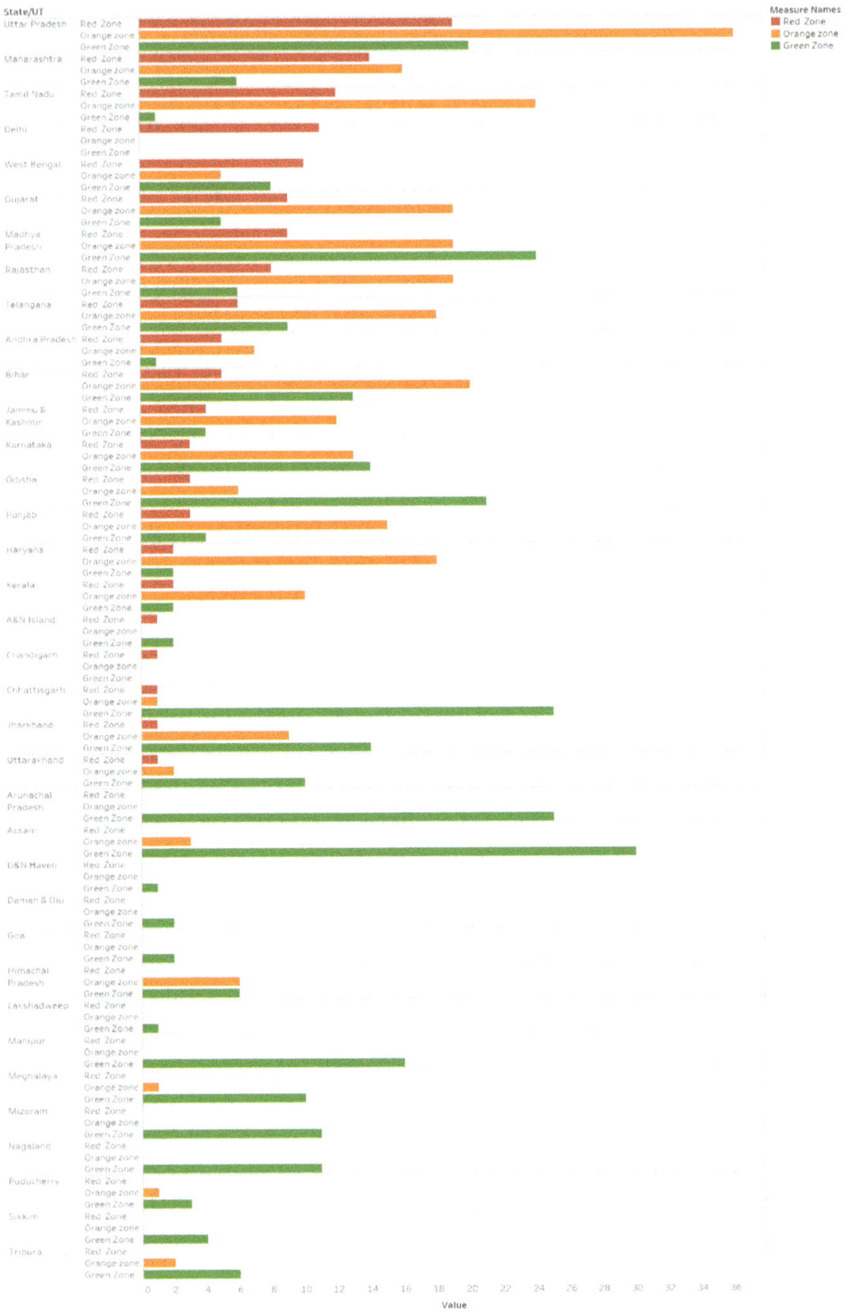

Fig. 8.5 Division of each state into red, green and orange zones

states including the seven sisters—Arunachal Pradesh, Assam, Manipur, Meghalaya, Mizoram, Nagaland and Tripura along with Goa, Daman & Diu, Dadra & Nagar Haveli, Himachal Pradesh, Lakshadweep and Sikkim which have no red zones.

8.3.1.6 Analysis of the Progression of COVID-19 in the Top States Over Time

This sub-section presents the growth of the total confirmed cases, recoveries and in the country over time. Figure 8.6 contains several line plots displaying the trends in the total confirmed cases for the most affected states [6]. Each line represents the trend in the total confirmed cases of a state over time, and the lines are coloured differently to represent different states. There has been a steady rise in the confirmed cases, especially since the lockdown was imposed on 25 March 2020. All the states show an upwards trend in the number of cases, and it does not look like any of these states are close to reaching a peak.

Figure 8.7 contains several line plots displaying the trends in the total deaths for the most affected states [6]. Each line represents the trend in the total deaths due to COVID-19 in a state over time, and the lines are coloured differently to represent different states. The deaths have steadily increased for each state, with the maximum number of deaths occurring in Maharashtra, followed by Delhi, where a sharp increase in the number of deaths has been observed starting from the end of May.

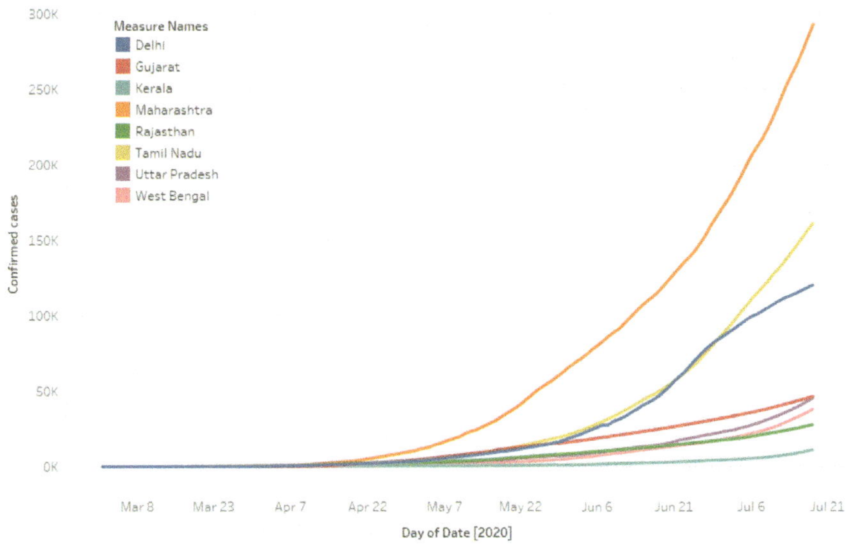

Fig. 8.6 Confirmed cases across top 10 states

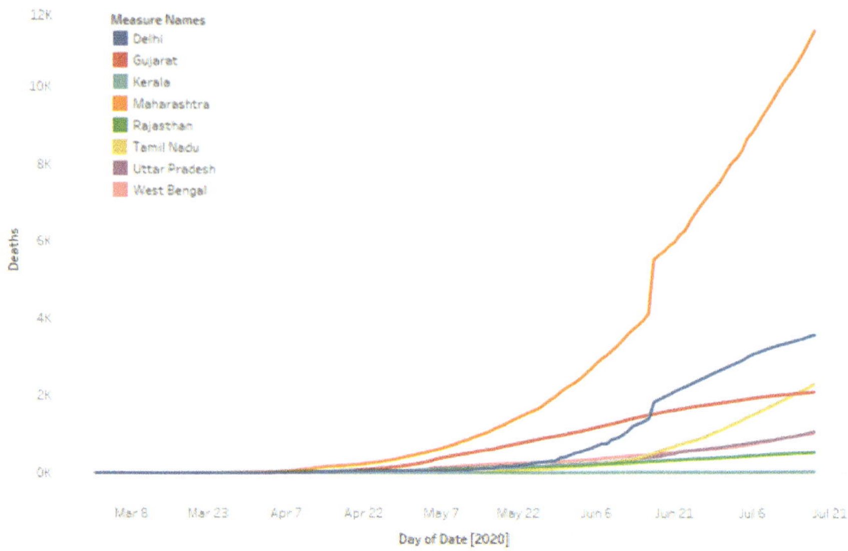

Fig. 8.7 Deaths across top 10 states

Figure 8.8 contains several line plots displaying the trends in the recoveries for the most affected states [6]. Each line represents the trend in the total recoveries from COVID-19 in a state over time, and the lines are coloured differently to represent different states. The graph shows a good sign, as recoveries are increasing steadily

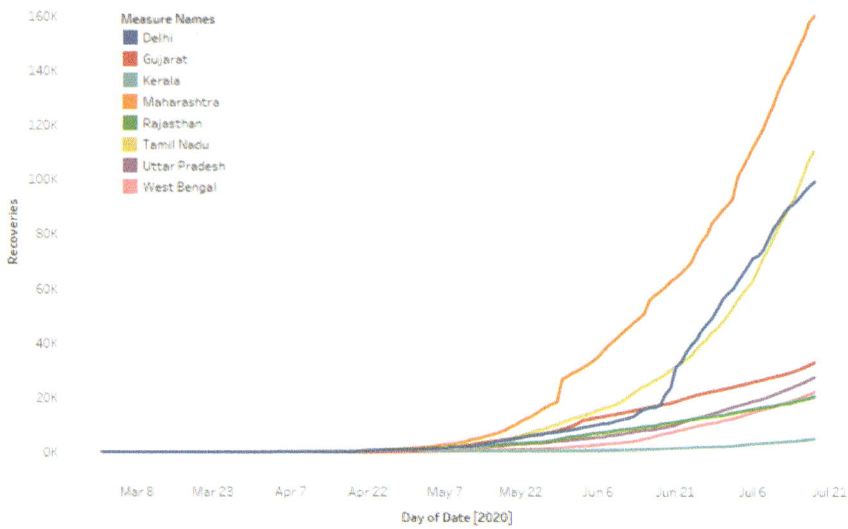

Fig. 8.8 Recoveries across the top states

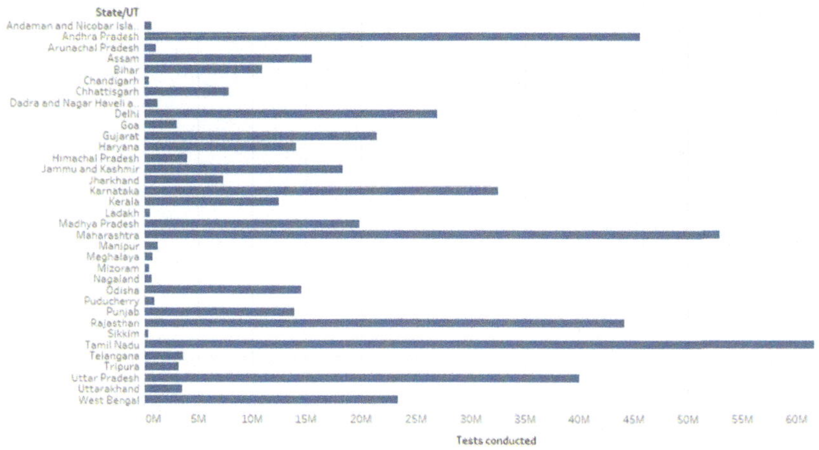

Fig. 8.9 Total tests conducted in each state

across all the states. Maharashtra, once again, has the highest number of recoveries followed by Tamil Nadu, Gujarat and Delhi. 49% of all cases in India have been cured of COVID-19.

8.3.1.7 Tests Conducted Across States

Let us look at how different states in India are faring with regards to the number of tests conducted. Figure 8.9 is a horizontal bar chart, where each bar represents a state and the length of the bar represents the total samples tested by the state [6]. Here, Tamil Nadu, Maharashtra and Andhra Pradesh are the top three states in terms of most samples tested for COVID-19. Therefore, it is not surprising that these three states also have high numbers of positive cases as well, since more testing will reveal more positive cases.

Figure 8.10 is an ordered bar chart, where each bar represents a state and the height of the bar represents the percentage of samples that have tested positive for COVID-19 by the state [6]. Telangana has 18.5% of all their samples testing positive, followed by Maharashtra at 16.6% and Delhi at 13.9%. The remaining states are less than 10%.

8.3.1.8 Comparison with Most Affected Nations of the World

India is currently the third worst affected nation in the world—preceded only by the USA and Brazil as shown in Table 8.2. Other top nations such as Spain, Russia and the UK are also included in this list as on 21 July 2020 [8].

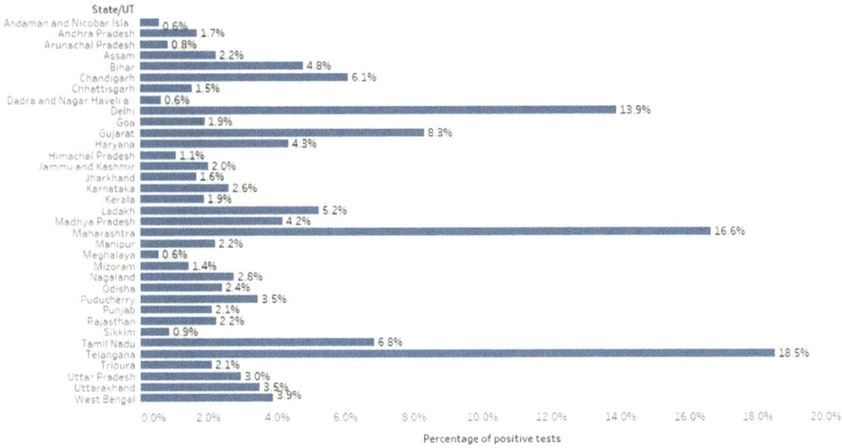

Fig. 8.10 Percentage of positive tests in each state

Table 8.2 Worst affected nations of the world

Country	Cases	Deaths	Region
USA	38,98,550	1,43,289	North America
Brazil	20,99,896	79,533	South America
India	11,19,412	27,514	Asia
Russia	7,77,486	12,427	Europe
Spain	3,07,335	28,420	Europe
UK	2,94,792	45,300	Europe

Figure 8.11 is a pie chart, which is normally used to visualize how a set of data is proportioned into different categories [9]. In this case, it is clearly seen that 25.54% of total confirmed COVID-19 cases are in the USA, followed by 14.5% in Brazil, and 7.49% in India, which places India at the third position in world's total COVID-19 cases.

Figure 8.12 is a horizontal bar chart, where each bar represents a country and the length of the bar represents the total cases (blue bar), total deaths (orange bar) and total recoveries (red bar). India and Russia have the lowest number of deaths among these nations [9]. It is important to note that no recovery data was available for Spain and UK.

A bar chart where each bar represents a country and the height of the bar represents the number of critical cases in that country as shown in Fig. 8.13 [9]. India has the second highest number of critical cases in the world, and only the USA has a higher number.

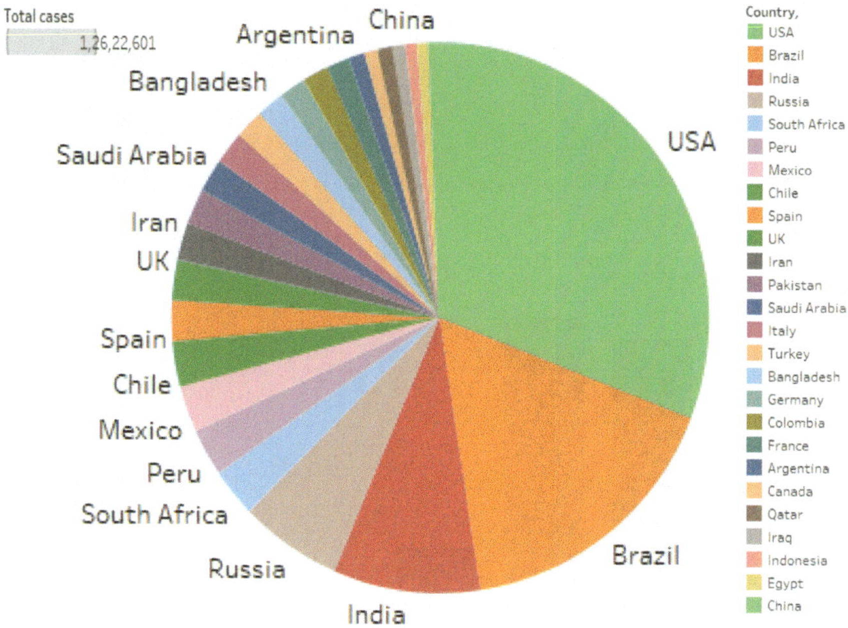

Fig. 8.11 Distribution of COVID-19 across the world

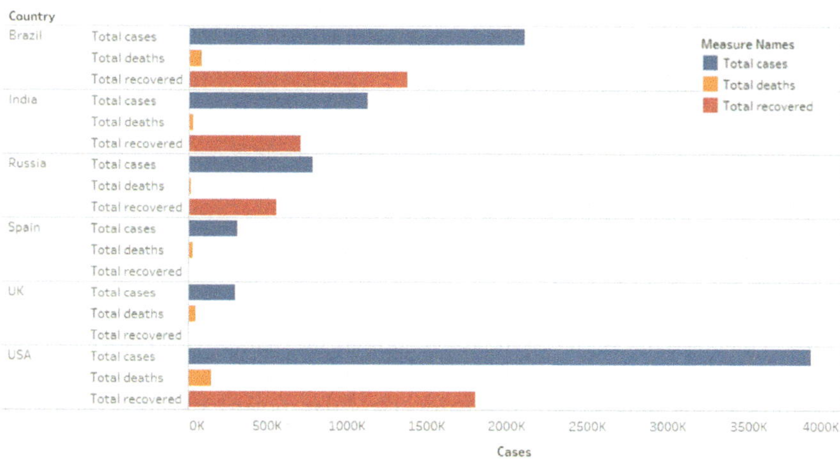

Fig. 8.12 Total cases, total deaths and total recoveries for worst affected countries

Figure 8.14 is a bar chart where each bar represents a country and the height of the bar represents the number of samples tested in that country per 1 million of the population [9]. India has conducted the fewest tests among these nations.

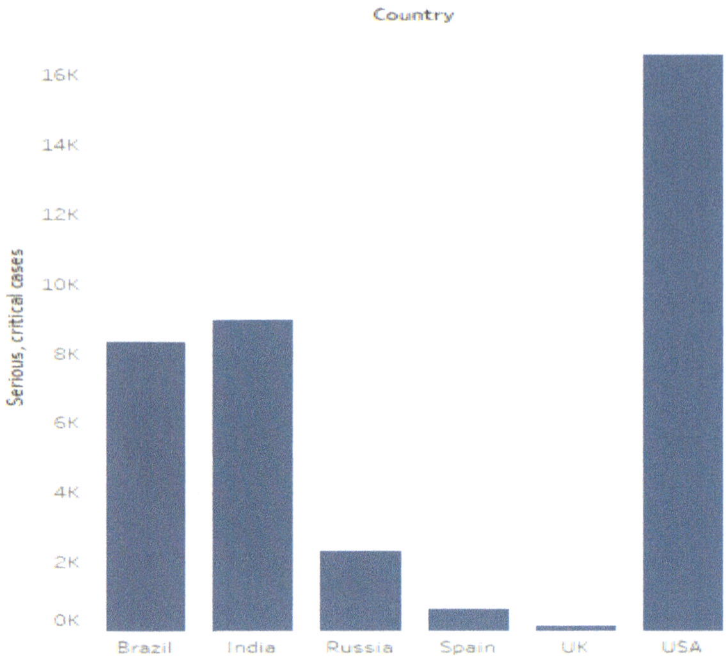

Fig. 8.13 Serious and critical cases for worst affected countries

It is very interesting to take a look at how the daily new cases have been progressing across these six worst affected nations.

(a) Daily cases in Brazil are presented in Fig. 8.15
(b) Daily cases in the USA are presented in Fig. 8.16
(c) Daily cases in Russia are presented in Fig. 8.17
(d) Daily cases in UK are presented in Fig. 8.18
(e) Daily cases in Spain are presented in Fig. 8.19.

The above graphs are all bar chart for different countries, with each bar representing a date and the height of the bar represents the number of COVID-19 cases added on that day. A line plot representing the 7-day moving average in blue has been added, to show the trend in daily new cases. The 7-day moving average, as described above, is calculated as the arithmetic mean of the daily cases added of 7 days.

The purpose of these graphs is to depict how cases have added up in different nations over time and to see which of these nations have managed to 'flatten the curve'. 'Flattening the curve' refers to changing the direction of the trend line from upwards to downwards. This would indicate that on average, every week has fewer added cases than the previous week. This is a good thing as it indicates that the spread of the virus is being contained and is the end goal of every nation. It is clearly evident from the above figures that Brazil and India are the only countries where the peak has not flattened, whereas for every other nation the curve has started flattening.

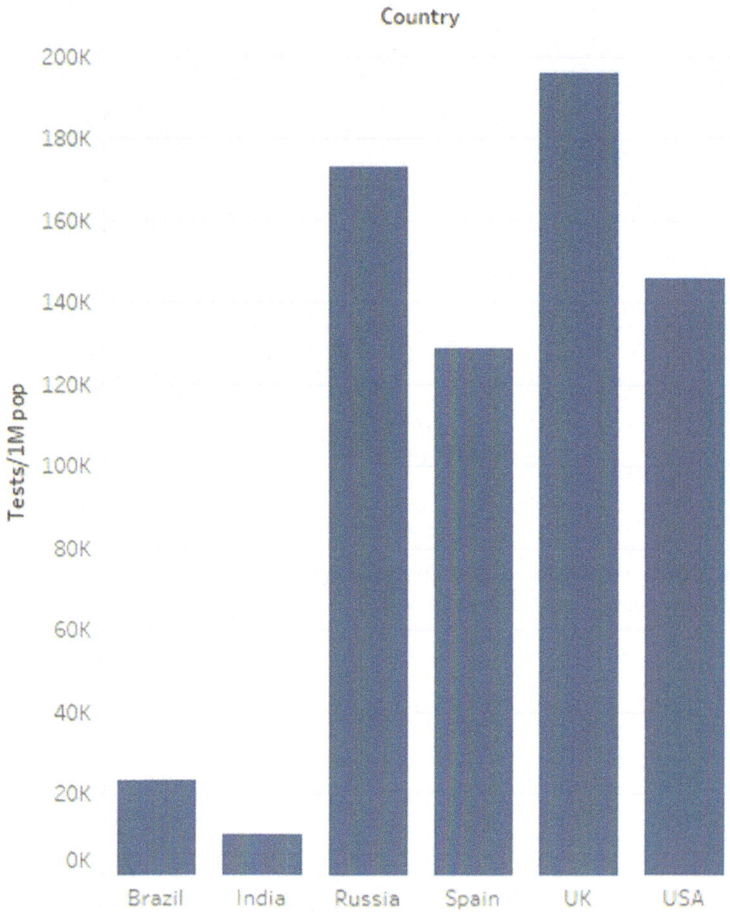

Fig. 8.14 Tests per 1 million population for worst affected countries

8.3.2 Healthcare Infrastructure in India

8.3.2.1 Population of India

This sub-section looks at the population of various states in India. Figure 8.20 is a horizontal bar chart where each bar represents a state and the length of each bar represents its population. India's population has reached 1,371,360,350. The highest populated state of India is Uttar Pradesh with estimated population of 237.9 million in 2019 which accounts to 17.35% of the whole population of the country. The second largest state by population is Bihar which overtakes Maharashtra with close to 125 million individuals. Delhi has the largest population of India as the union territory.

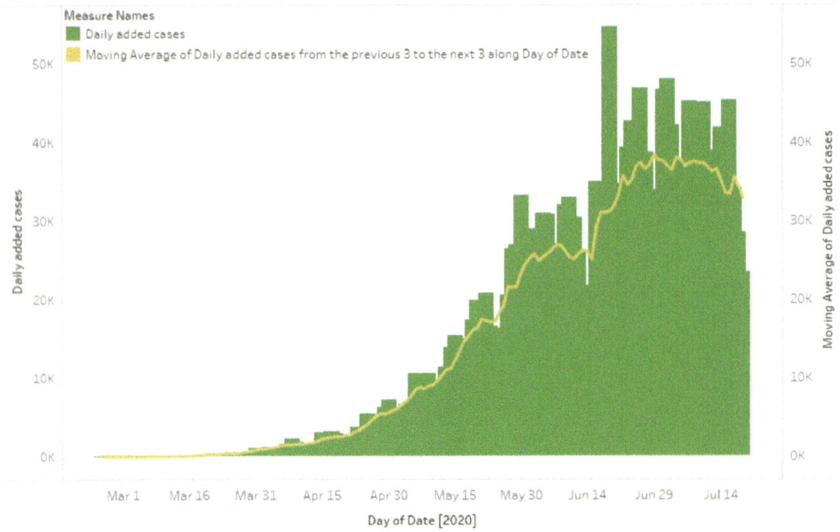

Fig. 8.15 Trends of daily added cases in Brazil and moving average of daily added cases from the previous 03 to next 03 along with day of date [15]

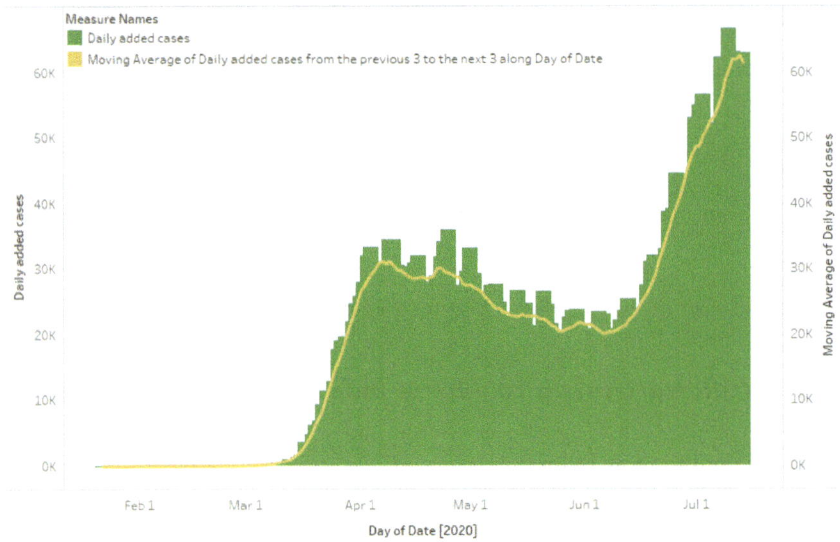

Fig. 8.16 Trends of daily added cases in the USA and moving average of daily added cases from the previous 03 to next 03 along with day of date [16]

c) Daily cases in Russia is presented in figure 17

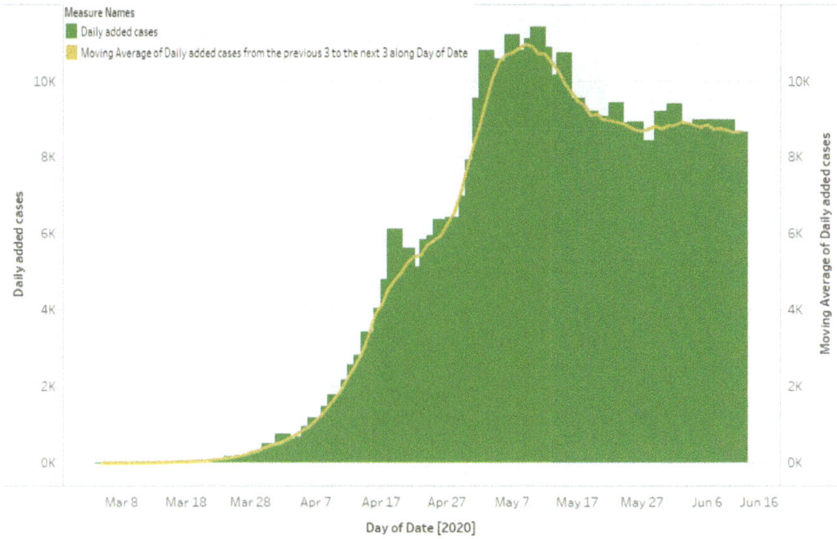

Fig. 8.17 Trends of daily added cases in Russia and moving average of daily added cases from the previous 03 to next 03 along with day of date [17]

d) Daily cases in UK is presented in figure 18

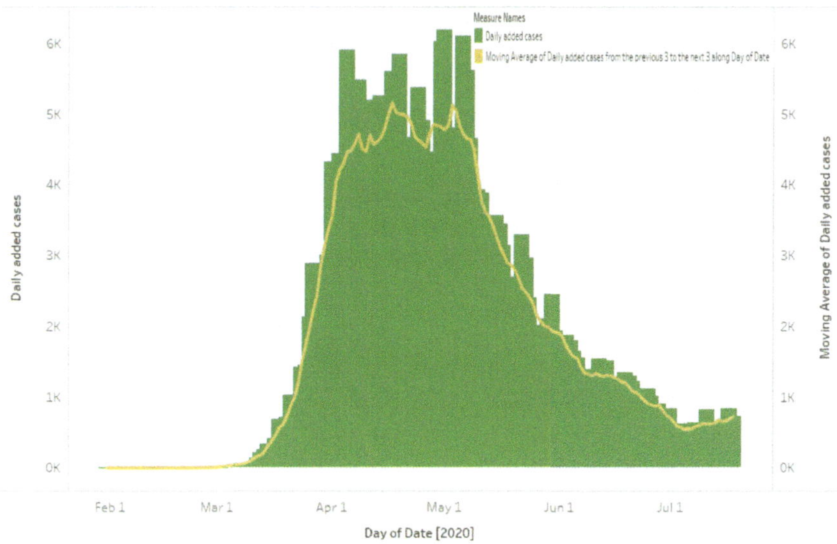

Fig. 8.18 Trends of daily added cases in UK and moving average of daily added cases from the previous 03 to next 03 along with day of date [18]

e) Daily cases in Spain is presented in figure 19

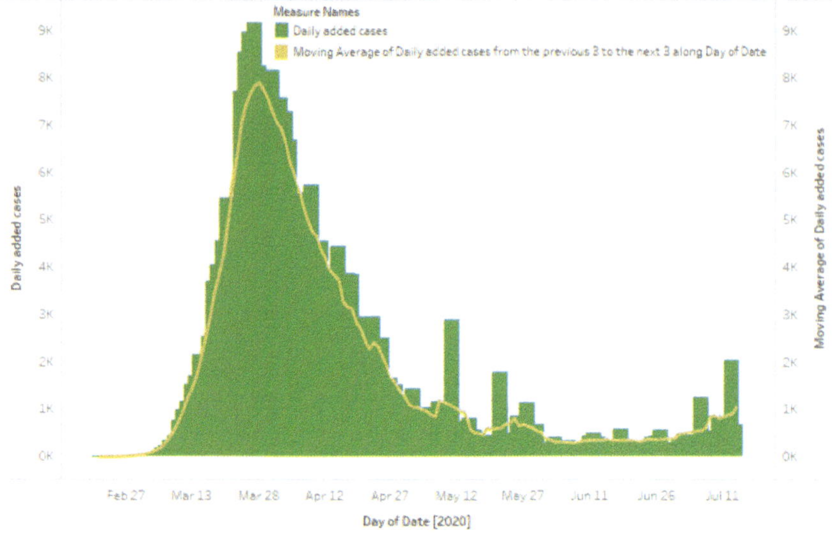

Fig. 8.19 Trends of daily added cases in Spain and moving average of daily added cases from the previous 03 to next 03 along with day of date [19]

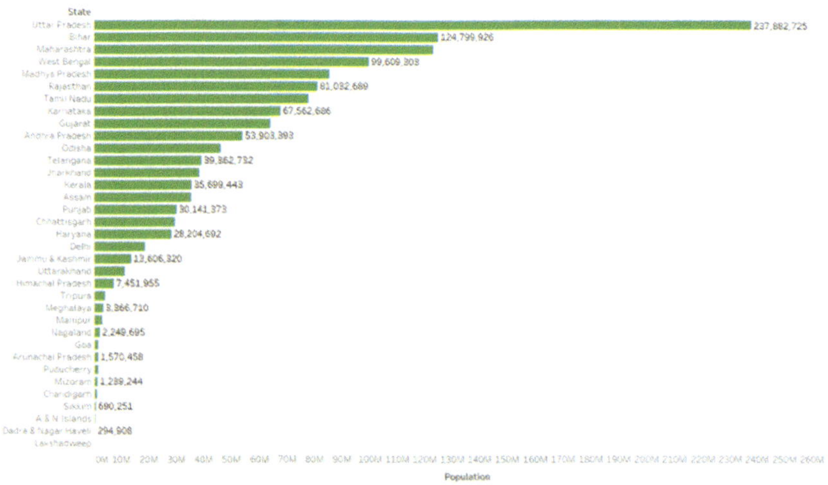

Fig. 8.20 Sum of population for each state

Uttar Pradesh, Maharashtra, Bihar, West Bengal and Madhya Pradesh are the five most populous states in India which makes almost half of India's population. Pakistan is the fifth most populated country in the world, and Uttar Pradesh has more population than Pakistan. Sikkim has the smallest state population, and Lakshadweep is the smallest union territory of India in terms of population.

8.3.2.2 Hospital Beds in India

This sub-section highlights the distribution of hospital beds across India. There is a total of 1,030,543 beds available in India across rural and urban government hospitals, medical colleges, etc. This is equivalent to 0.75 beds for every 1000 people belonging to the population of India. Figure 8.21 presents a box-plot, which is a visual summary of the data.

The following are the observations from the above box-plot:

1. The minimum number of beds is 240.
2. The maximum number of beds is 98,298.
3. The median is 15,443.
4. The first quartile is calculated as the number before which 25% of the observations lie. In this case, it is 3115.
5. Similarly, the third quartile is calculated as the number before which 75% if the observations lie. In this case, it is 42,565.
6. The data is widely dispersed and positively skewed.
7. The inter-quartile range (IQR) which is a measure of dispersion is calculated from the quartiles in the data. IQR for number of beds is 39,410, which indicates that there is a lot of variance in the data.
8. Karnataka (1,14,676) and Tamil Nadu (1,17,663) are the outliers in this dataset.

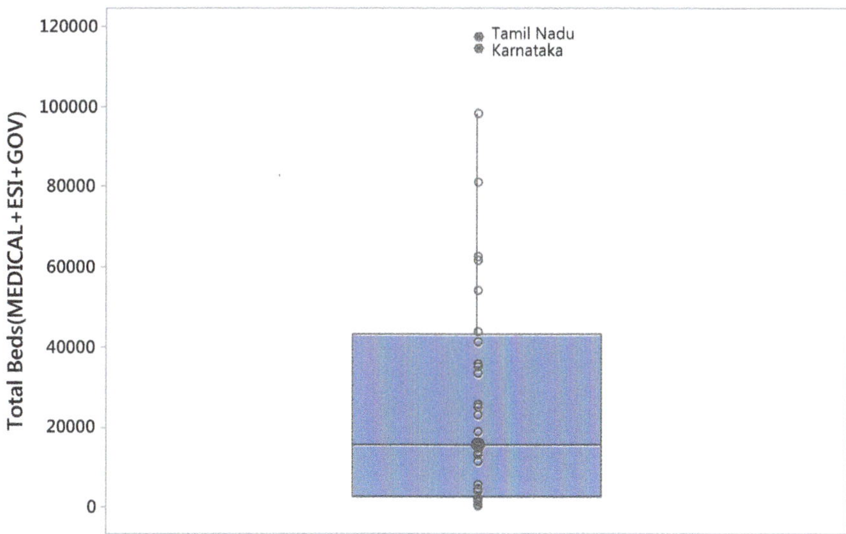

Fig. 8.21 Distribution of the total number of beds

Figure 8.22 is a horizontal bar graph where each bar represents a state and the length of each bar represents the ratio of beds to population of that state. The national average for beds to population of India (as of 2019) is 0.75 beds per 1000 people. As observed from the graph, some states lie below the national figure. These states are Bihar, Uttar Pradesh, Jharkhand, Maharashtra, Manipur, Chhattisgarh, Madhya Pradesh, Odisha, Haryana, Gujarat, Assam and Daman and Diu. Delhi has 1.88 beds per 1000 population, and the southern states of Kerala (1.72 beds per 1000) and Tamil Nadu (1.51 beds per 1000) have a higher number of beds available as well. Bihar has a significant shortage in the availability of hospital beds, with only 0.15 beds available per 1000 population.

India has an extremely low number of beds compared to other countries. This problem is escalated during a pandemic like COVID-19. It is estimated that anywhere between 5 and 10% of total patients require critical care in the form of lung support from a ventilator. The data for the availability of the number of ventilators is unavailable; nevertheless, a figure can be estimated from the number of available hospital beds—1,030,543 total hospital beds, out of which anywhere between 5 and 8% per cent are ICU beds (corresponding to 51,527 to 82,443 ICU beds) [10]. Assuming that 50% of these ICU beds are equipped with ventilators, therefore there are approximately 25,764 to 41,221 ventilators in the country. For the growing number of COVID-19 patients, more ventilators will be needed. This indicates that the growing demand for ventilators will surpass the short supply that is available.

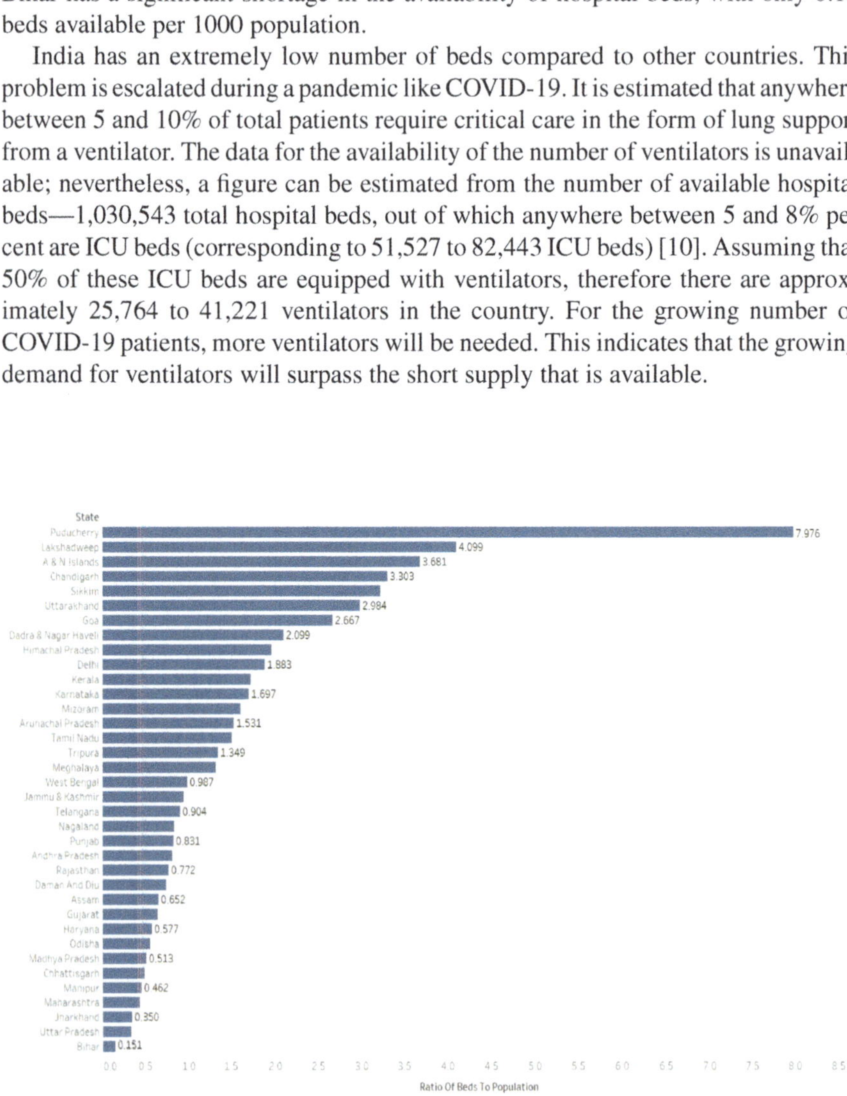

Fig. 8.22 Ratio of beds to population

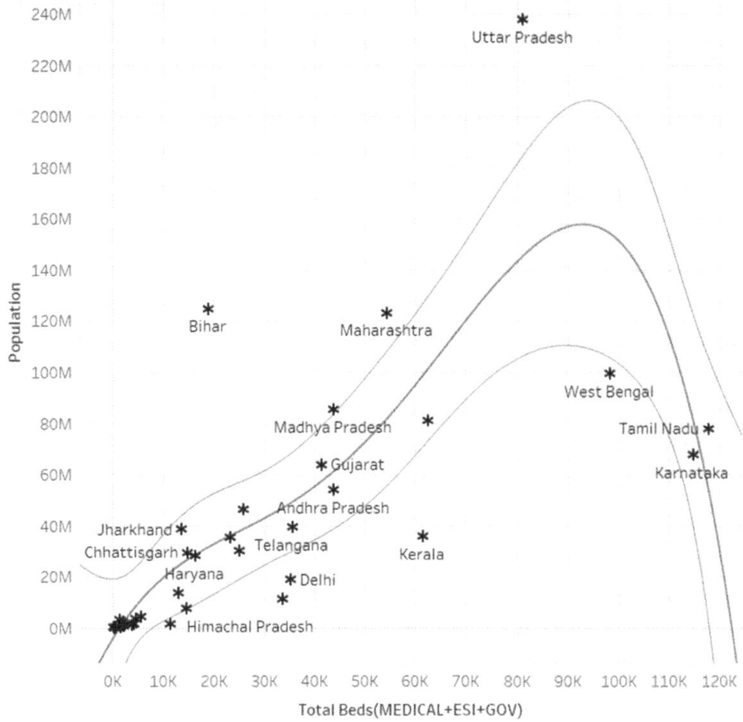

Fig. 8.23 Total beds versus population for each state

Figure 8.23 is a scatter plot which is normally used to observe if there is any relationship between two quantitative variables. Each point on the scatter plot represents a state, and its position is found by using the total beds in the state, the state population as the X coordinate and the state population as the Y coordinate.

The scatter plot in Fig. 8.23 shows a fourth degree polynomial curve with confidence bands fitted to the data with some outliers like Uttar Pradesh, Kerala and Bihar. A regression line is a statistical model where a polynomial trend equation is fitted to the data in order to make predictions. A polynomial regression model is computed for the sum of total beds, given the sum of population as described in Table 8.3. This model may be significant if the *p*-value is less than 0.05.

Individual trend lines (shown in Table 8.4).

The sign of the values of the parameters talks about the nature of the parameter. There are three parameters with negative values which means that the population decreases with increase of those parameters and two parameters with positive values which means population increase with their increase. Since the nonlinear regression takes infinite number of forms, it is impossible to form a single hypothesis for all parameters. That is why it is generally not advisable to interpret the significance of the parameters according to the *p*-values. All the coefficients of the model are not

Table 8.3 A computation of fourth degree polynomial regression model for quantitative variables (beds and population)

Model formula	(Total beds (Medical+ESI+Govt)^4+Total Beds (Medical+ ESI + Govt)^3 + Total Beds (Medical + ESI + Govt)^2 + Total Beds (Medical + ESI + Govt) + intercept)
Number of modelled observations	36
Number of filtered observations	0
Model degrees of freedom	5
Residual degrees of freedom (DF)	31
Sum squared error (SSE)	3.13748e+16
Mean squared error (MSE)	1.01209e+15
R-squared	0.6417
Standard error	3.18134e+07
p-value (significance)	<0.0001

significant since they all have *p*-values >0.05. We still continue with the model as the standard error of some of the parameters is comparatively high. The *R*-squared value is 0.64 which signifies that the model is a moderately strong fit for the data. This implies that 64% of the variance in total hospital beds of the states can be explained by the variances in the populations of those states.

8.3.2.3 Hospitals in India

This sub-section focuses on the number of hospitals in each state across India. Figure 8.24 is a box plot which is a visualization used to describe the distribution of the data.

From the above box-plot, it is observed that:

1. The minimum number of hospitals is 7.
2. The maximum number of hospitals is 3016.
3. The median is 484.
4. The first quartile is calculated as the number below which 25% of the observations lie. The value of the first quartile is 139.
5. The third quartile is calculated as the number below which 75% of the observations lie. The value of the third quartile is 1290.
6. The data is moderately dispersed and positively skewed.
7. The inter-quartile range (IQR), which is a measure of dispersion in the data calculated using the quartiles, is 1151.
8. Karnataka (3233) and Uttar Pradesh (4813) are the outliers in this dataset.

Figure 8.25 is a horizontal bar graph where each bar represents a state and the length of each bar represents the percentage ratio of hospitals to population of each state. It shows the number of hospitals in India across all the states. It can be observed

Table 8.4 Individual trend lines

Row	Column	*p*-value	DF	Term	Value	StdErr	*t*-value	*p*-value
Population	Total beds (Medical + ESI + Govt)	<0.0001	31	Total beds (Medical + ESI + Govt)^4	−1.05e−11	5.972e−12	−1.759	0.088
				Total beds (Medical + ESI + Govt)^3	1.922e−06	1.318e−06	1.4579	0.154
				Total beds (Medical + ESI + Govt)^2	−0.104036	0.0908978	−1.144	0.261
				Total beds (MEDICAL + ESI + GOV)	3243.12	2097.81	1.545	0.132
				Intercept	−4.2841e+06	1.13587e+07	−0.37	0.708

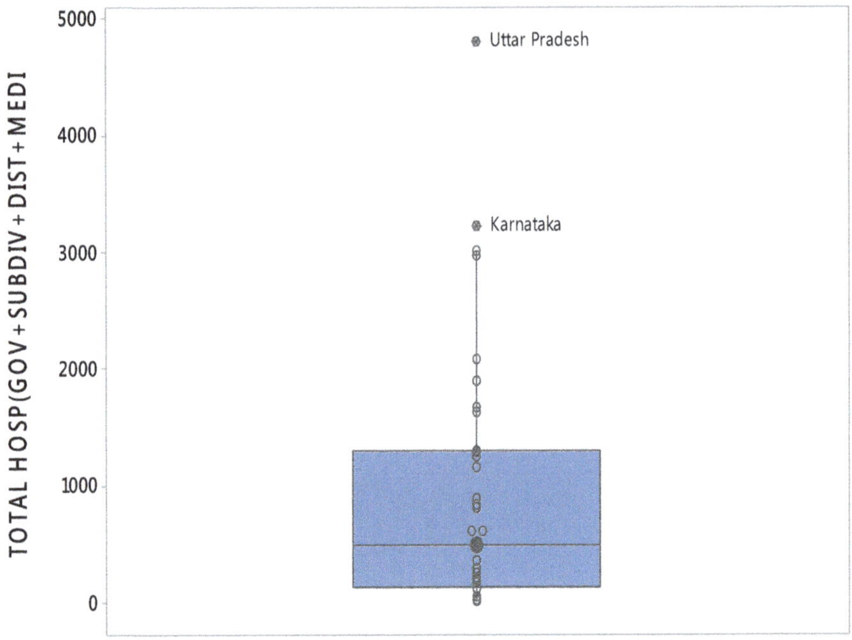

Fig. 8.24 Distribution of total hospitals across India

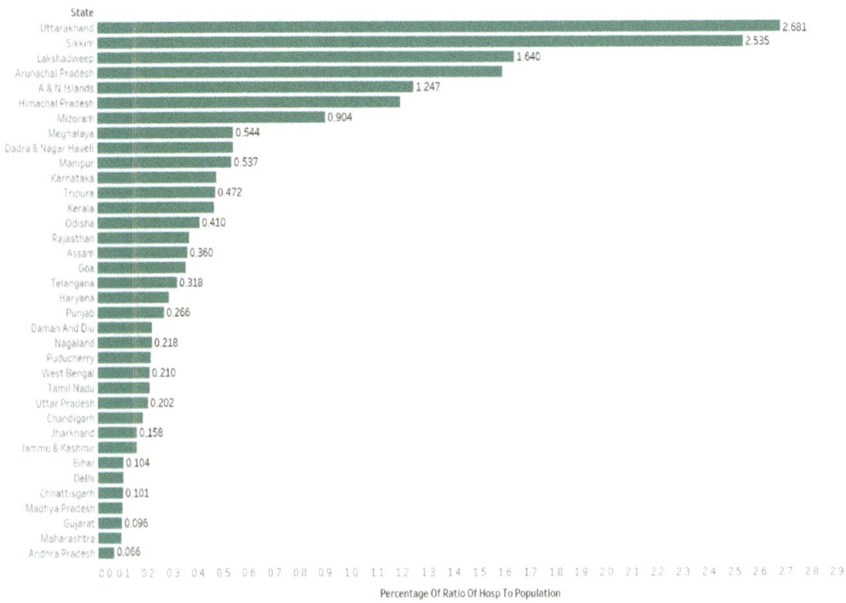

Fig. 8.25 Ratio of hospitals to population (state-wise)

that many states have fewer hospitals than the national average (0.24 hospitals per 10,000 population). These states are Andhra Pradesh, Maharashtra, Gujarat, Madhya Pradesh, Chhattisgarh, Delhi, Bihar, Jammu & Kashmir, Jharkhand, Chandigarh, Uttar Pradesh, Tamil Nadu, West Bengal, Puducherry, Nagaland and Daman & Diu. Andhra Pradesh has a significant shortage of government hospitals with just 0.066 hospitals availablc to 10,000 population. The capital city of Delhi has 0.10 hospitals per 10,000 population, whereas, Uttarakhand (2.68 hospitals per 10,000 population) and Sikkim (2.53 hospitals per 10,000 population) have higher availability of hospitals.

Figure 8.26 is a scatter plot which is normally used to observe if there is any relationship between two quantitative variables. Each point on the scatter plot represents a state and its position is found by using the total hospitals in the state as the X coordinate and the state population as the Y coordinate.

The above scatter plot also shows a third degree polynomial curve with confidence bands fitted to the data with some outliers like Himachal Pradesh, Uttarakhand, Maharashtra and Bihar. A polynomial regression line is a statistical model where a

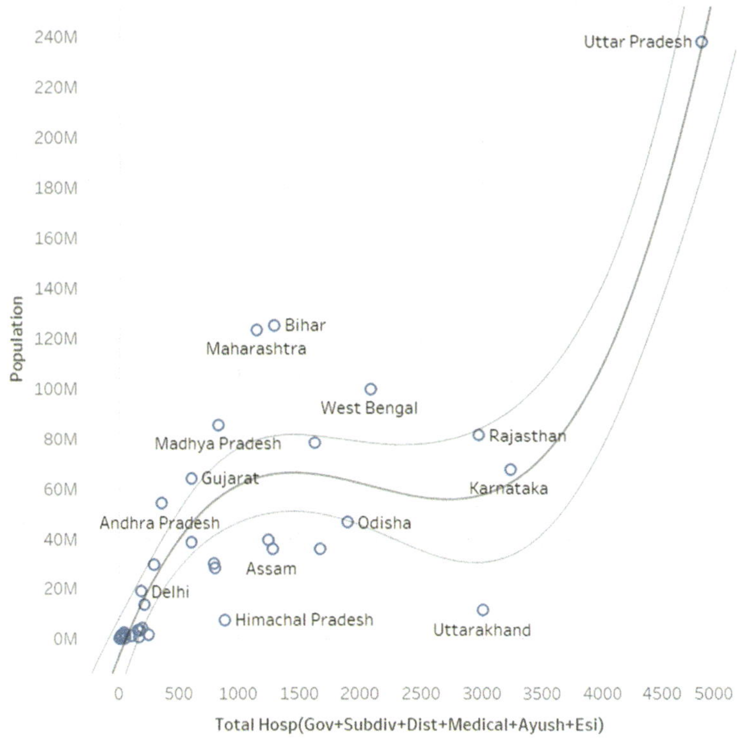

Fig. 8.26 Population versus total hospitals for each state across India

Table 8.5 A computation of third degree polynomial regression model for quantitative variables (hospitals and population)

Model formula	(Total Hosp (Gov + Subdiv + Dist + Medical + Ayush + Esi)^3 + Total Hosp (Gov + Subdiv + Dist + Medical + Ayush + Esi)^2 + Total Hosp (Gov + Subdiv + Dist + Medical + Ayush + Esi) + intercept)
Number of modelled observations	36
Number of filtered observations	0
Model degrees of freedom	4
Residual degrees of freedom (DF)	32
Sum squared error (SSE)	2.08582e+16
Mmean squared error (MSE)	6.51818e+14
R-squared	0.7618
Standard error	2.55307e+07
p-value (significance)	<0.0001

curve is fitted to the data in order to make predictions. Table 8.5 presents the details of the trend model.

Trend Lines Model

A polynomial trend model of degree 3 is computed for sum of population given sum of total Hospitals (Govt + Subdiv + Dist + Medical + Ayush + ESI). The model may be significant at $p \leq 0.05$.

Individual trend lines (as shown in Table 8.6).

The sign of the values of the parameters talks about the nature of the parameter. There are two parameters with negative values which means that the population decreases with increase in those parameters and two parameters with positive values which means population increase with their increase. The *p*-value of the trend model is <0.05. So, the null hypothesis is rejected, and it can be concluded that the model is significant. The *p*-value of the intercept is >0.05, so the null hypothesis is not rejected, and it can be concluded that its intercept is not significant. The *R*-squared value is 0.76 which signifies that the model is a moderately strong fit for the data. This implies that 76% of the variation in the number of hospitals in states can be explained by the variation in the state populations.

8.3.2.4 Clusters of Population-Wise Hospitals

Clustering is an unsupervised learning technique. Unsupervised learning is a machine learning technique which is used to look for patterns in a dataset. An unsupervised clustering technique called *k*-means clustering has been applied to the dataset of hospitals and population in order to look for patterns within the data. *K*-means clustering looks for clusters in the data by checking for similarities between nearby data

Table 8.6 Individual trend line

Row	Column	*p*-value	DF	Term	Value	StdErr	*t*-value	*p*-value
Population	Total Hosp (Gov + Subdiv + Dist + Medical + Ayush + Esi)	<0.0001	32	Total Hosp (Govt + Subdiv + Dist + Medical + Ayush + ESI)^3	0.0103519	0.0020246	5.113	<0.0001
				Total Hosp (Govt + Subdiv + Dist + Medical + Ayush + ESI)^2	−64.8548	13.4339	−4.82	<0.0001
				Total Hosp (Govt + Subdiv + Dist + Medical + Ayush + ESI)	122,890	21,993.6	5.5875	<0.0001
				Intercept	−7.34e+06	7.43828e+06	−0.987	0.330964

points and assigning them to clusters. These similarities are calculated based on the mathematical distance between the data points. The number of clusters is predetermined and can be adjusted depending on the results. The trial and error method is used to find the most suitable number of clusters. The steps used to apply *k*-means clustering to the infrastructure dataset are as follows:

Step 1: Fig. 8.27a shows the scatter plot of hospitals with respect to population to present the distribution of the data points. Once the fair understanding is established regarding the distribution of data, then *k*-means clustering algorithm is applied to this dataset.

Step 2: Fig. 8.27b represents the final step in *k*-means clustering. Here, K has been chosen to be 5. The algorithm has been applied, and different clusters have been formed on the basis of similarities and are highlighted in different colours. The Cluster 4 in brown colour has a strong case as these are the states with high

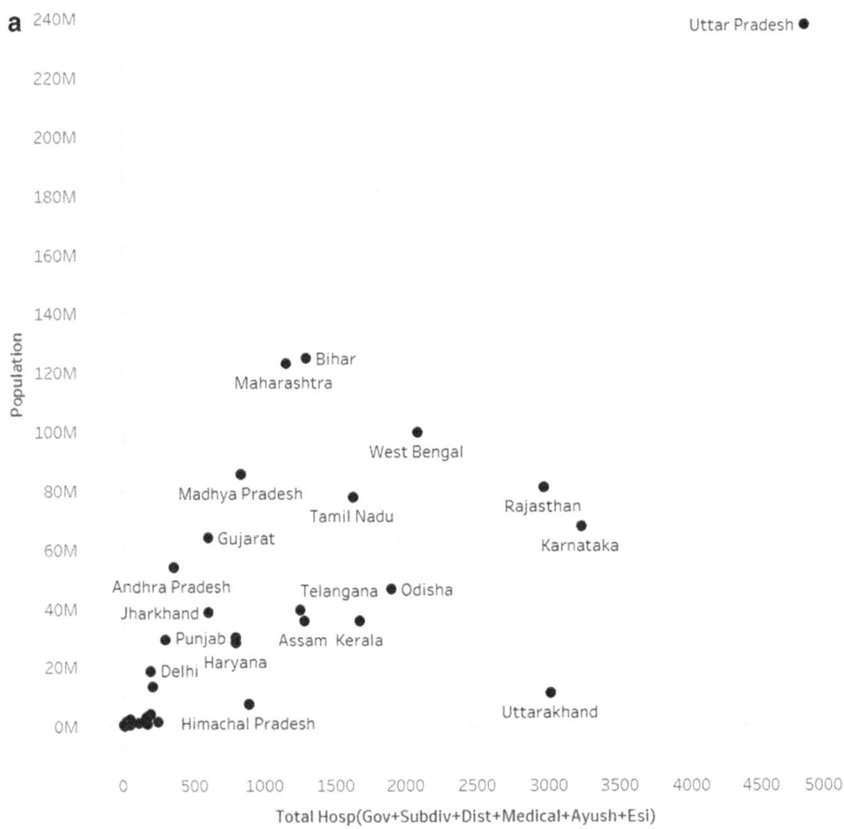

Fig. 8.27 **a** Step 1 of *k*-means clustering—scatter plot of hospitals with respect to population, **b** step 2 of *k*-means clustering—forming the clusters by calculating the distance

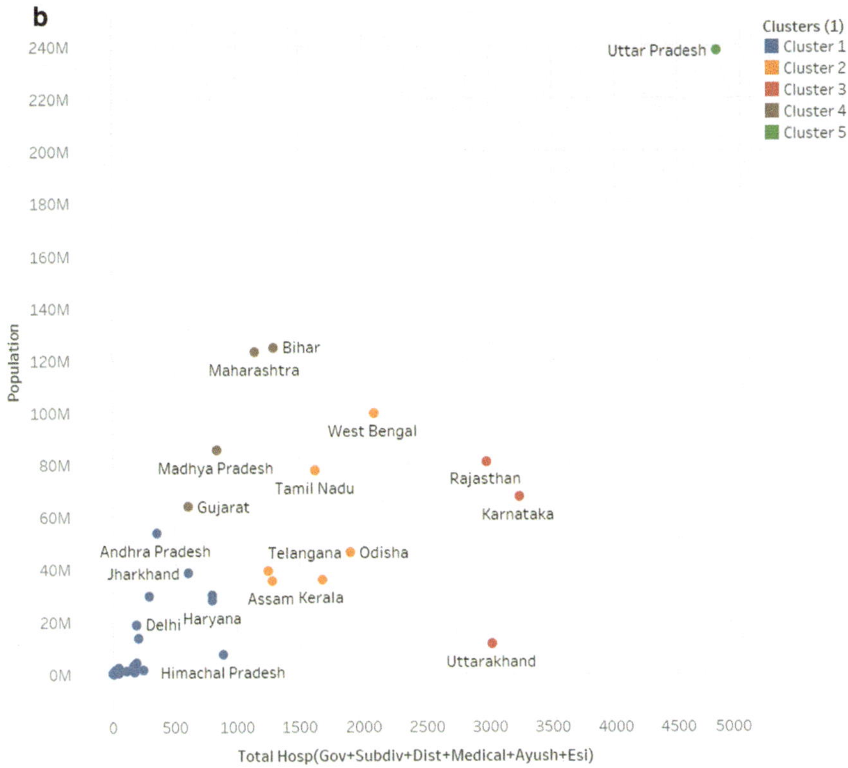

Fig. 8.27 (continued)

population compared to the number of hospitals available. The names of the states are Maharashtra, Bihar, Gujarat and Madhya Pradesh.

8.3.2.5 Analysis of Blood Banks and Dispensaries

This sub-section looks at the blood banks and medical dispensaries in each state and union territory of India. Figure 8.28 is a horizontal bar graph where each bar represents a state and the length of the bar represents the number of blood banks in the state. Maharashtra and Uttar Pradesh are the states with maximum number of blood banks. Lakshadweep and Dadra & Nagar Haveli have low counts. Blood banks could prove to be vital against COVID-19 if plasma therapy is used as a treatment for patients with more severe conditions.

Figure 8.29 is a bar figure where each bar represents a state and the height of the bar represents the number of dispensaries in the state. Sikkim (3952) has the maximum number of dispensaries followed by Uttarakhand (3728) and Bihar (2175). Delhi and Odisha have very few dispensaries. Dispensaries are important for those patients

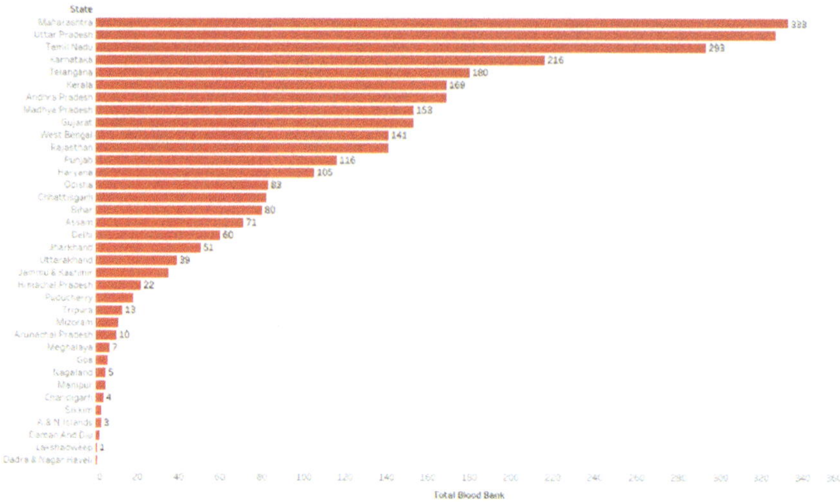

Fig. 8.28 Blood banks across India

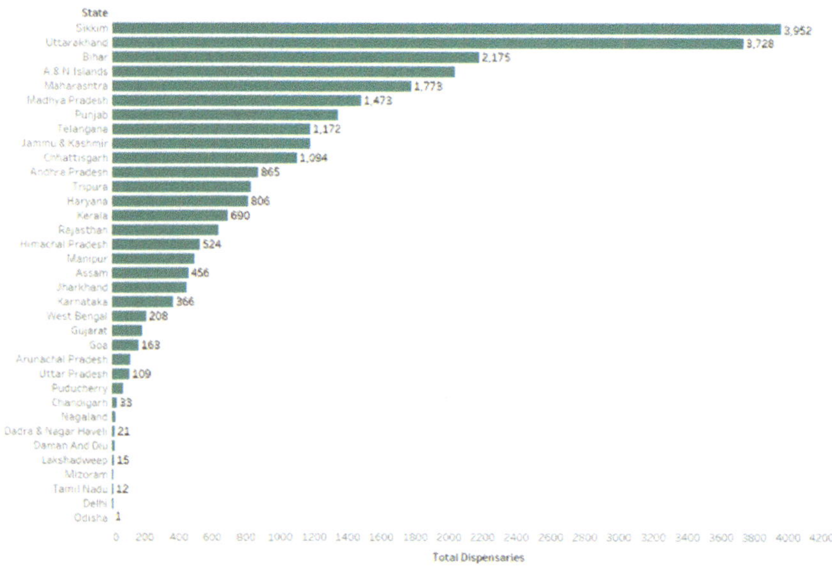

Fig. 8.29 Dispensaries across India

with mild symptoms that do not require hospitalization but still require medicines such as paracetamol and zinc.

8.3.2.6 Study of Healthcare Infrastructure of 03 States with Higher COVID-19 Cases

a. Details of Infrastructure of Maharashtra State are shown in Table 8.7. Maharashtra is the leading state of India with the highest number of confirmed COVID-19 cases. The population of the state is 8.97 per 100 of the total population, one of the top five most populated states in the country. It has 0.093 hospitals per 10,000 population which means that there are very few hospitals with respect to the population of the state. It also has low availability of hospital beds with respect to the population, which are 0.44 beds per 1000 population of the state. However, it has the maximum number of blood banks across the country. Excess facilities available via private healthcare institutions should be used by the government to combat the high number of cases. Hospital beds may also be set up in army camps.
b. Details of Infrastructure of Gujarat State are shown in Table 8.8.

Table 8.7 Healthcare infrastructure of Maharashtra

CHCs	361
No. of medical colleges	56
PHCs	1823
Population	1,23,44,223
Sub-centres	10,638
Total admission capacity	1865
Total beds	54,384
Total blood banks	333
Total dispensaries	1773
Total hospitals	1153

Table 8.8 Healthcare infrastructure of Gujarat

CHCs	363
No. of medical colleges	29
PHCs	1474
Population	6,38,72,399
Sub-centres	9153
Total admission capacity	198
Total beds	41,362
Total blood banks	153
Total dispensaries	182
Total hospitals	610

Table 8.9 Healthcare infrastructure of Tamil Nadu

CHCs	385
No. of medical colleges	50
PHCs	1421
Population	7,78,41,267
Sub-centres	8712
Total admission capacity	184
Total beds	1,17,663
Total blood banks	293
Total dispensaries	12
Total hospitals	1627

Gujarat is one of the states of India with the highest number of confirmed COVID-19 cases. The population of the state is 0.046 per 100 of the total population. It has 0.095 hospitals per 10,000 population of Gujarat which indicates that there is a shortage of hospitals with respect to the population of the state. It also has moderate availability of hospital beds with respect to the population, which are 0.64 beds per 1000 population of the state. Therefore, a higher number of healthcare facilities should be restored to keep up with the increasing number of COVID-19 patient

c. Details of Infrastructure of Tamil Nadu State are shown in Table 8.9.
Tamil Nadu has one of the highest numbers of confirmed COVID-19 cases in India. The population of the state is 0.056 per 100 of the total population. It has 0.20 hospitals per 10,000 population of Tamil Nadu which means there is a shortage of hospitals with respect to the population of the state. It also has moderate availability of hospital beds with respect to the population, which are 1.51 beds per 1000 of the population of the state. It has the third highest number of blood banks across India. It has better health infrastructure with respect to population as compared to the states of Maharashtra and Gujarat. However, more healthcare facilities should be restored to keep up with the increasing number of COVID-19 patients.

8.3.3 *Healthcare Human Resource in India*

8.3.3.1 Number of Doctors in India

Doctors play a crucial role in the pandemic as they treat the patients of the virus and their importance cannot be over-stated. Figure 8.30 is a box plot which is a visualization technique used to describe the distribution of the data. The boxplot of the total doctors was plotted according to their respective states.

The following are the observations from Fig. 8.30:

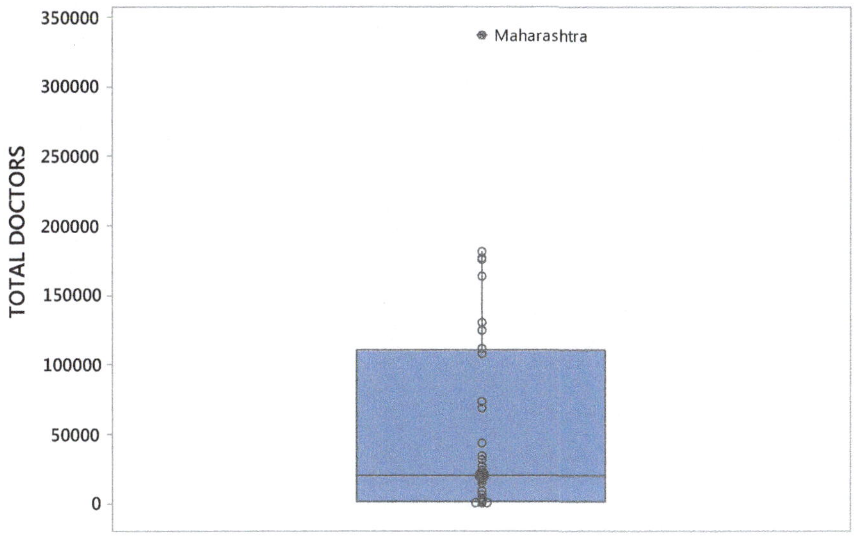

Fig. 8.30 Distribution of total doctors in India

1. The minimum number of doctors is 37.
2. The maximum number of doctors is 181,832.
3. The median is 20,125.
4. The first quartile is 109,777.
5. The third quartile is 1479.
6. There is one outlier and that is Maharashtra with number of doctors 337,530.
7. The column of doctors is highly positively skewed, and the figures are dispersed among themselves.

Figure 8.31 is a horizontal bar chart where each bar represents a state or union territory in India and the length of the bar represents the total number of doctors in that state. The figure shows that the total number of doctors in India is 2,058,204 which is equivalent to 1.5 doctors per 1000 population of the country.

The above figure highlights the availability of doctors in India across all the states. Uttar Pradesh (15.61) stands as an outlier with the highest number of doctors with respect to its population, followed by Goa (3.78) and Kerala (3.02). Many states lie below the national average of 1.5 doctors per 1000 members of the population. Daman & Diu has a severe shortage of doctors with just 0.20 doctors available for every 1000 members of their population. The national capital, Delhi, has 2.33 doctors available for every 1000 population.

Figure 8.32 is a scatter plot which is normally used to observe if there is any relationship between two quantitative variables. Each point on the scatter plot represents a state, and its position is found by using the total doctors of India as the X coordinate and the state populations as the Y coordinate. The scatter plot has a polynomial curve regressed in it with confidence bands. This represents the polynomial

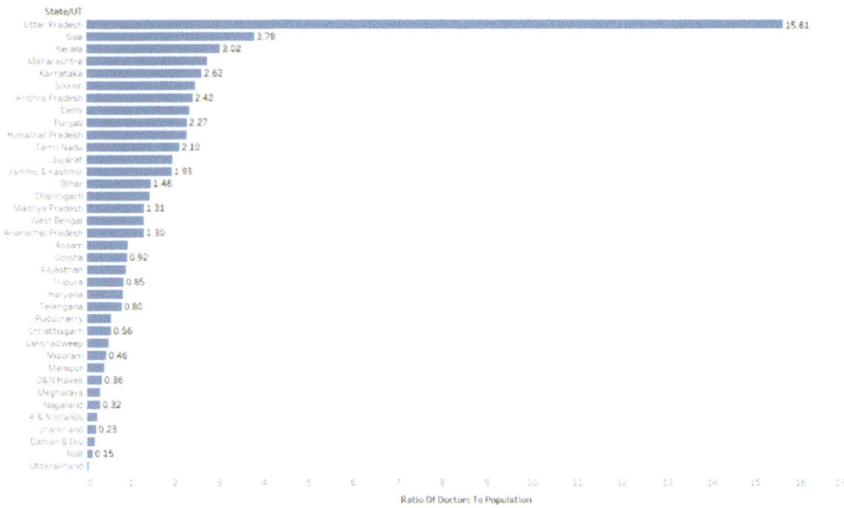

Fig. 8.31 Ratio of doctors to population

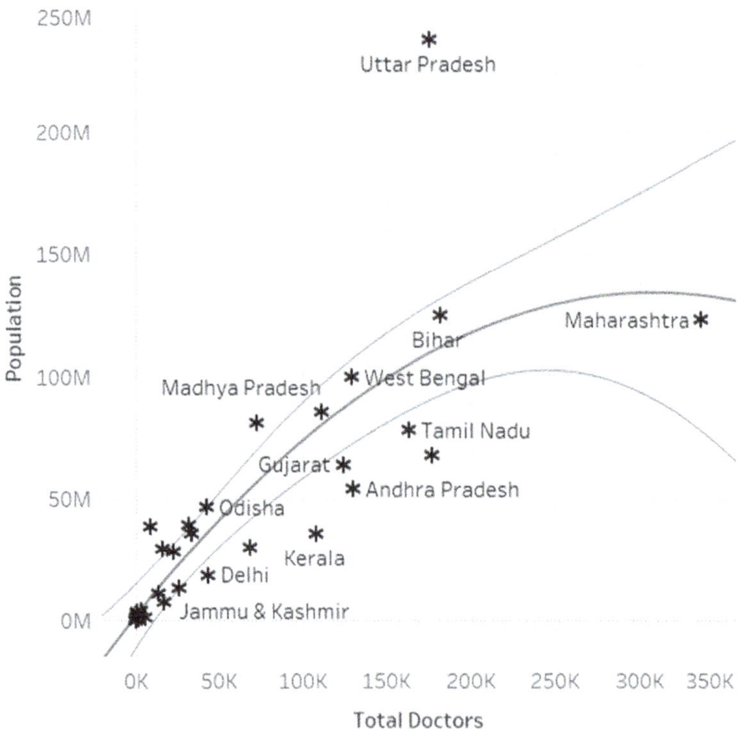

Fig. 8.32 Total doctors versus population

Table 8.10 Individual trend lines

Row	Column	*p*-value	DF	Term	Value	StdErr	*t*-value	*p*-value
Population	Total doctors	<0.0001	33	Total doctors^2	−0.001	0.0005651	−2.45	0.0195
				Total doctors	857.781	150.702	5.6918	<0.0001
				Intercept	1.65e+06	6.46e+06	0.2566	0.7990

regression model of the second degree that is fitted on the data. The outlier in this graph is Uttar Pradesh.

Trend Lines Model

A polynomial trend model of degree 2 is computed for population given total doctors. The model may be significant at $p \leq 0.05$.

Model formula	(Total Doctors^2 + Total Doctors + intercept)
Number of modelled observations	36
Number of filtered observations	0
Model degrees of freedom	3
Residual degrees of freedom (DF)	33
Sum squared error (SSE)	2.61258e+16
Mean squared error (MSE)	7.91691e+14
R-squared	0.701644
Standard error	2.8137e+07
p-value (significance)	<0.0001

Individual trend lines (as shown in Table 8.10).

The sign of the values of the parameters talks about the nature of the parameter. There is one parameter with negative values which means that the population decreases with increase in those parameters and two parameters with positive values which means population increase with their increase. The *p*-value of the trend model is <0.05. So, null hypothesis is rejected, and it can be concluded that the model is significant. The *p*-value of the intercept is >0.05, so we fail to reject the null hypothesis and conclude that its intercept is not significant. The *R*-squared value is 0.70 which signifies that the model is a moderately strong fit for the data. This implies that 70% of the variation in the number of doctors in India can be explained by the variation in the state populations.

8.3.3.2 Number of Nurses and Health Workers in India

Nurses and health workers may not get the same importance as doctors. Nevertheless, their availability in the states is crucial to a country whose healthcare system is under

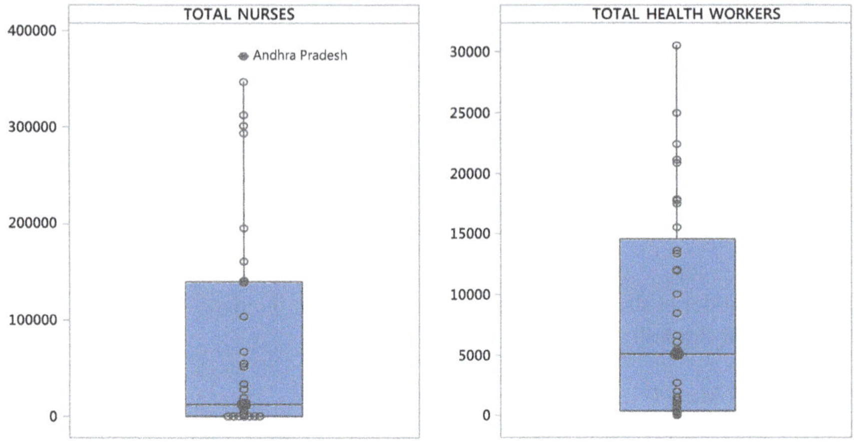

Fig. 8.33 Distribution of health workers and nurses

severe stress. Figure 8.33 is a combined box plot of health workers and nurses; this is a visualization technique used to describe the distribution of the data.

The following are the observation from the above figure:

1. The minimum number of health workers is 38.
2. The maximum number of nurses is 311,753 and that of health workers is 30,554.
3. The first quartile of nurses is 1123 and that of health workers is 554.
4. The median of nurses is 15,443 and that of health workers is 5215.
5. The third quartile of nurses is 138,985 and that of health workers is 14,548.
6. For nurses, the outliers are of Andhra Pradesh (373,536).
7. The column of nurses is highly positively skewed, and the figures are highly dispersed, whereas the column of health workers is positively skewed and the figures are dispersed among themselves.

Figure 8.34 consists of two horizontal bar charts where each bar represents a state or union territory of India. In the first figure, the length of the bar represents the percentage of the ratio of health workers to the population of that state. In the second figure, the length of the bar represents the percentage of the ratio of nurses to the population of that state. The total number of nurses in India is 2,973,462 which is equivalent to 0.217 per 100 population, and the number of health workers in India is 293,705 which is equivalent to 0.021 per 100 population of the country. From the combined bar plot of the ratio of health workers and nurses to 100 respective populations of the states, it can be concluded that Uttar Pradesh comes out on top which implies that Uttar Pradesh has the highest number of doctors, health workers and nurses with respect to its population. In the case of health workers, Lakshadweep (0.09) and Mizoram (0.08) are the second and third highest states after Uttar Pradesh (0.27). Similarly, in case of nurses, Kerala (0.844) and Andhra Pradesh (0.693) are the second and third highest states after Uttar Pradesh (1.226).

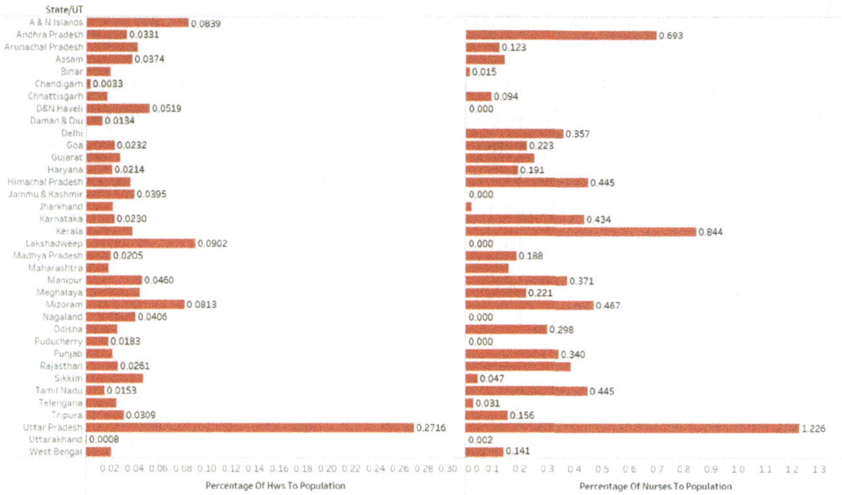

Fig. 8.34 Percentages of nurses and health workers to population

Figure 8.35 is a scatter plot which is normally used to observe if there is any relationship between two quantitative variables. Each point on the scatter plot represents a state, and its position is found by using as the total nurses and health workers in the state X coordinate and the state population as the Y coordinate. The outliers in both the figure are Uttar Pradesh. For the 'Total Nurses to Population' graph, the states of Andhra Pradesh, Tamil Nadu, Rajasthan, Kerala and Karnataka form a cluster at the right side—indicating an underlying similarity among these states. Uttar Pradesh has very few nurses with respect to population, thus resulting as an outlier.

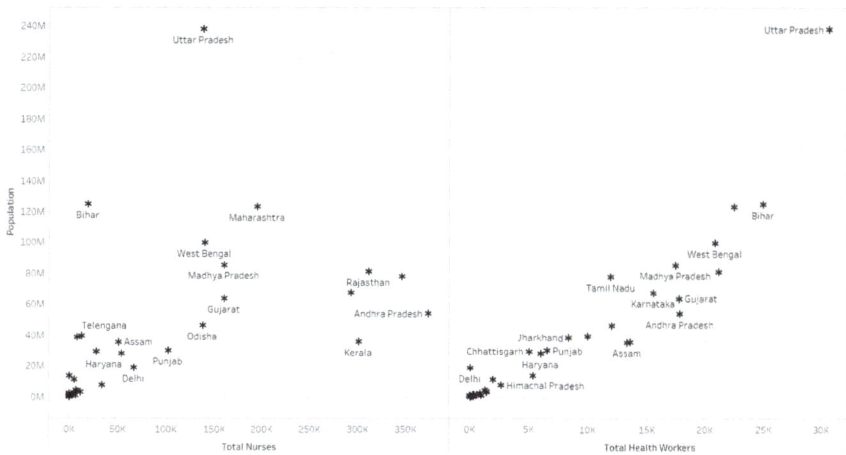

Fig. 8.35 Health workers and nurses versus population

8.3.3.3 Number of Pharmacists in India

Pharmacists provide medicines and additional services to the country, and they play a very important role in the healthcare system. It is interesting to look at the availability of pharmacists in the different states in India. Figure 8.36 is a box plot which is a visualization technique used to describe the distribution of the data. The boxplot of pharmacists was plotted according to their respective states.

The following are the observations from the figure:

1. The maximum number is 89,653.
2. The median is 12,591.
3. The first quartile range is 1087.
4. The third quartile range is 52,638.
5. Maharashtra is an outlier with value 233,337.
6. The column is highly positively skewed, and the figures are close to each other, i.e. not much dispersed.

Figure 8.37 is horizontal bar chart where each bar represents a state or union territory in India and the length of the bar represents the total number of pharmacists in that state. The state-wise bar plot of the pharmacists across India shows that Uttar Pradesh (0.7495 per 100 population) has the highest number followed by Chandigarh (0.37) and Maharashtra (0.1895).

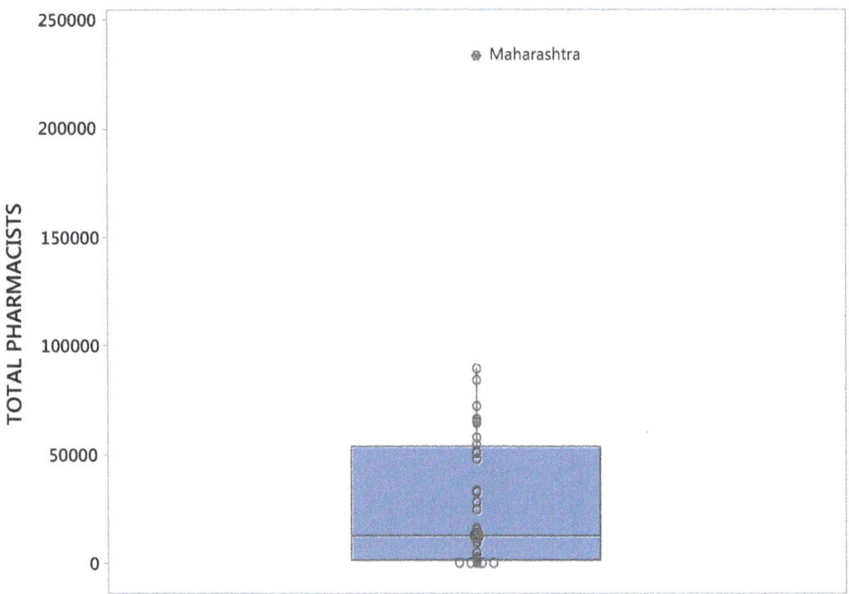

Fig. 8.36 Distribution of pharmacists across India

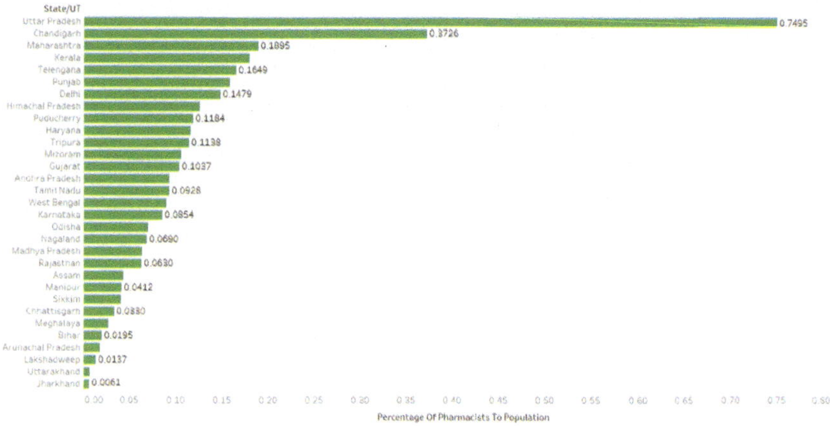

Fig. 8.37 Percentage of pharmacists to population

8.3.3.4 Admission Capacity of Medical Colleges for Different Courses

This section looks at different admission capacities of medical colleges for medical courses. This is done so that an estimate of the number of medical students in the country can be found. Medical students can aid doctors, nurses and health workers in the case that the healthcare system gets overloaded due to the pandemic. Figure 8.38 is a horizontal bar graph where each state is represented by a pair of different coloured bars. The purple bar represents the admission capacity of MBBS colleges, and the yellow bar represents the admission capacity of AYUSH colleges in each state. The length of the bar represents the respective admission capacities. Karnataka is followed by Maharashtra and Tamil Nadu which are the states having the highest medical college capacity. Maharashtra followed by Uttar Pradesh and Karnataka has maximum AYUSH courses capacity.

Figure 8.39 is a bar graph where each bar represents a state or union territory in India and the height of the bar represents the total admission capacity of post-graduate medical diploma courses in that state. The graph of diploma courses shows Maharashtra followed by Karnataka and Tamil Nadu as the states with the highest admission capacities for medical diploma courses. West Bengal, Odisha and Assam do not offer any such courses. The graph shows the capacities in decreasing order.

Figure 8.40 is a bar graph where each bar represents a state or union territory in India and the height of the bar represents the total admission capacity of post-graduate MD/MS courses in that state. The graph of MD/MS courses also shows Maharashtra followed by Karnataka and Tamil Nadu as the states with maximum admission capacity, whereas Sikkim and Meghalaya have the least. It is arranged in decreasing order of the figure.

Figure 8.41 is a bar graph where each bar represents a state or union territory in India and the height of the bar represents the total admission capacity of post-graduate DM/MCH courses in that state. The graph of DM/MCH courses shows

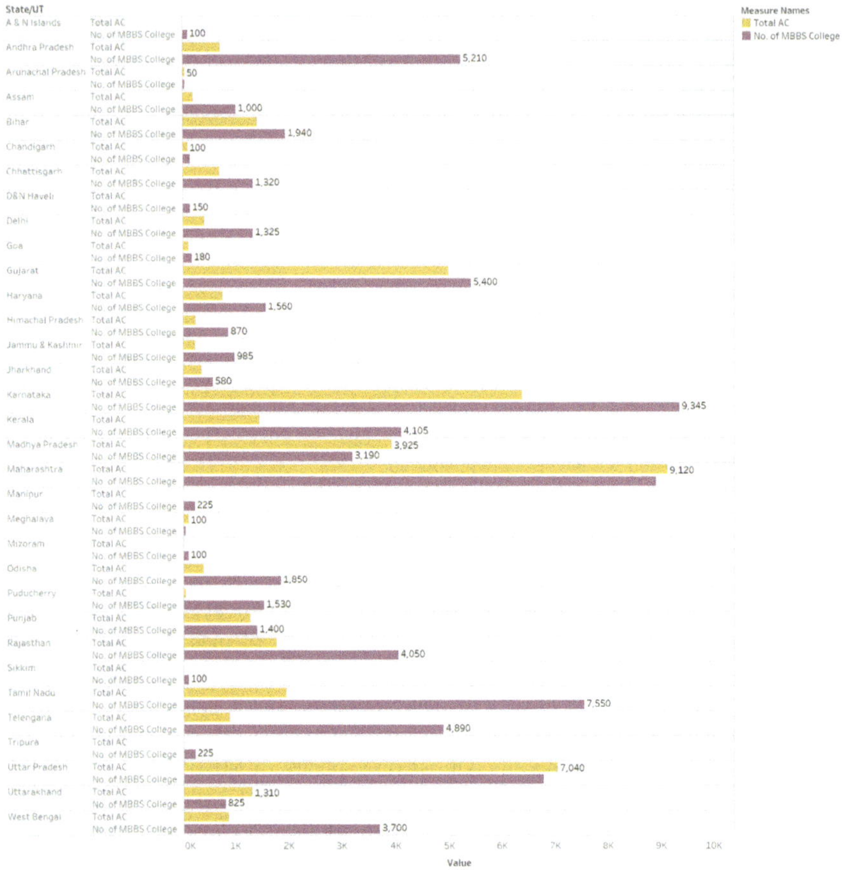

Fig. 8.38 Admission capacities of MBBS and AYUSH colleges

a decreasing order of DM/MCH admission capacity with Delhi having the highest capacity followed by Tamil Nadu, Maharashtra and Kerala. Jharkhand has the least capacity.

8.4 Conclusion

India has seen a steady increase in the number of COVID-19 cases added every day, and it does not look like it is going to improve. India also has a very high number of critical patients, when compared to five of the worst affected countries in the world. Maharashtra has the highest number of COVID-19 cases in India and makes up almost one-third of India's total cases. Tamil Nadu and Delhi are the next most affected states in India, but the situation is not nearly as bad as Maharashtra.

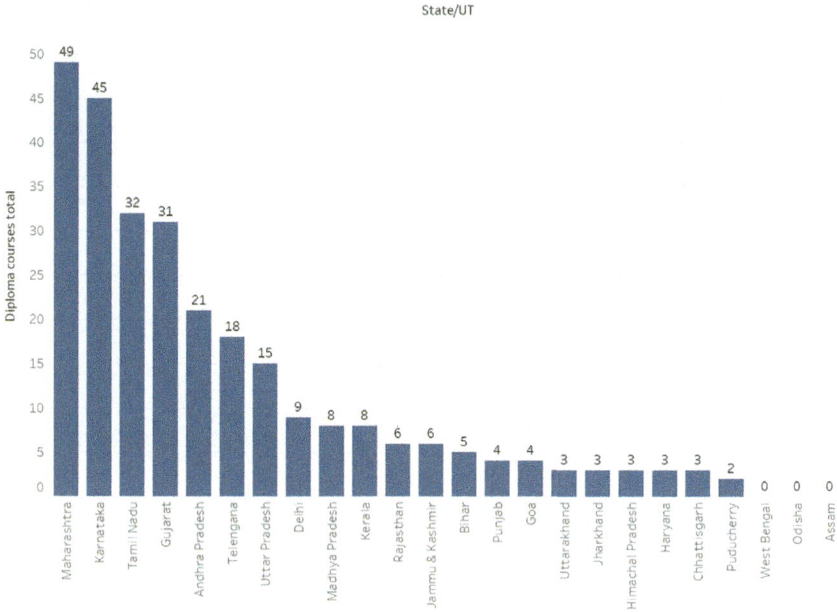

Fig. 8.39 Admission capacities in diploma courses

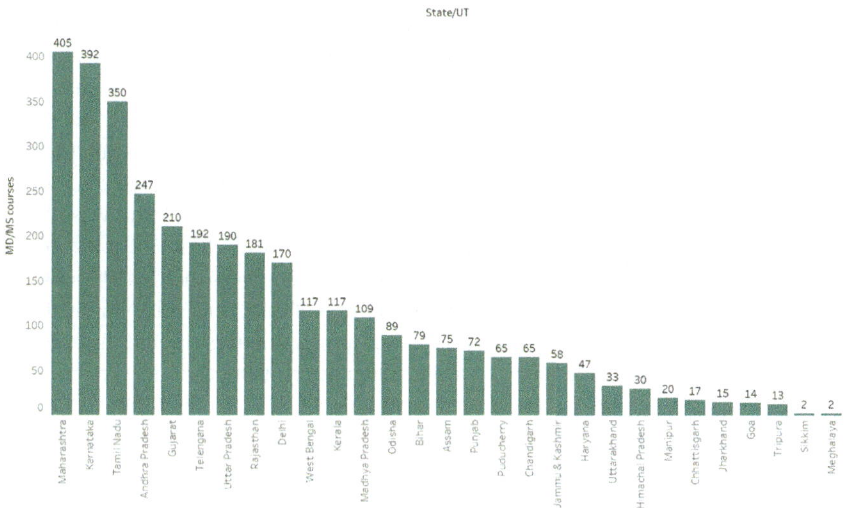

Fig. 8.40 Admission capacities in MD/MS courses

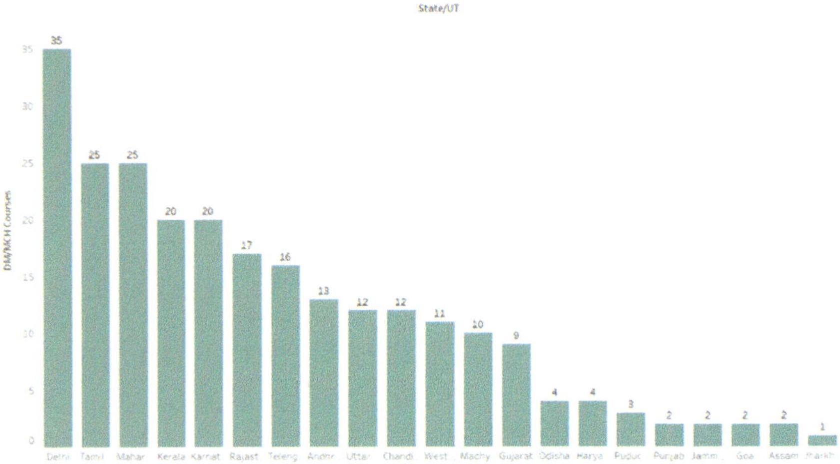

Fig. 8.41 Admission capacities in DM/MCH courses

In terms of testing, India has tested fewer samples than the more developed nations such as the US, UK and Russia. The developed nations, as shown above, have managed to flatten their curves (proving that the lockdown in these nations was effective) and look well on the way to recovery from COVID-19 (except Brazil). Brazil and India have not reached their peak yet as the number of cases continues to increase. This indicates that the lockdown has not been as effective in India as it has been in the more developed nations such as the USA, UK, Russia and Spain.

The healthcare infrastructure in India is very poorly equipped to handle the needs of its large population. It is high time to boost India's healthcare infrastructure and push for greater technological innovations in the healthcare sector. According to the WHO data, the general government expenditure on healthcare of India was 1% of the GDP in 2017, giving a rank of 165 out of 186 countries in terms of government expenditure on healthcare [11]. Other countries have spent more than 8–10% of GDP on health care and are still struggling with the burden of this pandemic, which makes a problem for the governments across the developing countries, including India. In order to cope with the inadequacy of the healthcare infrastructure, the government has plans in place to increase the current spending to 2.5% of the GDP within the next 2–3 years. The government has made some changes under Ayushman Bharat in order to strengthen the healthcare infrastructure in the form of the introduction of new tertiary institutes which has resulted in a further increase in the intake capacity of medical colleges. However, there is still significant room for improvement [12].

The severe shortage of human resources in health across different levels—between regions, between urban and rural areas, and between the private and public sector—may prove to be a catastrophe. There is, at the very least, a shortage of 90% of specialists in states like Chhattisgarh, Jharkhand and Rajasthan, while it is close to 86% in Uttarakhand and Odisha [13]. For example, the total population of Jharkhand was 38 million as per 2019 Indian population data, whereas the number of total required doctors in the system is only 8916. This is equivalent to only 0.23 doctors per 1000 members of the population—which is not desirable.

When it comes to the challenges facing the health infrastructure in India, the current state of the healthcare system in the country is not due to the lack of policies or managerial skills. Maharashtra, being the most affected state in India with COVID-19, has a moderate ratio of doctors to population, low number of health personnel, highest number of pharmacists with respect to other states and has good admission capacities for AYUSH colleges, medical colleges, diploma and MD/MS courses. However, Jharkhand has the lowest numbers in terms of admission capacities, pharmacists, heath workers and nurses and, most importantly, the lowest number of doctors with respect to population.

The scarcity of human resources appears to be a direct consequence of the absence of an integrated healthcare human resource development policy. According to WHO, India ranks 52nd out of the 57 countries facing a crisis in healthcare human resources [14]. The healthcare human resources have a major impact on the entire functioning of a modern integrated hospital. A shortage at any level of the healthcare personnel results in placing a heavier burden on the personnel who are present—who must work significantly harder and make up for the lack of numbers even in fair circumstances. This would, in turn, have a huge negative impact on the morale of the health workers. Nevertheless, health is a social phenomenon and hospitals are social institutions that should be looked after, strengthened and upgraded for the betterment of the society.

References

1. National Health Profile 2019
2. Population data of India state-wise 2019—http://statisticstimes.com/demographics/population-of-indian-states.php
3. Red, Orange and Green Zone data state-wise—https://static.mygov.in/rest/s3fs-public/mygov_158831498053877021.pdf
4. Medical Council of India—https://www.mciindia.org/CMS/
5. Ministry of Health and Family Welfare, India—https://www.mohfw.gov.in/
6. State COVID-19 data—https://www.kaggle.com/sudalairajkumar/covid19-in-india/data
7. State population 2020 estimates—http://www.indiaonlinepages.com/population/india-current-population.html
8. World COVID-19 data—https://www.worldometers.info/coronavirus/countries-where-coronavirus-has-spread/
9. In-depth world COVID-19 data—https://www.worldometers.info/coronavirus/#countries
10. https://japi.org/april2008/E-221.pdf
11. https://www.who.int/health-accounts/documentation/atlas.pdf?ua=1
12. https://www.expresshealthcare.in/blogs/covid-19-pandemic-long-term-impact-on-indian-healthcare/418670/
13. https://www.hindawi.com/journals/aph/2014/898502/
14. WHO, Primary Health Care: Report of the International Conference on Primary Health Care, Alma-Ata, USSR, Geneva, Switzerland, 1978
15. Brazil COVID-19 data—https://www.kaggle.com/unanimad/corona-virus-brazil?select=brazil_covid19.csv

16. USA COVID-19 data—https://www.kaggle.com/sudalairajkumar/covid19-in-usa?select=us_covid19_daily.csv
17. Russia COVID-19 data—https://www.kaggle.com/kapral42/covid19-russia-regions-cases?select=covid19-russia-cases.csv
18. UK COVID-19 data—https://www.kaggle.com/davehorton/uk-daily-confirmed-cases
19. Spain COVID-19 data—https://www.kaggle.com/danigarci1/covid19-in-spain?select=nacional_covid19.csv

Chapter 9
Estimating Cases for COVID-19 in India

9.1 Introduction

In the previous chapter, three datasets were analysed—COVID-19, healthcare human resources and healthcare infrastructure, to provide an insight into the situations of the various states and union territories of India with reference to corona cases. This information further is used as an input to predict the future. Predictions and forecasting are very important part of the study especially when dealing with sensitive issues like healthcare infrastructure and human resources during a pandemic. The goal at nation level is to flatten the daily added cases curve—as this indicates a decrease in the spread of the virus and more availability of beds and other healthcare facilities. To achieve this goal, all the rules and regulations for mitigating the virus should be strictly followed, as India is underprepared for handling the demands of the pandemic when it comes to healthcare infrastructure and human resources. The total susceptible cases that would be requiring ICU beds along with ventilators are estimated to be 322 million. As per the survey, the best-case scenario would have around 40,000 cases, most-likely scenario might have around 75,000 cases and a worst-case scenario projects 280,000 cases. The threat of community transmission of the virus is alarming, and it is clear that the distribution of the available hospital beds is not at all adequate for the number of affected patients.

The mild, severe and critical model, which is based on a study of 42,000 infections in China, was applied on the COVID-19 cases in India [1]. The result demonstrates that for patients below 60 years of age, the model shows 81% mild, 14% severe and 5% critical, whereas, for patients above 60 years of age, it is 70% mild, 20% severe and 10% critical [1]. In South Korea, the pandemic has passed its inflection point and the spread of the disease is currently under control. It was predicted that the situation would be under control by the end of March. In actuality too, the spread rate of virus has gradually decreased from 4.2 to 0.1 after control [2]—this is the success that

N. Sharma et al., *Open Data for Sustainable Community*,
Advances in Sustainability Science and Technology,
https://doi.org/10.1007/978-981-33-4312-2_9

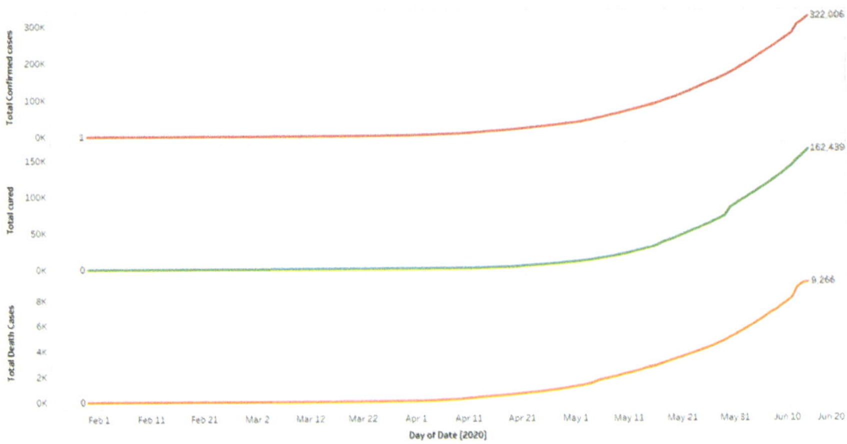

Fig. 9.1 Trend of total confirmed, recovered and death cases till 14th June 2020

India hopes to achieve as well. The government of India is now focusing on limiting the lockdown and concentrating on attacking the hotspots for COVID-19. After the Lockdown 4.0, India has entered the phase known as **Unlock 1.0**, which has resulted in a spike of **10,000+** cases per day.

India is a country with vast diversity, and it is necessary to learn about the increasing spread of COVID-19 in the different states of the country and also be able to predict the same. The current situation of India is shown in Fig. 9.1 with the cumulative number of infected cases and deaths from 30th January to 14th June 2020, whereas Fig. 9.2 presents the new cases reported. In this chapter, an attempt is made to forecast the number of cases of the country, using various algorithms.

9.2 Methodology

9.2.1 The Database

The database has variables such as Date, State, Confirmed, Recovered and Death Cases obtained from resources mentioned in Chap. 7 and also from Kaggle [3]. The extracted dataset consisted of cumulative numbers of COVID-19 cases, so separate dataset was created to store the data with respect to the date. These transformations on the data are performed using Microsoft Excel. Further, to convert this dataset into time-series data and applying statistical methods to it, R and python programming language are used.

The time-series data consists of four components:

Fig. 9.2 Daily new cases reported till May end

- Trend variations—They are the long-term, practically predictable pattern.
- Seasonal variations—They show repetitions after a specific period of time, and this can be days, weeks or months.
- Cyclical variations—They show cyclical patterns after a long period of time such as economic 'boom–bust' cycles.
- Random variations—They are truly random and exist in all time-series data, to some extent.

The COVID-19 data follows a steep, exponentially increasing trend but there is no seasonality present, which is clearly evident from Fig. 9.1, and applies to confirmed, recovered and the death cases.

9.2.2 The Models

The models that are built in this chapter are:

Polynomial Regression—It is simple to implement and gives good results for data having logistic curve
Long Short-Term Memory—It is a recurrent neural network that has proven to give accurate results in time-series forecasting

Autoregressive Integrated Moving Average—This type of model is highly used by statisticians to model various aspects of time-series data and analyse the trends in data to give accurate predictions. Another method known as exponential smoothing is also used in forecasting COVID-19 situations [4]

Prophet—It is forecasting model that works best with time series that have strong seasonal effects and several seasons of historical data.

9.3 Polynomial Regression

Polynomial regression is a type of regression analysis that analyses the relationship between the independent and dependent variable named X and Y, respectively. It is modelled in such a way that the dependent variable Y depends on nth degree polynomial in X. The objective of regression analysis is to model the expected value of Y depending on the value of independent variable X. In this model, X is a single independent variable that represents the number of days since 1st Jan 2020. The dataset consists of data from 1st January 2020 to 1st June 2020; thus, 1st Jan is Day 0 and 1st June is Day 152. Here, X is a vector of dimension (153, 1) which consists of the value of X for each day and Y is a vector of the same dimensions which consists of the number of newly reported cases on a particular day. The degree of the model (n) is 5.

9.3.1 Why Polynomial Regression Model?

After reviewing several hypotheses and mathematical models, it was concluded that the graph for the total number of cases against number of days, forms a logistic curve. A logistic curve rises very slowly in the beginning, then increases rapidly for over a certain range and flattens out after the point of inflection. The graph shown in Fig. 9.3 is the graph of the overall cases against the number of days in Italy; as it is seen, Italy is one of the nations that have passed its inflection point. Thus, the thought was to use polynomial regression to trace out such a function for India as well.

9.3.2 Polynomial Regression—Model A

There are two ways to train a polynomial regression model. The simpler way is to train the model to fit the curve of newly reported cases each day against the day since the first case was reported. The trained model can accurately predict the count of cases recorded each day till the desired date. This model performs very well when the count of cases recorded each day continues to increase. In the dataset used, the count

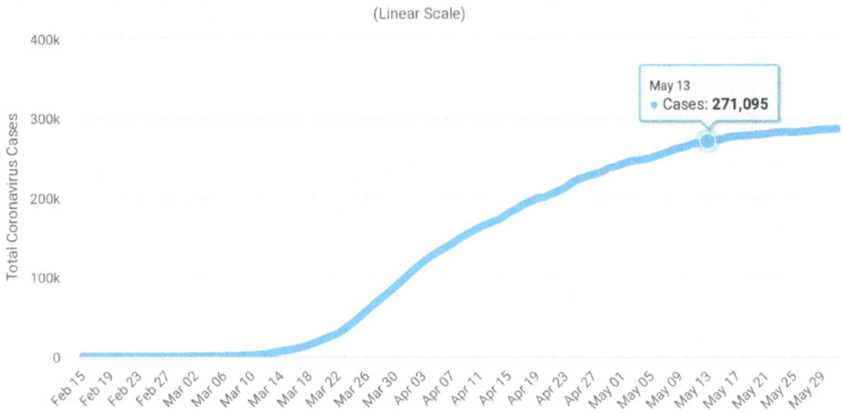

Fig. 9.3 Total coronavirus cases in Italy

of cases recorded is generally increasing continuously; however on certain days, it was observed to be slightly on the lower side. This phenomenon can be regarded as 'noise' in the dataset and might make it harder to train a model to make accurate predictions. However, the model adjusts automatically for the noise in the dataset and makes accurate predictions despite the presence of noticeable noise. The model shown in Fig. 9.4 is a model of type A and is capable of making accurate predictions which can be seen when the comparing the actual results to the predictions of the model.

As shown in Fig. 9.4, the blue points are the data points plotted using a scatter plot. The red curve is the model trained using the polynomial regression algorithm,

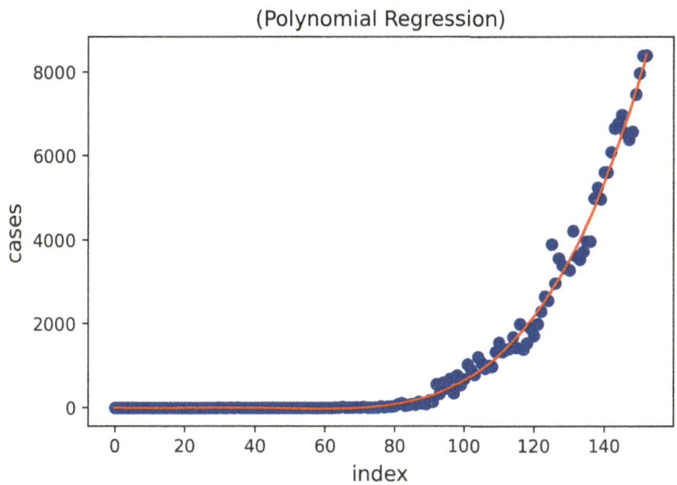

Fig. 9.4 Scatter plot of the number of cases

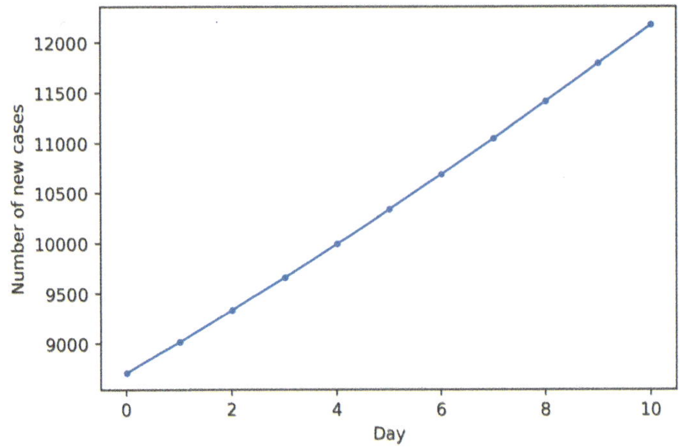

Fig. 9.5 Growth pattern in the number of new cases

and it can be seen that despite the noise in the dataset, the model adjusts very well and is not overfitting at all. Figure 9.5 shows the growth pattern in the number of new cases recorded.

Final predictions of Model A:
The total number of cases in India according to the polynomial regression model is **302,454** on 12th June 2020. The actual number of cases in India by 12th June 2020 was **297,535**.

9.3.3 Polynomial Regression—Model B

Another way to train a polynomial regression model to predict the number of cases of coronavirus is to train the model to trace the curve of cumulative number of confirmed cases rather than newly reported cases each day. This model can also accurately predict the number of confirmed cases, while adjusting very well to the noise in the dataset. A model of type B was trained on a dataset that contains the COVID-19 case data from 1st January 2020 to 17th June 2020 as Fig. 9.6. The polynomial degree of this model is also 5. The model predicts the number of confirmed cases till 30th June 2020.

As shown in Fig. 9.6, the blue dots are the data points plotted using a scatter plot. The red curve is the model trained using the polynomial regression algorithm. The model is a good fit to the training set. Now, let us see the predictions made by this model and the growth pattern depicted in Fig. 9.7.

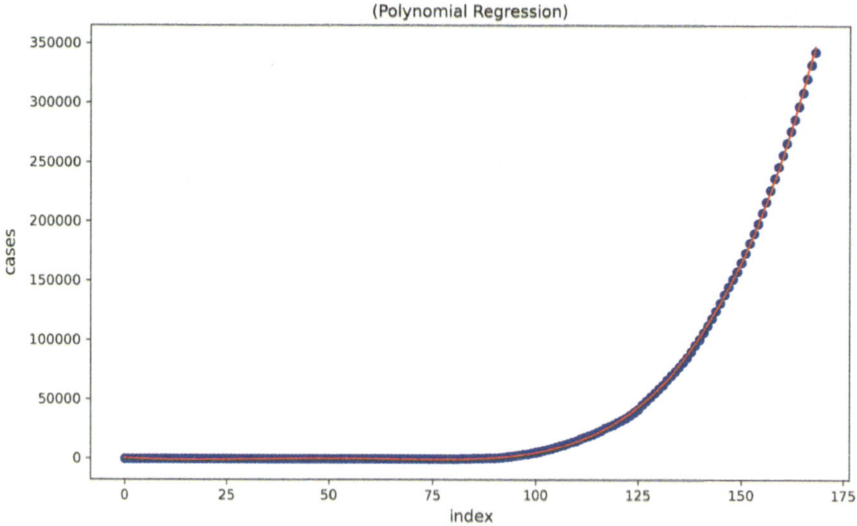

Fig. 9.6 Total number of confirmed cases against days

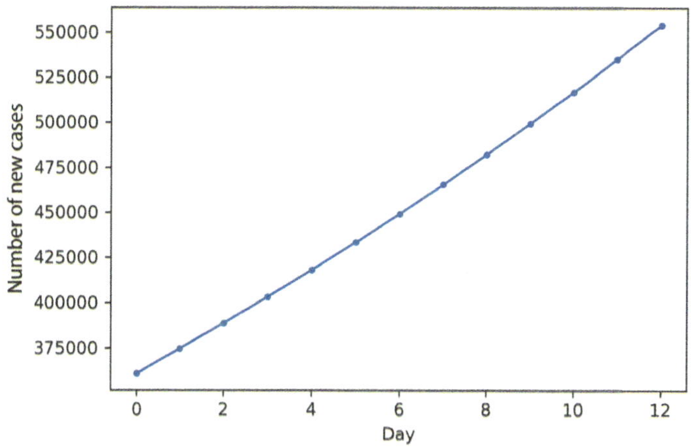

Fig. 9.7 Growth pattern in the number of new cases

The model predicts that the growth in the total number of confirmed cases will be linear. The X-axis in the graph represents the number of days since 17th June, and the Y-axis shows the total number of confirmed cases.

Final predictions:
The model predicts that the total number of confirmed cases in India by 30th June 2020 will be **554,063**. The actual number of cases in India by 30th June 2020 was **566,840**.

9.4 Long Short-Term Memory (LSTM)

LSTM stands for long short-term memory, and it is an artificial recurrent neural network architecture that is used in deep learning. Unlike the standard feedforward neural networks, LSTM has feedback loops, hence making it a useful tool for forecasting. It has the ability to process entire sequences of data. LSTM uses techniques like 'gating' to improve forecasting by disregarding previous inputs after a certain window. The LSTM model used in this chapter consisted of the following features:

- Number of hidden layers = 512
- Sequence length = 5
- Number of layers = 2
- Number of features = 1
- Dropout = 0.5.

9.4.1 Why LSTM?

LSTMs are considered most feasible for prediction tasks. Since the considered dataset is a time-series data, regressive models fail to consider the additional complexity of sequence dependency on input data added by the time-series data. LSTMs have proven to accurately recollect or overlook things and forecast the future values.

9.4.2 LSTM Model

Figure 9.8 shows the predictions for daily added cases made using the LSTM algorithm using the given code. The *X*-axis captures the dates, and the *Y*-axis captures newly added cases per day. The actual confirmed cases are marked with a blue line, and the predicted cases are marked in yellow. LSTM predicts an upwards trend in the daily added cases.

```
plt.plot(total_cases, label = “Historical Daily Cases”)
plt.plot(predicted_cases, label = “P + redicted Daily Cases”)
plt.legend()
```

Figure 9.9 shows the predictions for total confirmed cases made using the LSTM algorithm using the given code. The *X*-axis captures the dates, and the *Y*-axis captures the total confirmed cases per day. The actual confirmed cases are marked with a blue line, and the predicted cases are marked in yellow. LSTM predicts an upwards trend in the total confirmed cases as well.

```
plt.plot(total_cases, label = “Historical Daily Cases”)
plt.plot(predicted_cases, label = “P+redicted Daily Cases”)
plt.legend()
```

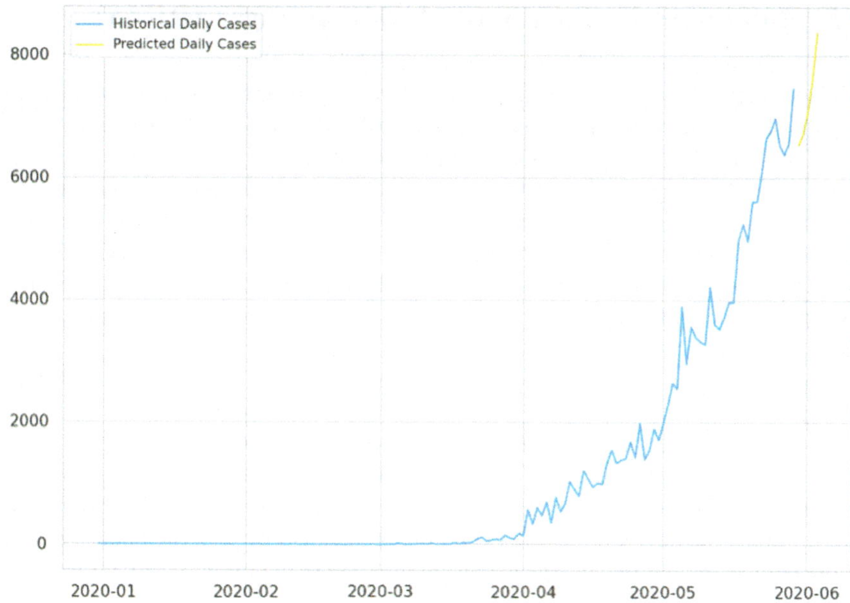

Fig. 9.8 New cases per day predictions

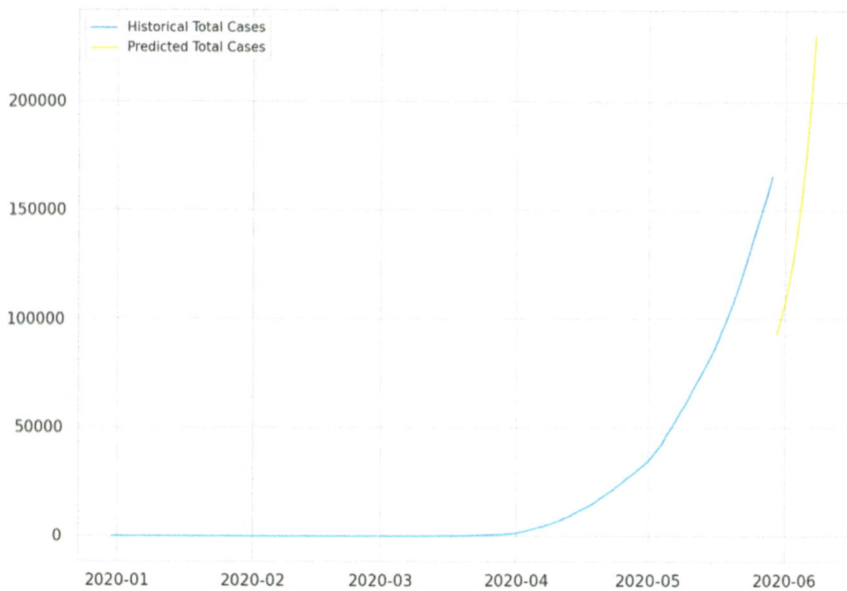

Fig. 9.9 Total cases per day prediction

9.5 Autoregressive Integrated Moving Average (ARIMA)

ARIMA stands for autoregressive integrated moving average and is just a generalization of the autoregressive moving average (ARMA) model in time-series analysis. These models are fitted to understand the data and also for predicting short-term future data points. The ARIMA model converts a non-stationary sequence to a stationary sequence by eliminating the trend in the non-stationary data using the differencing technique. However, after converting the data to a stationary series, an ARMA model is applied. Autoregressive model forecasts future values by regressing over past values. Moving averages model forecasts future values by averaging the previous values.

The time-series data available for building a model is non-seasonal and non-stationary. To determine the AR, differences and MA terms, visualizations such as autocorrelation function (ACF) and partial autocorrelation function (PACF) plots are used. To determine the degree of the AR component (denoted by 'p'), the indicator used is the number of significant spikes outside the shaded region in the PACF plot. To determine the degree of the MA component (denoted by 'q'), the indicator used is the number of significant spikes outside the shaded region in the ACF plot. The ARMA model after differencing is known as ARIMA (p, d, q).

The Akaike information criterion (AIC) is used to select the best performing model for forecasting. Mathematically,

$$\text{AIC} = \log\left(\frac{\sum_{i=1}^{T} e}{T}\right) + \frac{2p}{T} \tag{1}$$

The model which has the least AIC is selected as the best model. In order to study the stationarity of the series, the augmented Dickey–Fuller (ADF) test is performed. The null hypothesis is rejected, when the p-value is lower than the level of significance (which is taken as 0.05).

9.5.1 *Why Is ARIMA Preferred to Exponential Smoothing?*

Exponential smoothing is a method used for making forecasts and it does not take into consideration the correlation between the consecutive values of a time series. For making prediction intervals for the forecast, the forecast errors should not be correlated and should be normally distributed with zero mean and constant variance. ARIMA models use an explicit statistical model for the random component of a time series that gives nonzero autocorrelations in the irregular components.

9.5.2 ARIMA Model A

Figures 9.10 and 9.11 are ACF and PACF plots—they measure the autocorrelation and partial autocorrelation in the time-series data—and as explained above, these plots are used to determine the correct parameters to be used in the ARIMA model.

The model is trained with parameters ($p = 3, d = 2, q = 3$) and the model summary is given below:

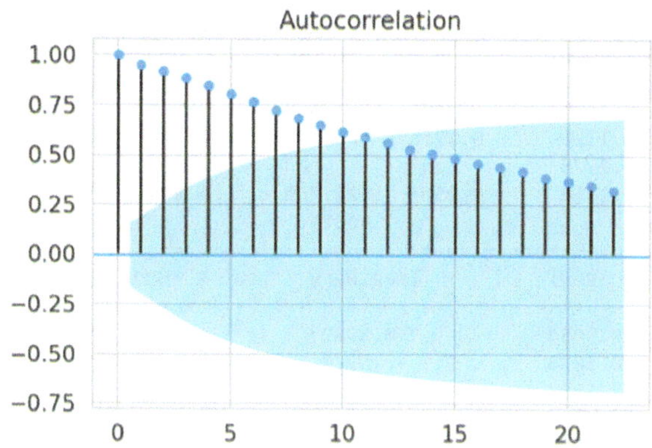

Fig. 9.10 Autocorrelation function plot for Model A

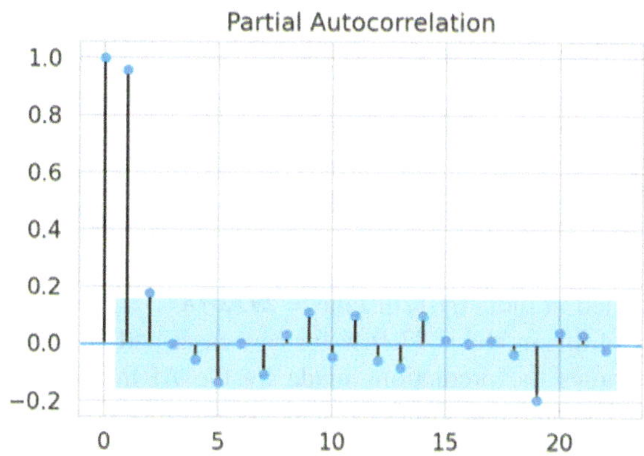

Fig. 9.11 Partial autocorrelation function plot for Model B

```
                              ARIMA Model Results
==============================================================================
Dep. Variable:                   D2.y   No. Observations:                  149
Model:                 ARIMA(3, 2, 3)   Log Likelihood               -1011.382
Method:                       css-mle   S.D. of innovations                nan
Date:                Sun, 14 Jun 2020   AIC                           2038.764
Time:                        18:50:15   BIC                           2062.795
Sample:                    01-02-2020   HQIC                          2048.527
                         - 05-29-2020
==============================================================================
                 coef    std err          z      P>|z|      [0.025      0.975]
------------------------------------------------------------------------------
const          0.0232      2.744      0.008      0.993      -5.355       5.401
ar.L1.D2.y    -1.4266        nan        nan        nan         nan         nan
ar.L2.D2.y    -1.3970        nan        nan        nan         nan         nan
ar.L3.D2.y    -0.4439        nan        nan        nan         nan         nan
ma.L1.D2.y     0.1320      0.125      1.057      0.290      -0.113       0.377
ma.L2.D2.y     0.2376        nan        nan        nan         nan         nan
ma.L3.D2.y    -1.1458      0.050    -22.769      0.000      -1.244      -1.047
                                    Roots
=============================================================================
                  Real          Imaginary           Modulus         Frequency
-----------------------------------------------------------------------------
AR.1           -0.5054           -0.8939j            1.0269           -0.3319
AR.2           -0.5054           +0.8939j            1.0269            0.3319
AR.3           -2.1363           -0.0000j            2.1363           -0.5000
MA.1           -0.4328           -0.7913j            0.9019           -0.3297
MA.2           -0.4328           +0.7913j            0.9019            0.3297
MA.3            1.0729           -0.0000j            1.0729           -0.0000
-----------------------------------------------------------------------------
```

Looking at the model the PACF and ACF plots and model summary, the parameters can be determined as:

- p (AR) = 3
- d (Differences) = 2
- q (MA) = 3.

Results:
Total cases predicted in India by 12th June = **293,078**.

The accuracy of this model is 97.14% for new cases and 98.58% for total cases.

Figure 9.12 shows the predictions made by the ARIMA model for the total confirmed cases. The blue line shows the actual observations, and the red line shows the predicted observations. The ARIMA model also predicts a continued increase in the total confirmed cases.

Table 9.1 presents a chart for predictions—made using the ARIMA model for total confirmed cases and daily added cases and Fig. 9.13 cross-validates the prediction.

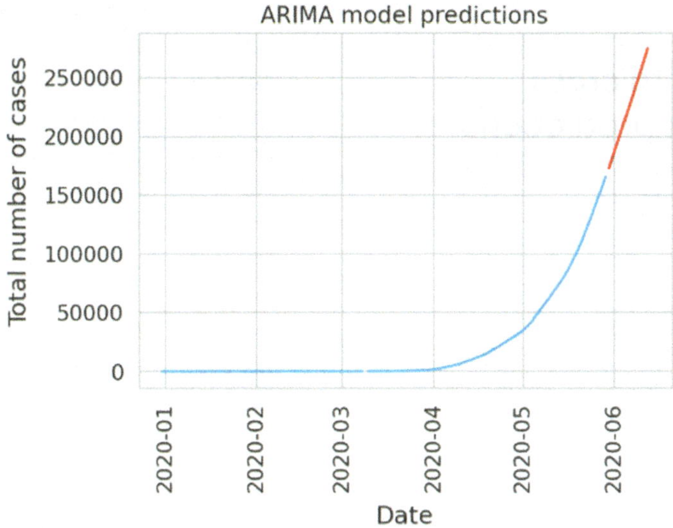

Fig. 9.12 Total cases prediction

Table 9.1 Predictions made using ARIMA model

Date	Actual new cases	Actual total cases	Predicted new cases	Predicted total cases
2020-06-01	7901	191,327	7905	189,119
2020-06-02	8390	198,706	8385	197,505
2020-06-03	8909	207,615	8494	205,999
2020-06-04	9304	216,919	8641	214,640
2020-06-05	9851	226,770	9065	223,704
2020-06-06	9887	236,657	9205	232,910
2020-06-07	9971	246,628	9346	242,256
2020-06-08	9983	256,611	9760	252,016
2020-06-09	9987	266,859	9910	261,926
2020-06-10	9985	276,583	10,055	271,980
2020-06-11	9996	286,579	10,454	282,434
2020-06-12	10,962	297,535	10,615	293,049

9.5.3 ARIMA Model B

Firstly, the ADF test is performed to check for the stationarity of the time series.

OUTPUT:

Dickey–Fuller = 4.649, Lag order = 5, **p-value = 0.99**
Alternative hypothesis: stationary.

Chart to cross check predictions

Total cases in india on
2020-06-01 : TC: 1,91,327
2020-06-02: TC: 1,98,706, NC: 8390
2020-06-03: TC: 2,07,615, NC: 8909
2020-06-04: TC: 2,16,919, NC: 9304
2020-06-05: TC: 2,26,770, NC: 9851
2020-06-06: TC: 2,36,657, NC: 9887
(The government relaxed the lockdown from this day)
2020-06-07: TC: 2,46,628, NC: 9971
2020-06-08: TC: 2,56,611 NC: 9983
2020-06-09: TC: 2,66,598 NC: 9987
2020-06-10: TC: 2,76,583 NC: 9985
2020-06-11: TC: 2,86,579 NC: 9996
2020-06-11: TC: 2,86,579 NC: 9996
2020-06-11: TC: 2,97,535 NC: 10962

	NewCases	TotalCases
2020-05-30	7567.752285	173366.752285
2020-05-31	7847.604103	181214.356388
2020-06-01	7904.927969	189119.284357
2020-06-02	8385.358900	197504.643257
2020-06-03	8494.083485	205998.726742
2020-06-04	8640.889870	214639.616612
2020-06-05	9064.927905	223704.544517
2020-06-06	9205.357371	232909.901888
2020-06-07	9346.289082	242256.190971
2020-06-08	9759.745710	252015.936680
2020-06-09	9909.700718	261925.637398
2020-06-10	10054.729912	271980.367310
2020-06-11	10454.027364	282434.394674
2020-06-12	10614.526377	293048.921051

Fig. 9.13 Cross checking of predicted cases

The *p*-value shows us that the stationarity of the time series is rejected and must be treated with differencing i.e. d = 2. After differencing, the ADF test is performed again.

OUTPUT:

Dickey–Fuller = -9.204, Lag order = 5, **p-value = 0.01**
alternative hypothesis: stationary.

This confirms that the time-series data is now stationary.

To find the optimal parameters of the prediction model, the ACF and PACF charts are plotted. From Figs. 9.14 and 9.15, the ACF shows one spike and PACF shows several significant spikes.

ARIMA (1, 2, 1) is selected based on the charts [5]. Many alternative models are also considered, and their AIC values are shown in Table 9.2.

A model with the least AIC value has well-behaved residuals. According to the table, the model ARIMA (1, 2, 3) is selected since it has the least AIC value. Using this, ARIMA (1,2,3) model observes 498,838.0 [95% PI(472,782.5, 524,893.5)] cumulative infected cases between by 24th June 2020, results are shown in Table 9.3 and plotted in Fig. 9.16.

The confirmed cases are forecasted for the next ten days till 24th June 2020. On 147th day, the count reaches 4,98,838; i.e. *the total number of confirmed cases reaches around 4,98,838 on 24th June 2020.* The figure above shows the predicted plots in blue with silver shade surrounding it denoting the confidence interval.

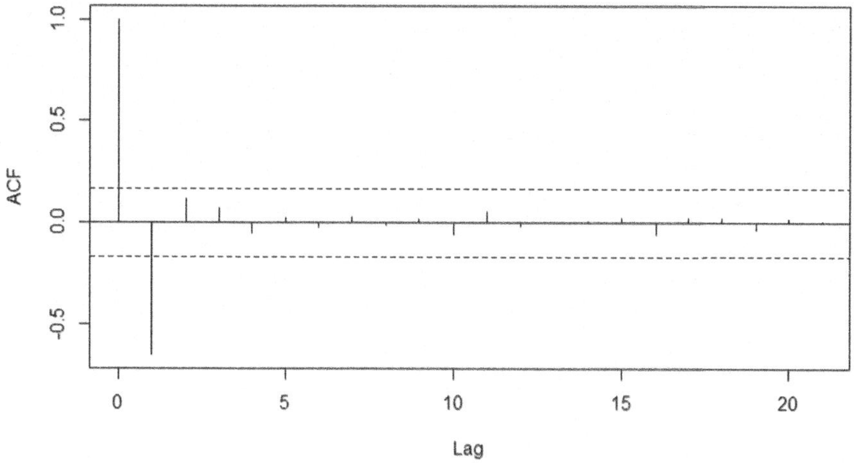

Fig. 9.14 Autocorrelation function plot for Model B

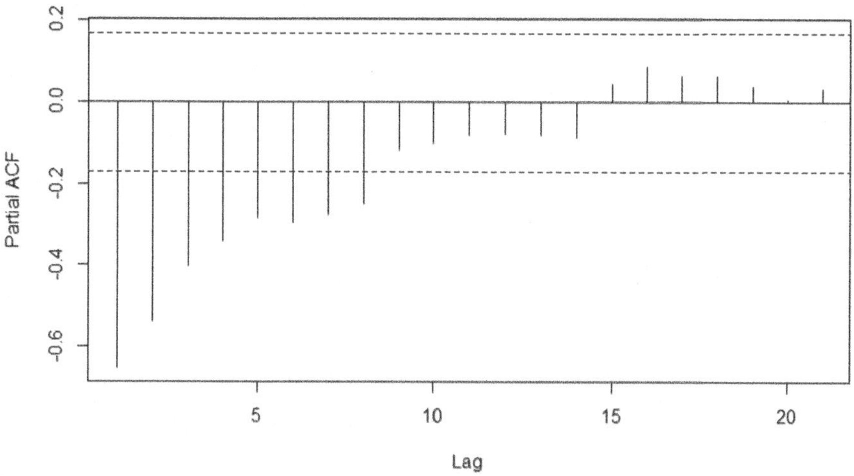

Fig. 9.15 Partial autocorrelation function plot for Model B

Table 9.2 ARIMA model parameters and their AIC value

ARIMA model	AIC value
ARIMA (0,2,2)	2320.45
ARIMA (1,2,2)	2304.77
ARIMA (0,2,1)	2323.14
ARIMA (1,2,1)	2322.71
ARIMA (1,2,3)	2299.18

Table 9.3 Forecasted points with 80 and 95% prediction interval

Point	Forecast	Lo 80	Hi 80	Lo 95	Hi 95
138	343,395.7	341,943.5	344,848.0	341,174.7	345,616.7
139	359,417.0	357,290.4	361,543.7	356,164.6	362,669.5
140	377,304.7	374,633.1	379,976.2	373,218.9	381,390.4
141	394,497.9	390,473.1	398,522.7	388,342.6	400,653.2
142	411,949.5	406,339.1	417,559.9	403,369.1	420,529.9
143	429,305.0	421,795.6	436,814.3	417,820.5	440,789.4
144	446,696.2	437,080.9	456,311.5	431,990.9	461,401.4
145	464,074.1	452,154.2	475,994.0	445,844.2	482,304.0
146	481,457.0	467,059.5	495,854.5	459,437.9	503,476.0
147	498,838.0	481,801.3	515,874.8	472,782.5	524,893.5

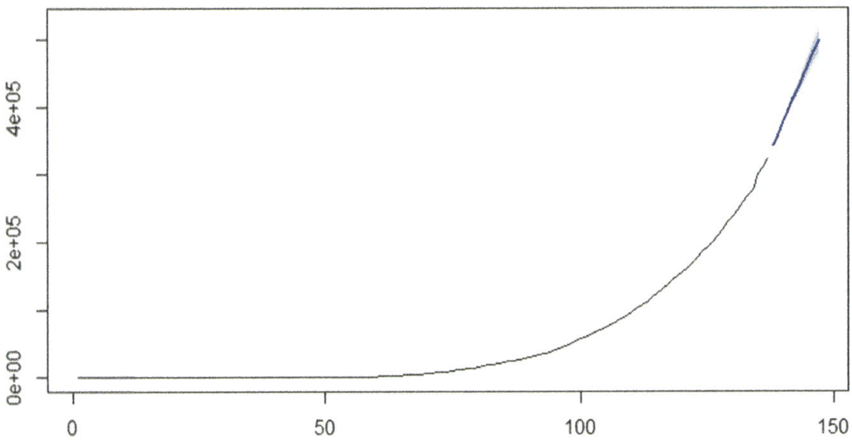

Fig. 9.16 Prediction points of the total number of confirmed cases as of 24th July 2020

9.5.4 ARIMA Model C

The procedure for Model B is repeated for Model C, using the recovered cases. The ADF test is performed to check for stationarity of the data with no differencing, and it returns the following result:

OUTPUT:

Dickey–Fuller = 3.8425, Lag order = 5, **p-value = 0.99**
Alternative hypothesis: stationary.

Next, with differencing = 2,
OUTPUT:

Dickey–Fuller = -9.7474, Lag order = 5, **p-value = 0.01**

Alternative hypothesis: stationary.

The first p-value indicates that the stationarity of the time series is rejected and must be treated with differencing, i.e. $d = 2$. After differencing, the ADF test is performed once again and it shows stationarity. To find out the parameters of the prediction model, the ACF and PACF charts are plotted. From Figs. 9.17 and 9.18, the ACF shows two spikes and PACF shows several significant spikes.

ARIMA (1, 2, 2) is selected based on the charts. Many alternative models are also considered, and their AIC values are shown in Table 9.4.

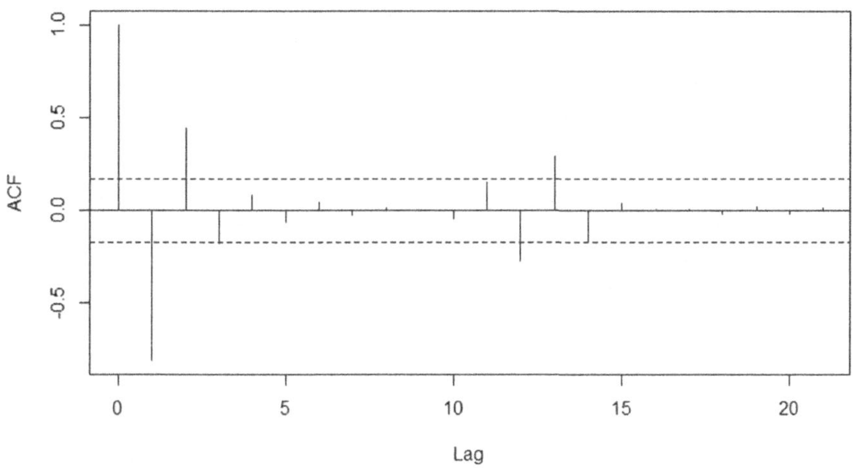

Fig. 9.17 Autocorrelation function plot for Model C

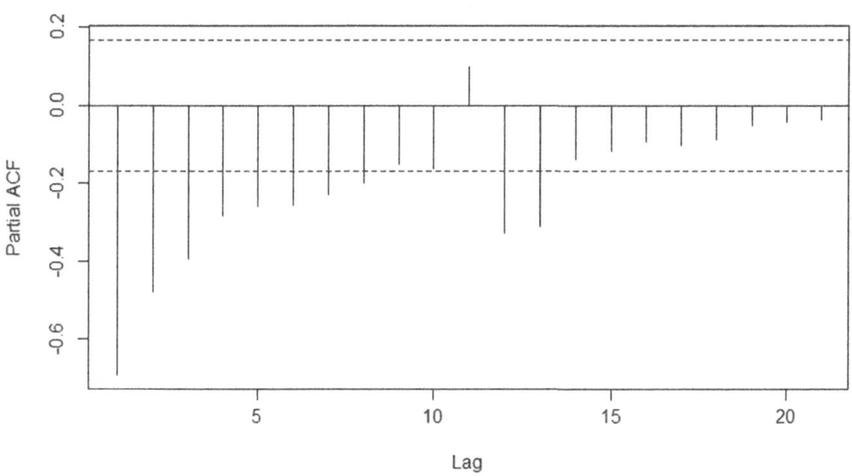

Fig. 9.18 Partial autocorrelation function plot for Model C

Table 9.4 ARIMA model parameters and their AIC values

ARIMA Model	AIC value
ARIMA (0,2,2)	2194.13
ARIMA (1,2,2)	2195.86
ARIMA (0,2,1)	2192.60
ARIMA (1,2,1)	2194.09
ARIMA (1,2,3)	2197.86

Table 9.5 Forecasted points with 80 and 95% prediction interval

Point	Forecast	Lo 80	Hi 80	Lo 95	Hi 95
138	168,984.3	167,958.9	170,009.7	167,416.0	170,552.6
139	175,529.6	173,821.8	177,237.4	172,917.7	178,141.5
140	182,074.9	179,661.0	184,488.8	178,383.1	185,766.7
141	188,620.2	185,455.7	191,784.7	183,780.5	193,459.9
142	195,165.5	191,202.0	199,129.0	189,103.8	201,227.2
143	201,710.8	196,899.9	206,521.7	194,353.2	209,068.5
144	208,256.1	202,550.7	213,961.5	199,530.4	216,981.8
145	214,801.4	208,156.0	221,446.8	204,638.1	224,964.7
146	221,346.7	213,717.4	228,976.0	209,678.7	233,014.7
147	227,892.0	219,236.6	236,547.4	214,654.7	241,129.3

A model with the least AIC value has well-behaved residuals. According to the table, we select the model ARIMA (0, 2, 1) since it has the least AIC value. Using this, ARIMA (0, 2, 1) model observe 227,892.0[95% PI (214,654.7, 241,129.3)] cumulative recovered cases, results are shown in Table 9.5 and plotted in Fig. 9.19.

The recovered cases are forecasted for the next ten days, till 24th June 2020. On 147th day, the count reaches 2,27,892; i.e. *the total number of recovered cases reaches around 2,27,892 on 24th June 2020*. The figure above shows the predicted plots in blue with a silver shade surrounding it which denotes the confidence interval.

9.5.5 ARIMA Model D

Model D is built using the same procedure as above, using the death cases.

Before starting the procedure, the time series must be made stationary. For investigating the stationarity of time series, we use ADF test, and results are shown below.

OUTPUT:

For $d = 2$:

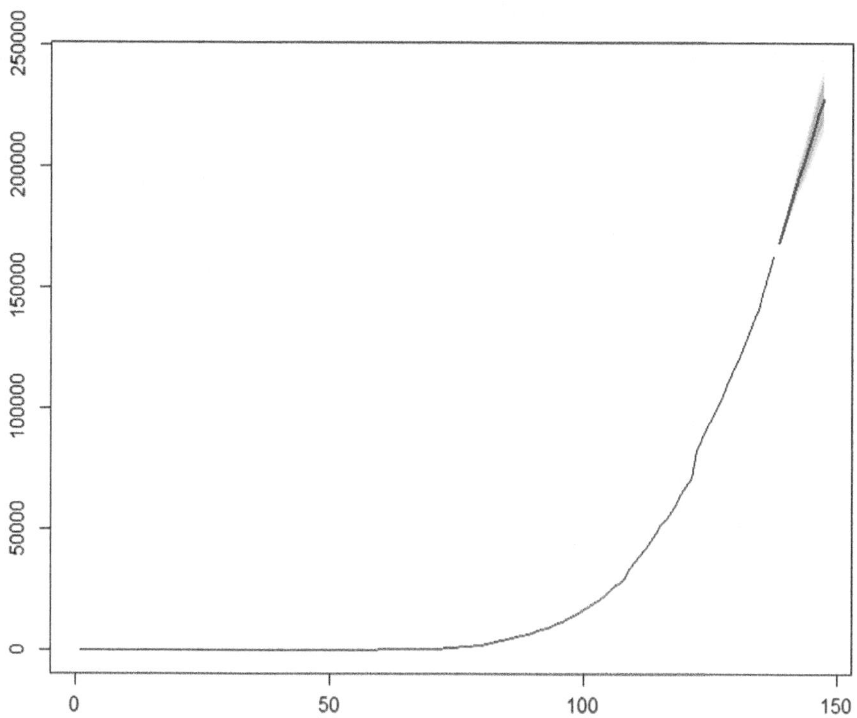

Fig. 9.19 Prediction points of the total number of recovered cases as of 24th July 2020

Dickey–Fuller = −11.428, Lag order = 5, **p-value = 0.01**
Alternative hypothesis: stationary.

The first p-value indicates that the stationarity of the time series is rejected and must be treated with differencing, i.e. $d = 2$. After differencing, the ADF test is performed once again and it shows stationarity. To find out the parameters of the prediction model, the ACF and PACF charts are plotted. From Figs. 9.20 and 9.21, the ACF shows one spikes and PACF shows no significant spikes.

ARIMA (0, 2, 1) is selected based on the charts. Many alternative models are also considered, and their AIC values are shown in Table 9.6.

A model with the least AIC value has well-behaved residuals. According to the table, we select the model ARIMA (1,2,3) since it has the least AIC value. Using this, ARIMA (1,2,3) model observes 13,620.0 [95% PI (13,062.313 −14,178.07)] cumulative death cases by 24th June 2020, and the results are shown in Table 9.7 and plotted in Figs. 9.21 and 9.22.

The death cases are forecasted for the next ten days, till 24th June 2020. On the 147th day, the count reaches 13,620; i.e. *the total number of death cases reaches around 13,620 on 24th June 2020*. The figure above shows the predicted plots in blue with silver shade surrounding it denotes the confidence interval.

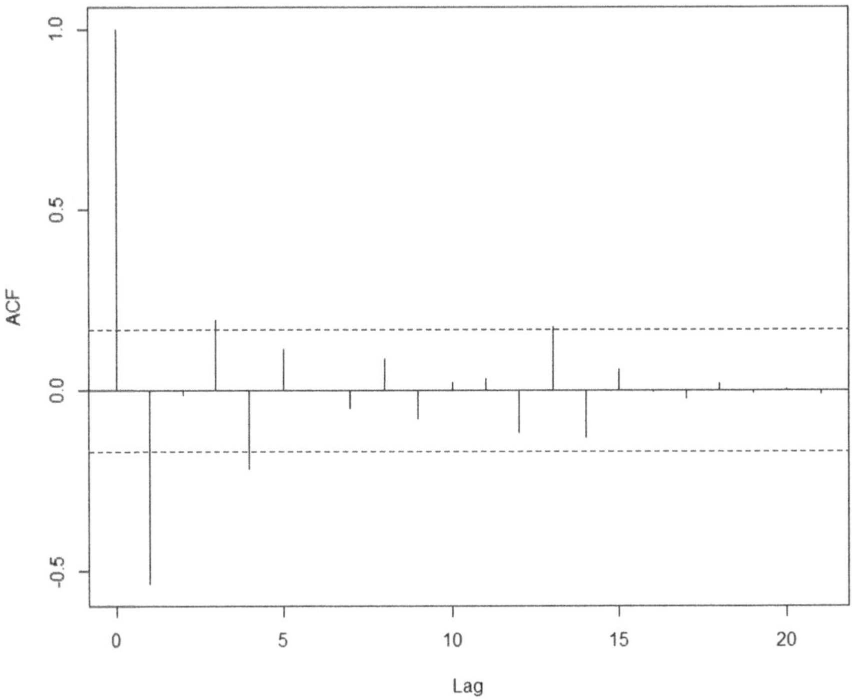

Fig. 9.20 Autocorrelation function plot for Model D

9.5.6 Validation of Models B, C and D

As of 15th June 2020, i.e. forecast point 138, there were **3,43,071** cumulative infected cases; **1,80,318** recovered cases; and **9918** cumulative deaths in India. Our forecasting results predict **3,43,395.7** [PI 95% (341,174.7, 345,616.7)] cumulative infected cases using ARIMA(1,2,3) model; cumulative recovered cases are predicted as **1,68,984** [PI 95% (167,416.0, 170,552.6)] using ARIMA(0,2,1) model; and cumulative deaths are predicted to be at **9670** [PI 95% (9580.734, 9759.39)] using ARIMA(1,2,3) model.

As of 16th June 2020, i.e. forecast point 139, there were **3,47,810** cumulative infected cases; **1,84,244** recovered cases and **10,015** deaths in India. Our forecasting results predicts **3,59,417** [PI 95% (356,164.6, 362,669.5)] cumulative number of infected cases using ARIMA(1,2,3) model; cumulative recovered cases are predicted as **1,75,529.6** [PI 95% (172,917.7, 178,141.5)] using ARIMA(0,2,1) model; and cumulative deaths are predicted as **10,141.5** [PI 95% (9996.051, 10,286.94)] using ARIMA(1,2,3) model.

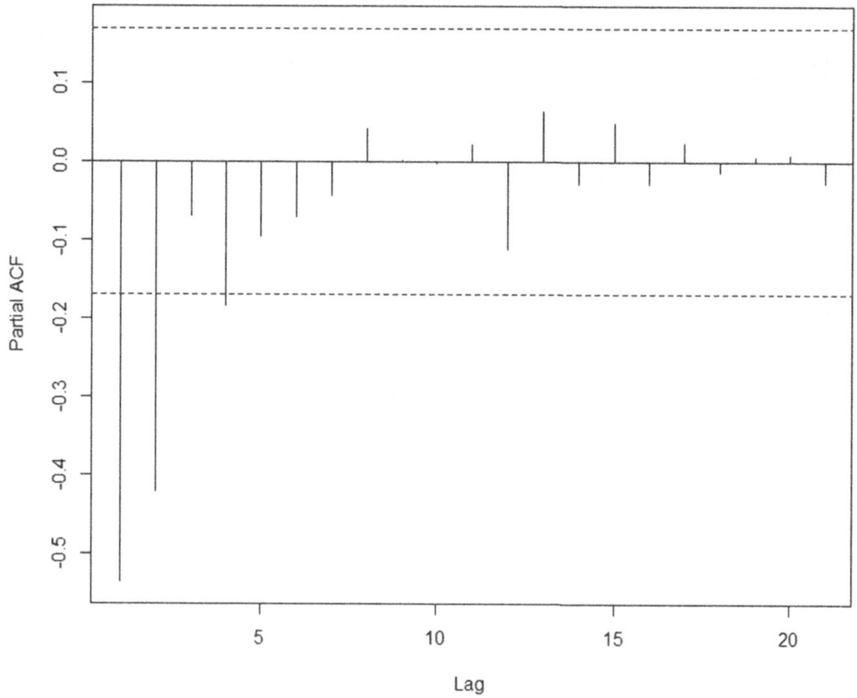

Fig. 9.21 Autocorrelation function plot for Model D

Table 9.6 ARIMA model parameters and their AIC value

ARIMA Model	AIC value
ARIMA (0,2,2)	1468.67
ARIMA (1,2,2)	1458.8
ARIMA (0,2,1)	1477.36
ARIMA (0,2,3)	1437.62
ARIMA (1,2,3)	1430.7

9.6 Prophet

Prophet follows the sklearn model API (sklearn is popular Python library used for machine learning techniques). It uses a class for forecasting time-series data based on an additive model where nonlinear trends are fitted with yearly, weekly and daily seasonality, including holiday effects. To predict using the prophet class, the fit method is used to fit the data and the predict method is used to predict for n days in the future. It is tolerant to missing data, outliers and shifts in trends. The prophet model is fully automatic. The input to prophet is a fixed format which is a data frame

Table 9.7 Forecasted points with 80 and 95% prediction interval

Point	Forecast	Lo 80	Hi 80	Lo 95	Hi 95
138	9670.1	9611.7	9728.4	9580.7	9759.4
139	10,141.5	10,046.4	10,236.5	9996.1	10,286.9
140	10,587.0	10,479.0	10,695.0	10,421.7	10,752.2
141	11,023.8	10,900.7	11,146.8	10,835.5	11,212.0
142	11,457.6	11,312.3	11,602.9	11,235.3	11,679.9
143	11,890.5	11,714.5	12,066.4	11,621.3	12,159.6
144	12,323.0	12,108.8	12,537.2	11,995.3	12,650.7
145	12,755.4	12,496.3	13,014.5	12,359.1	13,151.7
146	13,187.8	12,878.3	13,497.3	12,714.4	13,661.2
147	13,620.2	13,255.4	13,984.9	13,062.3	14,178.1

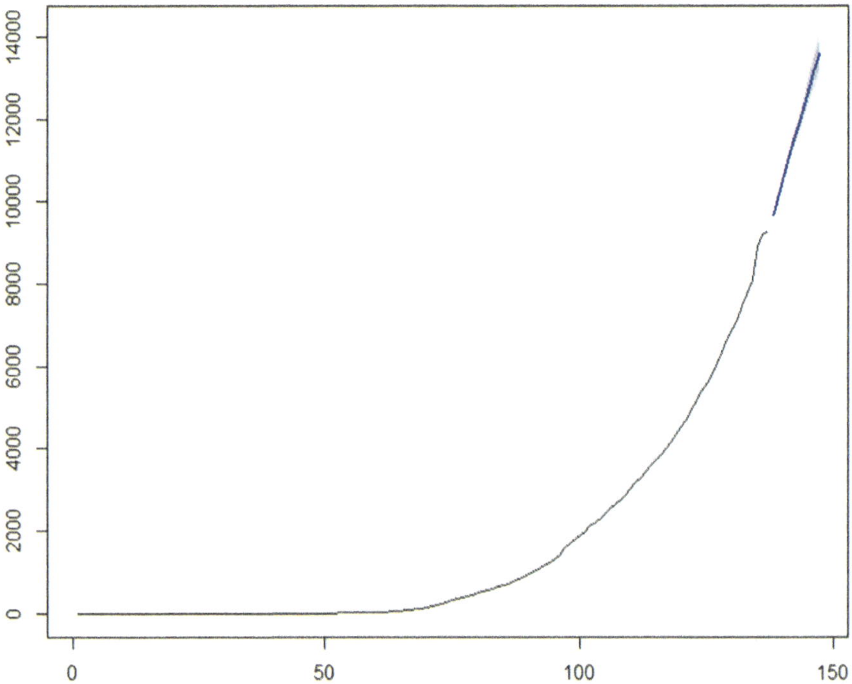

Fig. 9.22 Prediction points of the total number of death cases as of 24th June 2020

containing columns 'ds' and 'y', as shown as in Table 9.8, where 'ds' columns have the date stamp and y columns have the values that needs to be forecasted.

Table 9.8 Input data frame

ds	y
2019-12-31	0
2020-01-01	0
2020-01-02	0
2020-01-03	0
2020-01-04	0
2020-01-05	0

Table 9.9 Output data frame

Ds	yhat	yhat_lower	yhat_upper
2020-05-29	6926	6508	7325
2020-05-30	7081	6647	7510
2020-05-31	7303	6889	7714
2020-06-01	7477	7035	7903
2020-06-02	7600	7161	7996
2020-06-03	7727	7228	8166
2020-06-04	7889	7480	8339
2020-06-05	7095	7675	8532
2020-06-06	7251	7825	8679
2020-06-07	8743	8038	8919

9.6.1 Why Prophet?

Prophet does not require the user to have much prior knowledge or experience of forecasting. It is a highly optimized model provided by Facebook. Prophet utilizes Bayesian-based curve fitting method to forecast the time-series data and automatically finds the seasonal trends and makes prediction in a user-friendly visual representation.

The output data format includes columns 'ds', 'yhat', 'yhat_lower', 'yhat_upper' as shown in Table 9.9, where 'ds' is the date stamp, 'yhat' is the forecasted value, 'yhat_lower' is the lower limit of the forecasted value and 'yhat_upper' is the upper limit of the forecasted value.

9.6.2 Prophet Model

Total cases predicted in India by 12th June 2020 = **28,644** using prophet model as shown in Fig. 9.23.

```
new_cases_forecast = m.plot(forecast)
```

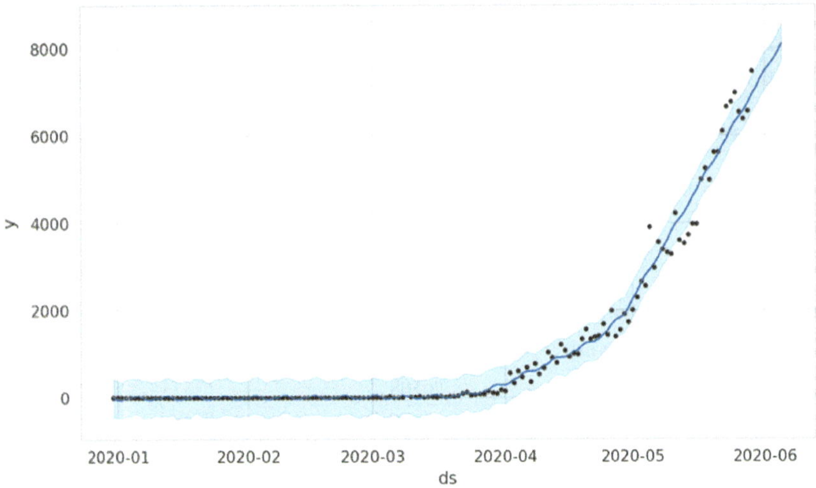

Fig. 9.23 New cases per day prediction using prophet

Figure 9.24 shows the overall trend plot, and it also captures the weekly seasonality of the new cases. This plot shows that the number of COVID-19 cases confirmed daily peaks on Fridays and is minimum on Saturdays. The reason for this pattern

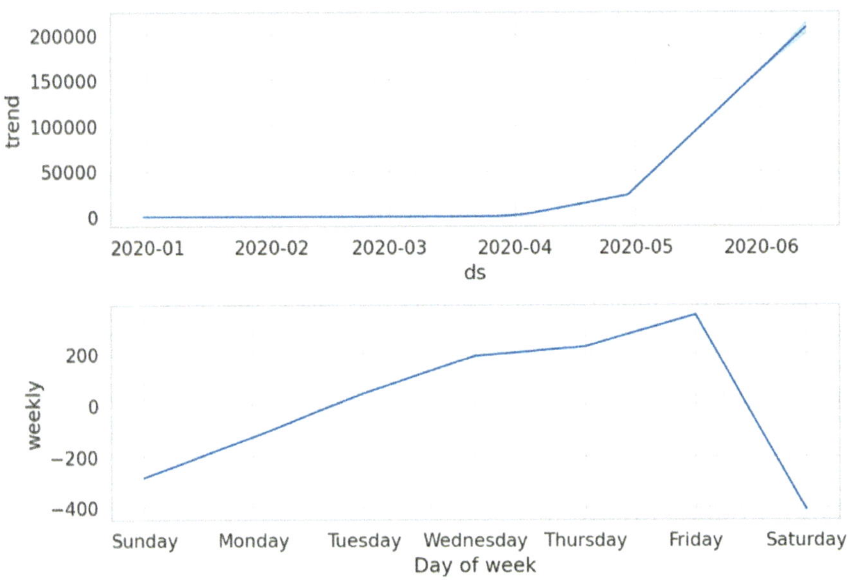

Fig. 9.24 Trend plot depicting weekly new cases

might be that the number of people going outdoors increased at the end of the week, on a Friday, thus increasing the number of cases reported. But after getting the news of the number of cases reported, people took precautionary measures and the number of people going outdoors decreased to a great extent. Another reason for this trend can be that the announcements made by the government were mostly during the weekend, thus resulting in a dip in the number of reported cases.

The accuracy of this model is 91.5% for new cases and 98.38% for total cases.

Table 9.10 presents a chart for predictions—made using the prophet model for total confirmed cases and daily added cases and Fig. 9.25 cross-validates the prediction.

Table 9.10 Predictions made using prophet

ds	Actual new cases	Actual total cases	Predicted new cases	Predicted total cases
2020-06-01	7901	191,327	7884	188,771
2020-06-02	8390	198,706	7997	196,768
2020-06-03	8909	207,615	8155	204,924
2020-06-04	9304	216,919	8322	213,246
2020-06-05	9851	226,770	8543	221,790
2020-06-06	9887	236,657	8683	230,474
2020-06-07	9971	246,628	8910	239,385
2020-06-08	9983	256,611	9128	248,513
2020-06-09	9987	266,859	9238	257,752
2020-06-10	9985	276,583	9360	267,112
2020-06-11	9996	286,579	9580	276,693
2020-06-12	10,962	297,535	9750	286,444

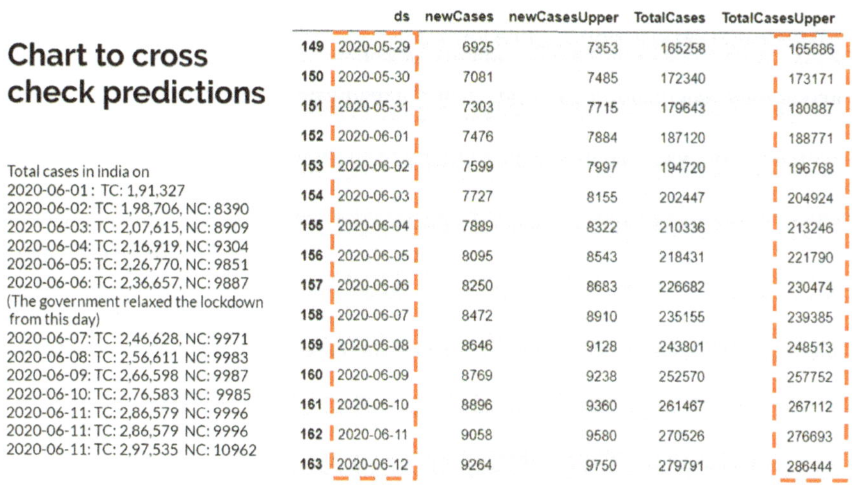
Chart to cross check predictions

Total cases in india on
2020-06-01 : TC: 1,91,327
2020-06-02: TC: 1,98,706, NC: 8390
2020-06-03: TC: 2,07,615, NC: 8909
2020-06-04: TC: 2,16,919, NC: 9304
2020-06-05: TC: 2,26,770, NC: 9851
2020-06-06: TC: 2,36,657, NC: 9887
(The government relaxed the lockdown from this day)
2020-06-07: TC: 2,46,628, NC: 9971
2020-06-08: TC: 2,56,611 NC: 9983
2020-06-09: TC: 2,66,598 NC: 9987
2020-06-10: TC: 2,76,583 NC: 9985
2020-06-11: TC: 2,86,579 NC: 9996
2020-06-11: TC: 2,86,579 NC: 9996
2020-06-11: TC: 2,97,535 NC: 10962

	ds	newCases	newCasesUpper	TotalCases	TotalCasesUpper
149	2020-05-29	6925	7353	165258	165686
150	2020-05-30	7081	7485	172340	173171
151	2020-05-31	7303	7715	179643	180887
152	2020-06-01	7476	7884	187120	188771
153	2020-06-02	7599	7997	194720	196768
154	2020-06-03	7727	8155	202447	204924
155	2020-06-04	7889	8322	210336	213246
156	2020-06-05	8095	8543	218431	221790
157	2020-06-06	8250	8683	226682	230474
158	2020-06-07	8472	8910	235155	239385
159	2020-06-08	8646	9128	243801	248513
160	2020-06-09	8769	9238	252570	257752
161	2020-06-10	8896	9360	261467	267112
162	2020-06-11	9058	9580	270526	276693
163	2020-06-12	9264	9750	279791	286444

Fig. 9.25 Cross-check the prediction done using prophet model

9.7 Conclusions

Multiple models, ranging from simple polynomial regression to forecasting models like LSTM, ARIMA, and prophet, were compared to predict the growth in the number of cases over some period in the near future. The predictions using polynomial regression and ARIMA models are very close to the actual number of cases (total and new cases daily) with ARIMA model A having the maximum accuracy of 98.58% and can be further optimized to model the data and get even more accurate results.

Analysis on the difference of rate of recovered cases to the rate of daily new cases, with any forecasting model, could potentially be used to forecast the date when the cases will plateau. When the difference between the growth rate and the recovery rate becomes zero, it can be deduced that the pandemic has plateaued; and when the difference is positive, it is indicative of the infection rate decreasing. Thus, a simple index is proposed which can help understand the situation in any pandemic:

$$\text{Pandemic State(PS)} = \text{(Rate of daily new cases)(Rate of recovery)}$$

This index can be interpreted as follows:

- Reduction and recovery, if $\text{PS} > 0$
- Plateauing, if $\text{PS} = 0$
- Increasing infection rate, if $\text{PS} < 0$.

The accuracy of the predictions is highly dependent on uncertainties such as political, social and environmental situations which greatly influence the spread of a pandemic. These predictions should be taken as a reference or guideline and not as prophecy, as these are drawn from the patterns in the data—such as the past number of cases—and do not consider any other factors that might hugely influence these numbers in the future.

With this ground situation, India has chosen to lift its lockdown, leaving it to the citizens to maintain physical distancing in crowded market places, malls, restaurants and religious congregations. The physical distancing norms were also in place during the entire duration of multi-phased lockdown. However, despite these measures, COVID-19 numbers increased exponentially from a few hundred on March 25, when the nationwide lockdown was enforced, to over 2.55 lakh when unlock 1.0 kicked off on June 8. Figures 9.26, 9.27 and 9.28 present the confirmed, recovered and death cases for top-5 affected stated of India, including the forecast of the number of cases [6].

These forecasts have been made using exponential smoothing and are made till 20th June 2020. The forecasts show a steep increase in the curve for most of the states.

- Maharashtra has the maximum number of confirmed, recovered and death cases. The forecast shows the number of confirmed cases could probably increase to 117,420, the recovered cases to 59,256 and death cases to 4423 as of 20th June 2020.

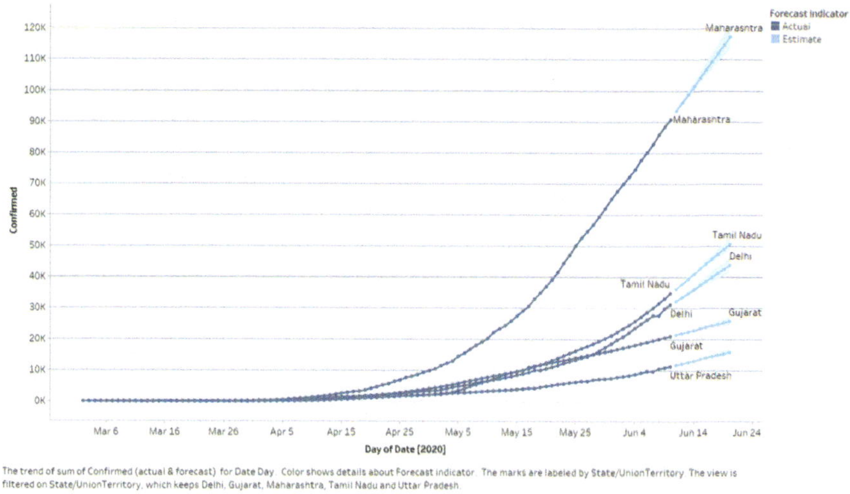

Fig. 9.26 Top 5 COVID-19 affected states with forecasts of total confirmed cases

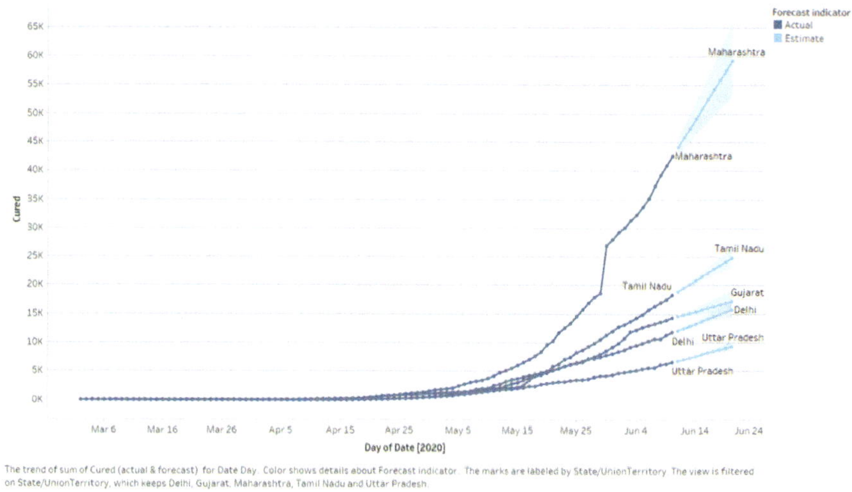

Fig. 9.27 Top 5 COVID-19 affected states with forecasts of total recovered cases

- Tamil Nadu has the second highest number of confirmed cases and recovered cases. Its forecast shows that the number of confirmed cases could probably increase to 50,711, the recovered cases to 24,857 and death cases to 491 as of 20th June 2020. It has the lowest number of deaths compared to the other four states.
- Delhi has the third highest number of confirmed and death cases and fourth highest recovered cases. Its forecast shows that the number of confirmed cases could

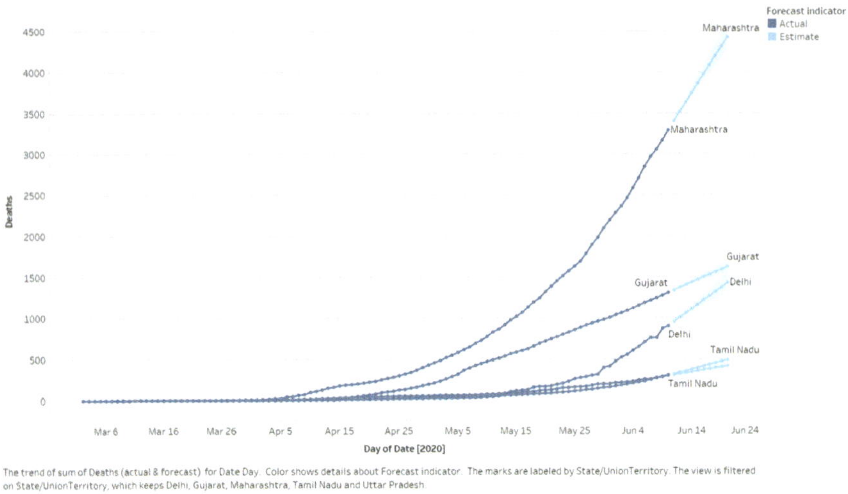

Fig. 9.28 Top 5 COVID-19 affected states with forecasts of total death cases

probably increase to 43,959, the recovered cases to 15,791 and death cases to 1433 as of 20th June 2020.

- Gujarat has the fourth highest number of confirmed, second highest number of death cases and third highest recovered cases. Its forecast shows that the number of confirmed cases could probably increase to 25,894, the recovered cases to 17,235 and death cases to 1629 as of 20th June 2020.
- Uttar Pradesh the fifth highest number of confirmed, recovered and death cases. The forecast shows the number of confirmed cases could probably increase to 15,941, the recovered cases to 9381 and death cases to 419 as of 20th June 2020. It showed a sudden increase in the number of cases after the Unlock 1.0.

References

1. https://imgk.timesnownews.com/site/times-fact/COVID-Report-June-14.pdf
2. Li L, Yang Z, Dang Z, Meng C, Huang J, Meng H, Wang D, Chen G, Zhang J, Peng H, Shao Y (2020) Propagation analysis and prediction of the COVID-19. https://www.sciencedirect.com/science/article/pii/S2468042720300087
3. The time-series data on cumulative cases of COVID-19 in India: https://www.kaggle.com/sudalairajkumar/covid19-in-india/data?select=covid_19_india.csv
4. Gupta R, Pal SK (2020) Trend analysis and forecasting of COVID-19 outbreak in India. MedRxiv
5. https://people.duke.edu/~rnau/411arim3.htm
6. https://www.covid19india.org/

Chapter 10
Multifacet Impact of Pandemic on Society

10.1 Introduction

COVID-19, an uninvited guest, makes its way by posing a great global health concern known as 'pandemic'. This unprecedented visitant, 'coronavirus', also pretenses a question before us, 'What is the new normal we all are heading to?' Global health authorities advise less or no human intervention to prevent the further spread of the virus. This situation led many nations to enforce a lockdown. However, as World Health Organization (WHO) predicts coronavirus to be with us for a long time, some nations have decided to relax containment measures. The nations across globe have a difficult task before them to strike a balance between reopening the economy and preventing a second wave of infections. The objective is to combat the unfortunate impacts on the economy, which needs to be restarted to avoid further destruction.

The International Monetary Fund (IMF) estimates that the global economy growing at −3% in 2020 is an outcome "far worse" than the 2009 global financial crises. IMF also calls it 'Worst Crisis since the Great Depression' that occurred a century ago. During such a rare time, the economy is one of the problems that every country is grappling with and has been affected the most because of restrictions due to lockdown [1]. The fall of rupee, an Indian currency in last few months is shown in Fig. 10.1.

The revival of the economy becomes even more difficult due to the effects of labour mobility on the economy. No sectors have remained untouched from the harsh impact of the virus. During these times, the demand for essential commodities has also proved difficult to be fulfilled on time. The existing approach of supply chain management failed to meet the demand during such crucial times. In this chapter, an improvised network of the 'Decentralized Supply Chain Management' has been proposed that aims at overcoming the existing problems in the management of the chain of supplies. The framework consists of four flexible layers that incorporate

N. Sharma et al., *Open Data for Sustainable Community*,
Advances in Sustainability Science and Technology,
https://doi.org/10.1007/978-981-33-4312-2_10

Fig. 10.1 Fall of rupee

Data Management, Automation, Real-Time Data Visibility, Real-Time Tracking, and Quality Control.

The chapter is organized as follows: next section reviews the literature; Sect. 10.3 investigates the impact of lockdown on the economy from various perspectives. Section 10.3.2 discusses the layoffs of employees during the time of the pandemic. Sections 10.3.3 and 10.3.4 describe how the migration of labourers has impacted the economy. In Sect. 10.4, we focus on supply chain management and propose an effective model that will prove to be effective during unprecedented times.

10.2 Literature Review

The highest double-digit decline was also recorded in India during the initial phase of the nationwide lockdown. Exports declined by 34.57% in March 2020 as lockdown restrictions made it difficult for the smooth functioning of the supply chain management. Experts have termed this the highest drop in exports in recent times. Except for exports of iron ore, which grew 58.43%, all other commodity groups registered negative growth in March 2020 compared to March 2019. According to the Federation of Indian Export Organisations (FIEO), recovery is only expected after the second quarter of the 2021 financial year. A decline in imports was also seen which accounted for a drop of over 28% in March 2020. The merchandise trade deficit narrowed to $9.76 billion in March 2020 from $11 billion in March 2019. Fall across categories is depicted in Table 10.1 which shows the percentage of decline in exports in March 2020 [2].

Top commodities like engineering goods saw over a 42% drop in shipments. Exports of other commodities that register billions of dollar worth of exports every month also declined anywhere between 20% and 41%. Among major import

Table 10.1 Fall across categories

Commodity	Export in March 2020	Decline %
Engineering goods	$5.44 BN	42.32
Petroleum products	$2.49 BN	31.12
Gems and jewellery	$2.02 BN	41.05
Organic/Inorganic chemicals	$1.56 BN	32.88
Drugs and pharmaceuticals	$1.55 BN	22.80

commodities, there was a 15% decline in petroleum and crude, 29.09% drop in electronic goods, 31.72% declines in electrical and non-electrical machinery, 53.46% drop in pearl, precious and semi-precious stones and 62.64% drop in gold imports.

According to Xu [3], managing supply chain disruptions revolves around two goals: first, to thoroughly understand the potential of identified risks and second, to increase the capacity of the supply chain—within reasonable limits. The author further highlights the need of taking into account the vulnerabilities of more comprehensive scenario planning to model the dynamics and consequences of high-impact risks, like the 2003 Severe Acute Respiratory Syndrome (SARS) outbreak in China. Few days after the SARS outbreak in China, Chung et al. [4] highlighted the need for adopting effective contingency plans for the management of supply chains. Despite highlighting this need, supply chain management in most countries has failed to fulfil the demand of people during the ongoing COVID-19 pandemic. This emphasizes on being resilient in managing chains of supply.

The coronavirus pandemic affects the global and local economies at a larger scale. Supply availability in global supply chains has been drastically reduced and misbalanced with the demands. To overcome this extraordinary situation, it clearly shows a need for progressing the supply chain resilience research and practices [5]. A critical step in the risk management process is risk assessment, shown in Fig. 10.2.

Supply chain resilience increases as capabilities increase and vulnerabilities decrease and vice versa shown in Fig. 10.3a. (1) Excessive vulnerabilities relative to capabilities will result in excessive risk. (2) Excessive capabilities relative to vulnerabilities will erode profitability. (3) Thus, supply chain resilience should be balanced, as shown in Fig. 10.3b [6].

Let us try to visualize the same with an example. There is a small perturbation introduced in the system leading to event (2). This will lead price to decrease. The system is dynamic and tries to auto correct itself due to some existing resilience in the system. This will lead towards event (1). Events (1) and (2) will happen serially trying to counter each other's effect trying to achieve a locally stable state. Measuring this oscillating behaviour needs a new approach in the new normal.

To fulfil the demand for essential commodities, especially during unprecedented times, the supply chain needs to be managed effectively [7]. It is now evident that supply chain is a network of facilities that performs the functions of procurement of material, the transformation of material to intermediate and finished products, and distribution of finished products to customers. Often, organizational barriers between

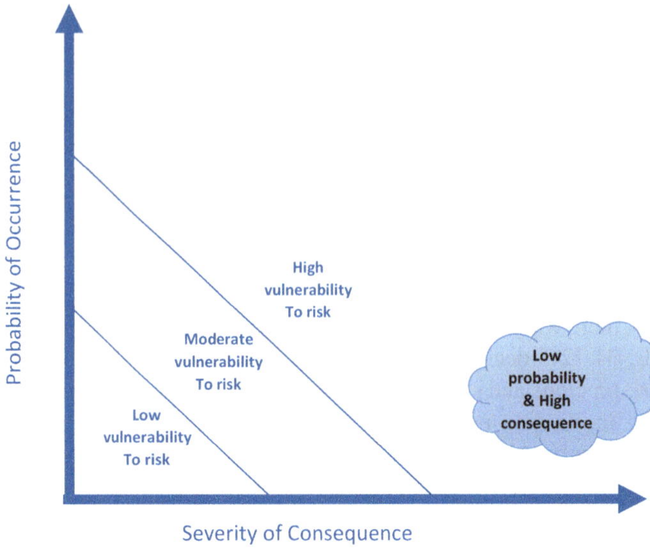

Fig. 10.2 Risk assessment of supply chain management

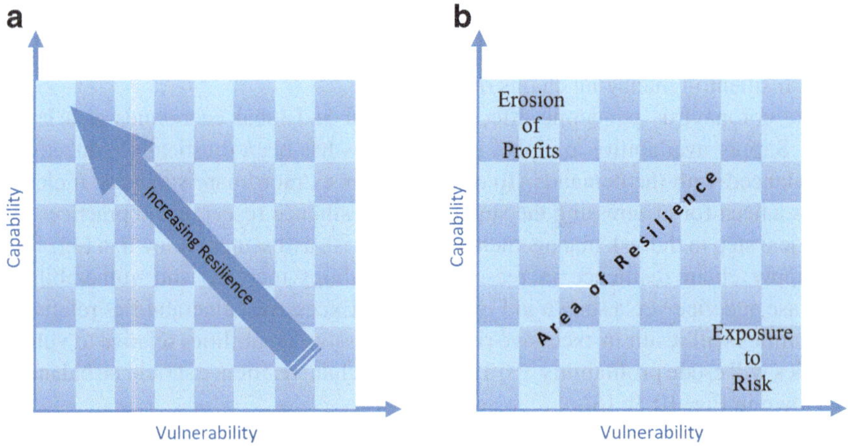

Fig. 10.3 **a** Supply chain resilience increases as capabilities increase and vulnerabilities decrease. **b** Balanced supply chain resilience decrease

these facilities exist and information flows can be restricted such that complete centralized control of material flows in a supply chain may not be feasible or desirable. Consequently, most companies use decentralized control in managing the different facilities at a supply chain [8].

Biswas [9] throws light on a nearly two-decade-old initiative to reach fresh produce directly to consumers in Maharashtra, India—bypassing the local markets. Decentralization and direct-to-home delivery being the essential features of this alternative market channel will remain valuable even after the lockdown, when efforts to avoid crowding in the wholesale markets are likely to continue. The government and Maharashtra State Agricultural Marketing Board (MSAMB) identify farmer groups and Farmer Producer Companies (FPCs) and form clusters; local bodies choose the market sites and link the markets for direct delivery to co-operative housing societies. Large numbers of vegetable growers in Maharashtra have got into direct selling to consumers. Direct marketing, although not a new concept in perishables, had thus far failed to get momentum among local farmers. Now, with the retail stores and local markets being shut, rural entrepreneurs have stepped into supply the demand in the cities.

Dukach [10] highlights the vulnerability of medical products to get affected during transportation. Damage and delay are the biggest challenges to overcome in supply of medical equipment. In-transit circumstances like rough handling, temperature and humidity fluctuations may result in failure of some fragile electronic medical equipment's causing it to become inoperable. Pharmaceutical products require utmost care to be taken since it has been observed that sudden temperature changes during transportation may lead to decomposition of medicine. Products like drug becomes unconsumable if it gets accidentally exposed to drastic temperatures. IoT sensors provide manufacturers and distributors with real-time visibility into the location and condition of their in-transit goods, giving them the information, they need to identify and eliminate delays and damages if any. This multisensor tracking system can function over a year on a single charge and monitor not just the location of medical device shipments, but also their temperature, humidity, shock levels, and other key metrics. All of this vital supply chain data is sent to the manufacturer in real time over the cellular network, triggering immediate notifications if shipments are likely to be late or are experiencing potentially damaging temperature or shock levels.

In a study by IIM Lucknow [11], Joseph Mathew with two more researchers states that new trends in healthcare supply chain like RFID technology, use of analytics and virtual centralization of supply chains prove to be helpful attaining inventory visibility and accurate counts at every stage of the supply chain and also helps reducing chain shrinkage, shipping errors and wastage. Especially for pharmaceutical products, Khouki et al. [12] state that improvement in inventory management systems in hospitals is very significant for optimized sharing of life-saving surgical kits in emergency.

10.3 Economy During the Time of Pandemic

According to Agence France-Presse (AFP), some economists bet on a solid rebound once the worst has passed and hope about a "V-shaped" recovery while some expect a

"U-shaped" recovery proceeding slowly, in terms of opening up the economy again. However, in doing so there may be few setbacks. The coronavirus pandemic has hit like a worldwide hurricane, and the entire economics of the supply chain has been greatly affected by it. To bring economy back on its foot is to rejig the supply chain machinery and fulfil the demand for essential commodities.

10.3.1 Impact of Lockdown on Banking Sector and Insurers

Though the lockdown was necessary and inevitable to prevent the faster spread of novel coronavirus (COVID-19) and to save the lives of people of the country, it is going to affect the various sectors of our economy severely [13]. The banking and non-banking finance companies (NBFCs) which are the backbone of India's economy are no exception to this. Even if the lockdown is lifted partially, operations of several companies in specific sectors will not see an immediate revival in their functioning in terms of the economy, one reason being the migration of labourers that took place recently, hampering the production to start. It is observed that in sectors like aviation, real estate, consumer durables, etc., the product demand in the market has reduced, movements of people and products are not free. Due to constraints, people are not able to adapt to the new normal. Retail trade logistics is struggling because of the complete disruption of supply chains from China, as a result of which they are way behind on project schedules and could face payment issues from their customers.

Furthermore, there is a fear that if the lockdown gets extended, there can be an increase in non-performing assets (NPA) and non-performing loans (NPL). The lockdown would certainly impact the moneylenders from MSME (Micro, Small and Medium Enterprises). The banking sector is even more concerned about the recovery of debts, as reports by Moody's Investors Service and Credit Rating Information Services of India Limited (CRISIL) predict that the repaying capacity of borrowers is going to be negative from stable. In the study of 35 sectors, both from manufacturing and services, there is a sharp variation in resilience in a post-COVID-19 landscape. They are categorized into three classes:

High Resilience Category: Pharmaceuticals, telecom, Fast Moving Consumer Goods (FMCG), fertilizers, oil refineries, power and gas distribution centres, and transmission are among the sectors comprising nearly 44% of the debt.

Moderate Resilience Category: Another 52% debt is seen in sectors like—automobile manufacturing companies, power generation units, road and construction.

Least Resilient Category: Around 4% of debt in sectors such as airlines, gems and jewellery, auto dealers, and real estate—given the discretionary nature of goods and services and weak balance sheets.

According to the guidelines issued by Reserve Bank of India (RBI), the banks and other NBFCs are bound to extend the moratorium on their loans for another three months, in addition to the moratorium declared for the previous quarter as well [14]. In another statistic, banking system credit grew by 6.5% year-on-year on the fortnight that ended on May 22, which is down by almost 50% from its value a year ago, reflecting the impact of nationwide lockdown imposed during the end of March [15].

10.3.2 *Layoffs in Various Industries*

According to estimates released on 31 March 2020 by the National Sample Survey and the Periodic Labour Force Survey suggests that more than 136 million non-farm jobs are in immediate danger, and also International Labour Organization (ILO) had estimated that the coronavirus outbreak could threaten more than 25 million jobs worldwide. According to the layoff survey carried out in 2020, it was observed that in the first week of May 68% of employers had started or been planning to lay off their employees. Among the surveyed organizations, it was also observed that:

1. 73% of employers have plans to decrease the salary of employees.
2. 57% of employers say that this layoff is temporary.
3. 21% of employers plan for permanent layoffs for at least 2 years.

What is even more bizarre is the way companies in India, to recover from the financial loss are imposing harsh decisions on their employees—one of the fastest-growing companies fire 3700 workers on 14 May 2020, in just a 2-min video call. Another social commerce start-up laid off 150 executives' just days after assuring them their jobs were not at risk. No prior notices were issued to the employees making it even worse for them. Data of layoffs reported in various industries between 11 March and 26 May 2020 is shown in Table 10.2.

As depicted in Table 10.2, no sector has remained unaffected by the harsh impact posed on the economy. The government had suspended all ride-hailing services along with other means of transport during the enforcement of the lockdown in March. With all travel modes being prohibited, the tourism economy has been undergoing a drastic fall. Several companies in food delivery services have fired their employees, despite the government's appeal to not do so. Real estate is another sector that has been suffering a severe setback, to only increase further uncertainty due to the migration of labourers. Sectors like entertainment, tourism, and fitness are undergoing critical salary cut-offs and even employee layoffs. It is perceived that not all industries will take the same time span to overcome the problem of economic disruption.

10.3.3 *Migration and Livelihood*

Since the economic distribution is greatly diverse in India, migration is an ancient common strategy adopted for a livelihood by workers in the labour category. In most of the cases, migration in search of employment is directed towards the urban area, clustered around metropolitans. Statistics reveal that around half of the population in

Table 10.2 Layoffs reported between 11 March–26 May 2020

Industry	Layoffs	% of Employees
Transportation	6700	25
	982	17
	406	30
Travel	1900	25
	900	25
	400	33
	300	25
Consumer	1000	50
	1000	17
	900	30
Retail	2800	44
	669	100
	250	50
Food	1300	50
	400	44
Fitness	784	98
	700	35
Real estate	600	35
	550	N/A
	375	15
Recruiting	400	39
	300	30
Marketing	400	10
	300	50
Logistics	349	18
	300	18
Entertainment	500	45
Finance	460	30
Healthcare	340	53
Other	260	10

urban localizations are migrants, of which around 20% are from another state. Some observations made in the 2011 census regarding trends in migration are insightful to understand demographics in already established metros or in new urban cities is presented in Fig. 10.4.

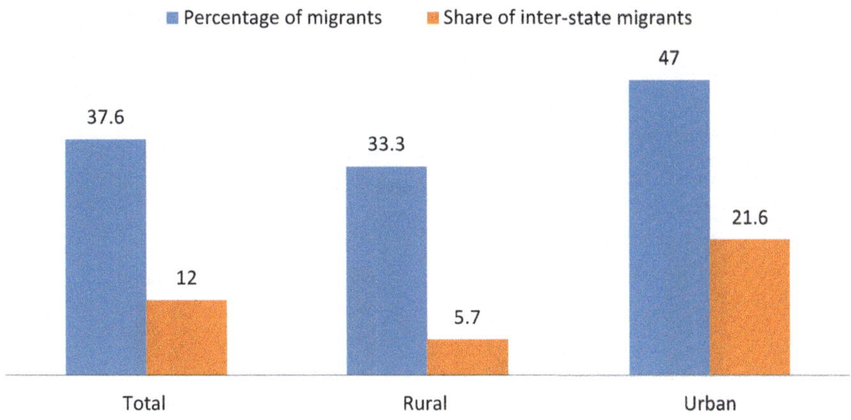

Fig. 10.4 Migration intensity and share of interstate migrants in rural and urban areas, India, 2011

10.3.4 Reverse Migration

Migrant labourers have a great contribution to economic growth, innovation, skill development, and entrepreneurship development in the process of nurturing the economy. But, migrant workers are undermined in various policies and programs about urban development and economic planning. Not realizing that migrant labourers are stakeholders in the urban economy has harshly impacted the agricultural sector and manufacturing sector during the time of the pandemic. Migrant labourers play an important role in keeping the enterprises active, especially where production and sales have seen a sudden drop.

The first case of COVID-19 surfaced in India on 30 January 2020, and following the outbreak, the lockdown in the entire country was announced on 24 March for a period of 21 days. Borders were sealed, transportation got stopped, factories, shops, restaurants and all type of the economic activities were shut, barring only the essential services. This proved to be a nightmare for hundreds of thousands of migrant workers, who lost their livelihoods overnight and became homeless. The immediate challenges faced by these migrant workers were related to food, shelter, loss of wages, fear of getting infected and anxiety. As a result, thousands of them started fleeing from various cities to their native places. This phenomenon is termed as reverse migration.

Temporary shutting down of industries has led to a loss of livelihood for those who either work on short-term contracts or those who are without any job contracts. This includes several employees from different industries. The transportation sector is also badly affected. This results in the cut down of the jobs in the market and also makes it difficult for job creation. This will result in pay-cuts and delayed increments. Figure 10.5 presents the percentage of migrants to the total population and also the percentage share of interstate migrants to the total migrants for top cities of India. Figure 10.6 presents the stock and category of migrant workers in India.

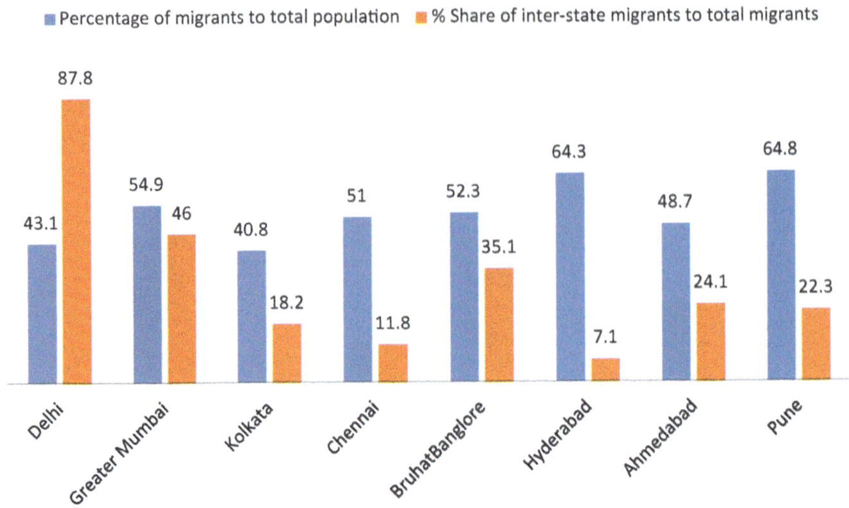

Fig. 10.5 Migration intensity and share of intersate migrants for top Indian cities

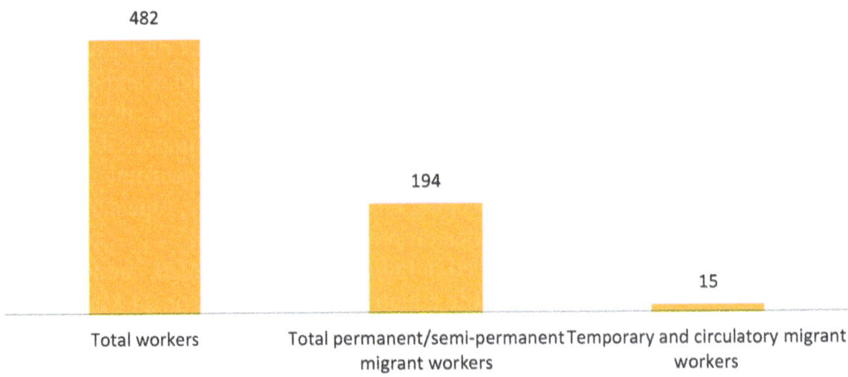

Fig. 10.6 Stock of migrant workers (in million), India

10.4 Supply Chain Management

The cycle of demand and supply chain in the economy is the strength that needs to be harnessed to its full potential.

10.4.1 Demand of Essential Commodities

In order to meet the increased demand in the country, every stakeholder in the supply chain needs to be empowered. With the lockdown been enforced in most countries, meeting the demand and supply for essential commodities is of the utmost importance. Food, a basic necessity, managed to create the maximum panic as the lockdown was announced. Individuals started rushing to grocery stores, super-markets and other such places to indulge in panic buying which to some extent led to the shortage of commonly consumed food items. Thus, there arises a necessity to strike a balance between maintaining social distancing and recovering from the threat posed on the economy.

In countries like India, movement restrictions have also hampered the supply of Personal Protective Equipment (PPE) kits as well as medicines to the healthcare centres. Mobility being a key bottleneck as trucks struggled to get past state borders has greatly hampered the supply of essential commodities. Manufacturers of testing kits and PPE began facing difficulties in transporting the urgent medical materials to the hospitals. India may have the capacity to mass manufacture rapid antibody diagnostic kits to test for COVID-19, but shortage of crucial elements such as the synthetic copies of the virus and its antibodies leave it dependent on other countries to assemble these test kits in a bid to scale up testing. Several companies are working on these rapid tests, but it will take around some 1–2 months for these to come out, because it will take time to develop the reagents, viral protein and antibodies. Lack of crucial components, delay in overseas regulatory nods hinder the progress on testing kits to be deployed at various locations. Global supply chain management has failed to fulfil the demand and supply in the crucial times of pandemic. Considering the restrictions on maintaining minimum human contact and prohibitions implicated at sensitive areas, both wholesale and retail supply chains were not able to function smoothly. Therefore, decentralization of supply chain process is the need of the hour.

10.4.2 What Are Decentralized Supply Chains?

In the emergency situation wherein limitations of centralized and global markets are highlighted, decentralized and local markets standout with their flexibility. Decentralized supply chain acts as the bridge between the nearest source or producer and the end user. It can be referred to as the network of material flow from the vendor to the customer, which means the forward flow and when the flow is from the customer to the vendor, it is called as the reverse flow. To facilitate and regulate the flow of material within a specified area, the network involves runtime database. These chains are not limited to one particular category of product, but can also include service utilities like hospitality and health care.

10.4.3 Framework of Decentralized Supply Chain

The proposed framework comprises four layers. Each layer represents one component of the network. These four layers are storage houses, integrated distribution centres or hubs, distribution agents and clusters of end users, respectively. Each component in this chain is meant to perform specific role allocated to it. They are as given below:

10.4.3.1 Storage House

A primary block in any supply chain is the producer or manufacturer. Storage houses, shown by S1 and S2 in Fig. 10.7, are the first element of proposed supply chain. Since the network focuses on supply of food, medicines and emergency safety kits on priority, the storage houses are centres where these essential commodities are collected, stored and preserved if necessary. Local producers, manufacturers and farmers are the sources of goods for the storage. Considering that in emergency situations, product warehouses have key significance to avoid wastage of any perishable products. The storage houses are equipped with sensors to monitor and record data about temperature, humidity, availability of stock, etc. and share it with manufacturers and distribution centres. This ensures data visibility about status of storage house and alerts other elements to arrange goods accordingly.

It is necessary to categorize and segregate perishable goods and medicines for their preservation against ambient conditions. Therefore, a Quality Assurance Department

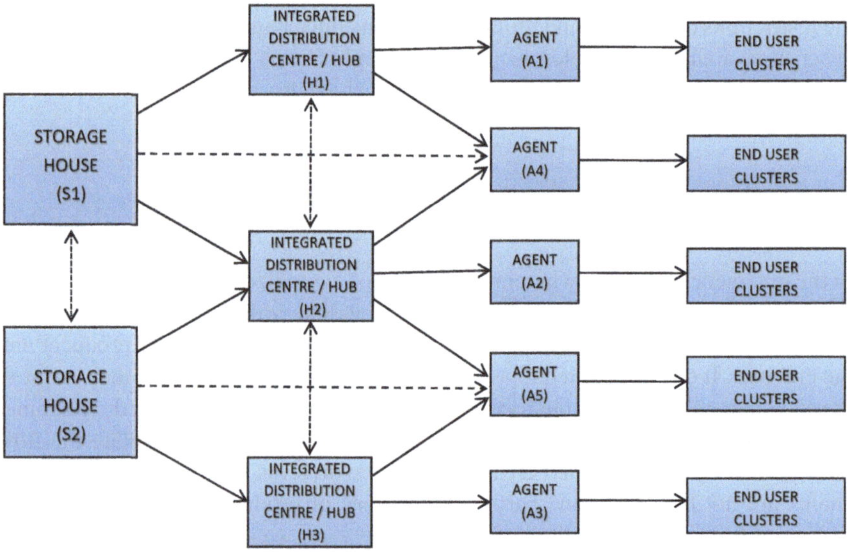

Fig. 10.7 Framework of decentralized supply chain management

(QAD) is associated with each storage house. Storage houses supply the commodities to integrated distribution centres or hubs. In certain circumstances, a storage house can itself act like a hub. Also, mutual sharing of products in two storage houses to tackle occasions like shortage is also feasible.

10.4.3.2 Integrated Distribution Centre or Hub

The hubs are secondary component of the chain. They collect essential goods from storage houses and distribute them among delivery agents. Along with distribution, they also perform quality check of the products to check if the product has been damaged during transportation. Hubs can approach to both, storage house or any other hub nearby for refilling stock. The flexibility of network enables hub to communicate with storage house, another hub and delivery agent at a time to minimize damage and delay to the essentials being delivered.

10.4.3.3 Agent

The penultimate component of the supply chain is the delivery agent. These agents are any third-party individual or transport agency which is responsible for delivery of products from hub to the end customer. Every agent is allocated with one cluster to which it serves the delivery within shortest time possible.

10.4.3.4 End User Cluster

A city is divided into the region in which the proposed supply chain is expected to function. The city is divided into several imaginary clusters which include definite number of customers or end users. Each cluster is assigned with an agent. Hence, all the customers within one cluster are served by the same delivery agent.

These clusters are dynamic and can be restructured as and when required.

10.4.4 Technologies Used

Interdependency of various layers of the model can be made efficient by using some of the emerging technologies, namely advanced robotics, Internet of Things (IoT), Big Data, smart warehousing, 3D printing, block chain and so on. To avoid bottlenecks and speed up the process in the supply chain network, technological advancements play a vital role.

Real-time data collection is of the primary importance that needs to be addressed for the proper management of the supply chain network. It can be done using the IoT system. IoT-based devices that make use of sensors and actuators help in the exchange of data between different layers of the network. In addition to it, sensors are also used to monitor and track the flow of essential goods. It becomes easier to get timely information about the location of the goods, storage of the goods, thereby estimating the time required to deliver the product at a specific location. Real-time inventory monitoring can be achieved using various smart warehousing techniques.

During the time of the medical emergency, where social distancing is of the utmost importance, advanced robotics can be used to handle the materials at warehouses and loading and unloading points. Automated Guided cart Vehicle (AGC); the low-cost Automated Guided Vehicle System (AGVs) can easily integrate into the production environment. To bridge the gap between probability and reality in the supply chain, software assistance tools like Enterprise Resource Planning (ERP), Warehouse Management System (WMS), Transportation Management System (TMS) are used. Algorithms like shortest job first (SJF) and banker's algorithm are used to avoid extremities like deadlock and starvation. Big Data analytics is yet another technology which is playing an instrumental role in improving supply chain management. It has had a good impact on all facets of the supply chain network which includes accurate planning and scheduling, inventory management, real-time decision-making, demand analysis, order optimization, improving delivering time, and improving responsiveness.

Healthcare workers rely on personal protective equipment to protect themselves and their patients from being infected and infecting others. 3D printing can be used to manufacture gloves, medical masks, respirators, goggles, face shields, gowns and aprons to increase the supply of these essential commodities against shortage.

10.4.5 Quality Assurance

Trade plays a vital role in ensuring the availability of essential medicines, protective gear, vegetables and fruits. Monitoring the quality of these products is as important as dispensing them on time. At times manufacturers ramp up the production neglecting its quality. It emphasizes the need for monitoring the quality of the products being delivered. Various initiatives are being taken to supply the best quality products [16].

Indian Council of Agricultural Research has come up with an initiative of manufacturing 'Solar-Powered Vegetable Vending Vans' that keeps vegetables fresh for two days with the help of evaporating cooling technology. The hi-tech vans designed by The Indian Institute of Horticultural Research (IIHR) is one of the steps in achieving quality control.

During the time of the lockdown, farmers are left stranded with fruits and veggies lying in the fields. To avoid damage to fruits and to ensure long usability, fresh fruits can be converted to dry fruits. This can be achieved by using a drying system. With an efficient drying system, farmers would be able to dry their farm produce using appropriate temperature and humidity control systems. Drying inhibits the growth of bacteria, yeasts and mould through the removal of water [17].

In the proposed framework, storage houses and integrated distribution centres will have consist of the Quality Assurance Department (QAD) where the quality of products will be monitored and accordingly proceed for the delivery. Various techniques, viz. image processing and IoT-based monitoring systems will be incorporated for monitoring the quality of pharmaceuticals and food products.

Kausar et al. [18] rightly emphasizes that the pharmaceuticals need to be stored under conditions that prevent contamination and, as far as possible, deterioration. The conditions under which pharmaceutical products are manufactured and stored can have a major impact on their quality. High temperature and relative humidity (RH) are the most important factors involved in drug degradation. Many vehicles with an inbuilt IoT-based monitoring system are used for the delivery of these drugs. Moreover, the quality of the drugs is also verified at the storage house and the hub.

10.4.6 Salient Features

During a medical emergency, the supply chain has to be flexible and should adopt the restrictions that it has laid. The layers of the proposed model are complementary to each other and can share tasks among them. For example, an agent can occasionally serve and deliver products to a user, not necessarily from the cluster assigned to him. Also, suppliers can directly hand over specific perishable products to the agent, skipping one of the steps—it need not be transported for further storage and quality checking in the hub. Thus, the elements participating in the supply chain can interchange their roles as and when required. This flexibility reduces the latency of procedure; in case of supplying medical commodities where slight delay can lead to permanent loss.

The use of IoT sensors provides transparency about real-time inventory data. During transportation, any adverse change in temperature, humidity, exposure to light, or leakage of air, the ambiance of storage containers can be detected with the help of sensors. Even a slight change in such parameters can lead to the decomposition of sensitive products like medicines, packaged foods, vegetables and delicate safety kits. Unexpected events such as vehicle failure or road blockage can cause a delay in the delivery. Timely notifications about such events can be given on the seller's and customer's dashboard that is connected to the IoT network. This will not only help the seller to prevent further damage to the product by taking necessary actions but also helps the customers to track their orders. The model discussed here possesses this provision to ensure minimum damage and delay.

10.5 Conclusion

The current situation suggests that the recovery rate of the pandemic is a slow process. The pandemic has pushed the global economy into recession. Some nations have decided to relax containment measures. This comes after the unfortunate impacts on the economy which has to restart to avoid further destruction. The immigration of labourers has had a huge impact on the economy. Due to many restrictions laid down, the functioning of the global supply chain failed as the demands could not be met on time. Hence, it suggests resuming an old initiative—'Decentralized Supply Chain Management' which also proves to be useful when the normal times return. The model proposed will ensure the proper functioning of the network. The network consists of various layers that help to bridge the gap between the producers and the end users, and it aims to meet the demands on time. Quality monitoring is also carried out during the delivery of the commodities. During the times when social distancing is of the utmost importance, using various emerging technologies helps in not only achieving prompt results but also providing better results with less human intervention.

References

1. Agence France-Presse, Washington (2020) V-shaped or U-shaped? Economists divided on virus recovery, Indian Express
2. ENS Economic Bureau, New Delhi (2020) Exports decline by 34.57% in March as COVID-19, lockdown snap supply chains, Indian Express
3. Xu J (2008) Managing the risk of supply chain disruption: towards a resilient approach of supply chain management. ISECS Int Colloquium Comput Commun Control Manage
4. Chung A, Gilliland M, Wiggins D (2003) SARS highlights need for supply chain contingency plans, Gartner
5. Ivanov D, Dolgui A (2020) Viability of intertwined supply networks: extending the supply chain resilience angles towards survivability. A position paper motivated by COVID-19 outbreak. Int J Prod Res
6. Pettit TJ, Fiksel J, Croxton KL (2010) Ensuring supply chain resilience: development of a conceptual framework. J Bus Logistics
7. Mentzer JT, DeWitt W, Keebler JS, Min S, Nix NW, Smith CD, Zacharia ZG (2001) Defining supply chain management. J Bus Logistics
8. Lee HL, Billington C (1993) Material management in decentralized supply chains. Oper Res Soc Am
9. Biswas PS (2020) Direct farm-to-kitchen: relief, with promise. Indian Express
10. Dukach D (2018) Real world and data driven answers to supply chain problems, "Medical Design and Outsourcing" dated 9 Oct 2018
11. Mathew J, John J, Kumar S (2013) New trends in healthcare supply chain
12. Khouki S, Bojji C, Bensouda Y (2019) A review of medical distribution logistics in pharmaceutical supply chain. Int J Logistics Syst Manage 34(3)
13. Singh J, Bodla BS (2020) Covid-19 pandemic and lockdown impact on India's banking sector: a systemic literature review. In: Singh RP, Singh A, Kumar R (eds) COVID-19 pandemic: a global challenge, pp 21–32. ISBN 978-93-86695-28-4
14. COVID lockdown impact: banks, NBFC's urge RBI to extend moratorium by another three months. Financial Express, 5th May 2020
15. Lockdown impact: bank credit growth plunges 50% in fortnight to May 22 by Abhijeet Lele in Business Standard on 6th June 2020
16. Satish Kumar BS (2020) Solar-powered vegetable vending vans come in handy during lockdown. The Hindu, 06 Apr 2020. https://www.thehindu.com

17. Kiran P, Sonam S, Ashwini T, Pratidnya M (2016) Solar powered automatic fruit drying system. IJARECE
18. Shafaat K, Hussain A, Kumar B, Ul Hasan R, Prabhat P, Yadav VK (2013) An overview: storage of pharmaceutical products. World J Pharm Pharm Sci

The manufacturer's authorised representative in the EU is Springer Nature Customer Service Centre GmbH, Europaplatz 3, 69115 Heidelberg, Germany. If you have any concerns regarding our products, please contact ProductSafety@springernature.com

Printed and bound by CPI Group (UK) Ltd, Croydon, CR0 4YY
30/09/2026
02225289-0008